HEAVY METAL CONTAMINATION OF WATER AND SOIL

Analysis, Assessment, and Remediation Strategies

HEAVY METAL CONTAMINATION OF WATER AND SOIL

Analysis, Assessment, and Remediation Strategies

Edited by
Elham Asrari, PhD

Apple Academic Press

TORONTO   NEW JERSEY

Apple Academic Press Inc.	Apple Academic Press Inc.
3333 Mistwell Crescent	9 Spinnaker Way
Oakville, ON L6L 0A2	Waretown, NJ 08758
Canada	USA

©2014 by Apple Academic Press, Inc.

First issued in paperback 2021

Exclusive worldwide distribution by CRC Press, a member of Taylor & Francis Group
No claim to original U.S. Government works

ISBN 13: 978-1-77463-328-1 (pbk)
ISBN 13: 978-1-77188-004-6 (hbk)

Library of Congress Control Number: 2013954919

Library and Archives Canada Cataloguing in Publication

Heavy metal contamination of water and soil: analysis, assessment, and remediation strategies/edited by Elham Asrari, PhD.

Includes bibliographical references and index.
ISBN 978-1-77188-004-6
1. Heavy metals--Environmental aspects--Case studies. 2. Water--Pollution--Case studies. 3. Soil pollution--Case studies. 4. Water--Analysis--Case studies. 5. Soils--Analysis--Case studies. 6. Water quality--Case studies. 7. Soils--Quality--Case studies. 8. Soil remediation--Case studies. 9. Water--Purification--Case studies. I. Asrari, Elham, writer of introduction, editor of compilation

| TD196.M4H42 2013 | 628.5'2 | C2013-907166-0 |

Apple Academic Press also publishes its books in a variety of electronic formats. Some content that appears in print may not be available in electronic format. For information about Apple Academic Press products, visit our website at **www.appleacademicpress.com** and the CRC Press website at **www.crcpress.com**

ABOUT THE EDITOR

ELHAM ASRARI, PhD

Dr. Elham Asrari is an assistant professor in the Civil Engineering Department at Payame Noor University, Iran. She is a researcher and author, having published numerous peer-reviewed articles in the fields of air pollution, water and wastewater pollution. She received her MSc in Civil and Environmental Engineering from Mazandran University, Iran, and her PhD in Environmental Sciences from Pune University, India.

CONTENTS

Acknowledgment and How to Cite.. *xi*

List of Contributors.. *xiii*

Introduction..*xxi*

Part I: Introduction

1. **Heavy Metals in Contaminated Soils: A Review of Sources, Chemistry, Risks and Best Available Strategies for Remediation**......... 1

    Raymond A. Wuana and Felix E. Okieimen

Part II: Heavy Metal Contamination

2. **Leaching Behavior of Heavy Metals and Transformation of Their Speciation in Polluted Soil Receiving Simulated Acid Rain** 53

    Shun-an Zheng, Xiangqun Zheng, and Chun Chen

3. **Spatially Explicit Analysis of Metal Transfer to Biota: Influence of Soil Contamination and Landscape** 69

    Clémentine Fritsch, Michaël Coeurdassier, Patrick Giraudoux, Francis Raoul, Francis Douay, Dominique Rieffel, Annette de Vaufleury, and Renaud Scheifler

4. **Heavy Metal Contamination of Soil and Sediment in Zambia**........... 109

    Yoshinori Ikenaka, Shouta M. M. Nakayama, Kaampwe Muzandu, Kennedy Choongo, Hiroki Teraoka, Naoharu Mizuno, and Mayumi Ishizuka

5. **Human Exposure Pathways of Heavy Metals in a Lead–Zinc Mining Area, Jiangsu Province, China**.. 129

    Chang-Sheng Qu, Zong-Wei Ma, Jin Yang, Yang Liu, Jun Bi, and Lei Huang

Part III: Analysis and Assessment of Heavy Metal Contamination

6. **Integrated Assessment of Heavy Metal Contamination in Sediments from a Coastal Industrial Basin, N. E. China** 157

    Xiaoyu Li, Lijuan Liu, Yugang Wang, Geping Luo, Xi Chen, Xiaoliang Yang, Bin Gao, and Xingyuan He

7. A Determination of Metallothionein in Larvae of Freshwater Midges
    (*Chironomus riparius*) Using Brdicka Reaction 183
    Ivo Fabrik, Zuzana Ruferova, Klara Hilscherova, Vojtech Adam, Libuse Trnkova,
    and Rene Kizek

8. Multivariate Statistical Assessment of Heavy Metal Pollution
    Sources of Groundwater Around a Lead and Zinc Plant 201
    Abbas Ali Zamani, Mohammad Reza Yaftian, and Abdolhossein Parizanganeh

9. Assessment of Heavy Metal Contamination of Agricultural Soil around
    Dhaka Export Processing Zone (DEPZ), Bangladesh:
    Implication of Seasonal Variation and Indices 221
    Syed Hafizur Rahman, Dilara Khanam, Tanveer Mehedi Adyel,
    Mohammad Shahidul Islam, Mohammad Aminul Ahsan,
    and Mohammad Ahedul Akbor

Part IV: Remediation of Heavy Metal Contamination

10. Phytoremediation of Heavy Metals: A Green Technology 249
    P. Ahmadpour, F. Ahmadpour, T. M. M. Mahmud, Arifin Abdu, M. Soleimani,
    and F. Hosseini Tayefeh

11. Assessment of the Efficacy of Chelate-Assisted Phytoextraction of
    Lead by Coffeeweed (*Sesbania exaltata Raf.*) 269
    Gloria Miller, Gregorio Begonia, Maria Begonia, Jennifer Ntoni, and Oscar Hundley

12. Sustainable Sources of Biomass for Bioremediation of Heavy
    Metals in Waste Water Derived from Coal-Fired Power
    Generation 285
    Richard J. Saunders, Nicholas A. Paul, Yi Hu, and Rocky de Nys

13. Characterization of the Metabolically Modified Heavy Metal-Resistant
    Cupriavidus metallidurans Strain MSR33 Generated for Mercury
    Bioremediation 305
    Luis A. Rojas, Carolina Yáñez, Myriam González, Soledad Lobos,
    Kornelia Smalla, and Michael Seeger

14. A Ferritin from *Dendrorhynchus zhejiangensis* with Heavy Metals Detoxification Activity 329

Chenghua Li, Zhen Li, Ye Li, Jun Zhou, Chundan Zhang, Xiurong Su, and Taiwu Li

Author Notes 345

Index 349

ACKNOWLEDGMENT AND HOW TO CITE

The chapters in this book were previously published in various places and in various formats. By bringing them together here in one place, we offer the reader a comprehensive perspective on recent investigations into heavy metal contamination of water and soil. Each chapter is added to and enriched by being placed within the context of the larger investigative landscape. Specifically:

- Chapter 1 explains how remediation of heavy metal contaminated soils is necessary to make the land resource available for agricultural production, enhance food security, and scale down land tenure problems arising from changes in the land use pattern.
- Chapter 2 considers the leaching behavior and transformation of heavy metals when influenced by acid rain.
- Chapter 3 studies the relationship between landscape influence and trace metal concentrations in animals and in soils
- Chapter 4 provides basic information on the accumulation and transportation of studied pollutants to environmental conservation.
- Chapter 5 discusses the importance of human health risk assessment as a tool for estimating the nature and probability of adverse health effects in humans: one that can be used towards sustainability development
- Chapter 6 focuses on one of the most important subjects discussed in this book: the presence of metal pollution in coastal sediment. Heavy metal contaminations in sediment could affect the water quality; quantifying and explaining the spatial distribution of heavy metal contaminants can help control sediment chemistry and identify the potential ecological risks of heavy metals.
- Chapter 7 discusses the use of metallothionein (MT), a low molecular mass protein, as a tool for the assessment of heavy metal environmental pollution.
- Chapter 8 details several studies that monitor heavy metals in groundwater, studies that are important for their implications on public health.
- Chapter 9 argues for the adoption of an effective effluent management strategy, one that moves towards control over enhanced metal levels with recycling of effluents for toxic metal separation and soil remediation and reclamation.
- Chapter 10 considers phytoremediation as a cost effective and environmentally friendly technology for the remediation of heavy metals.

- Chapter 11 also focuses on phytoremediation: the study establishes an optimal time frame for harvesting Sesbaniaexaltata after chelate amendment, thereby limiting the likelihood of exposure of heavy metals to grazing animals.
- Chapter 12 investigates the biosorption of heavy metals, which is considered a practical method of wastewater bioremediation.
- Chapter 13 summarizes the use of bioremediation to remove mercury from polluted areas.

We wish to thank the authors who made their research available for this book, whether by granting permission individually or by releasing their research as open source articles. When citing information contained within this book, please do the authors the courtesy of attributing them by name, referring back to their original articles, using the credits provided at the beginning of each chapter.

LIST OF CONTRIBUTORS

Arifin Abdu
Department of Forest Production, Faculty of Forestry, Universiti Putra Malaysia, 43400 UPM Serdang, Selangor, Malaysia

Vojtech Adam
Department of Chemistry and Biochemistry, and Department of Animal Nutrition and Forage Production, Faculty of Agronomy, Mendel University of Agriculture and Forestry, Zemedelska 1, CZ-613 00 Brno, Czech Republic

Tanveer Mehedi Adyel
Department of Environmental Sciences, Jahangirnagar University, Dhaka 1342, Bangladesh

F. Ahmadpour
Environment and Energy Department, Islamic Azad University, Science and Research Branch, Tehran, Iran

P. Ahmadpour
Department of Forest Production, Faculty of Forestry, Universiti Putra Malaysia, 43400 UPM Serdang, Selangor, Malaysia

Mohammad Aminul Ahsan
Analytical Research Division, Bangladesh Council of Scientific and Industrial Research (BCSIR) Laboratories, Dhaka 1205, Bangladesh

Mohammad Ahedul Akbor
Analytical Research Division, Bangladesh Council of Scientific and Industrial Research (BCSIR) Laboratories, Dhaka 1205, Bangladesh

Gregorio Begonia
Plant Physiology/Microbiology Laboratory, Department of Biology, P.O. Box 18540, College of Science, Engineering and Technology, Jackson State University, 1000 Lynch Street, Jackson, Mississippi 39217, USA

Maria Begonia
Plant Physiology/Microbiology Laboratory, Department of Biology, P.O. Box 18540, College of Science, Engineering and Technology, Jackson State University, 1000 Lynch Street, Jackson, Mississippi 39217, USA

Jun Bi
State Key Laboratory of Pollution Control and Resource Reuse, School of the Environment, Nanjing University, Nanjing, China

Chun Chen
Agro-Environmental Protection Institute, Ministry of Agriculture, Tianjin, People's Republic of China, Key Laboratory of Production Environment and Agro-Product Safety, Ministry of Agriculture, Tianjin, People's Republic of China, and Tianjin Key Laboratory of Agro-Environment and Agro-Product Safety, Tianjin, People's Republic of China

Xi Chen
State Key Laboratory of Desert and Oasis Ecology, Xinjiang Institute of Ecology and Geography, Chinese Academy of Sciences, Xinjiang, China

Kennedy Choongo
Department of Biomedical Studies, School of Veterinary Medicine, University of Zambia, P.O. Box 32379, Lusaka, Zambia

Michaël Cœurdassier
Department of Chrono-Environment, UMR UFC/CNRS 6249 USC INRA, University of Franche-Comté, Besançon, France

Rocky de Nys
School of Marine and Tropical Biology & Centre for Sustainable Tropical Fisheries and Aquaculture, James Cook University, Townsville, Australia

Annette de Vaufleury
Department of Chrono-Environment, UMR UFC/CNRS 6249 USC INRA, University of Franche-Comté, Besançon, France

Francis Douay
Université Lille Nord de France, Lille, France and Laboratoire Génie Civil et géoEnvironnement (LGCgE), EA 4515, Lille, France

Ivo Fabrik
Department of Chemistry and Biochemistry, Mendel University of Agriculture and Forestry, Zemedelska 1, CZ-613 00 Brno, Czech Republic

Clémentine Fritsch
Department of Chrono-Environment, UMR UFC/CNRS 6249 USC INRA, University of Franche-Comté, Besançon, France

Bin Gao
College of Resources Science and Technology, Beijing Normal University, Beijing, China

Patrick Giraudoux
Department of Chrono-Environment, UMR UFC/CNRS 6249 USC INRA, University of Franche-Comté, Besançon, France

Myriam González
Laboratorio de Microbiología Molecular y Biotecnología Ambiental, Departamento de Química and Center for Nanotechnology and Systems Biology, Universidad Técnica Federico Santa María, Valparaíso, Chile

Xingyuan He
State Key Laboratory of Forest and Soil Ecology, Institute of Applied Ecology, Chinese Academy of Sciences, Liaoning, China

Klara Hilscherova
Research Centre for Environmental Chemistry and Ecotoxicology, Faculty of Science, Masaryk University, Kotlarska 2, CZ-611 37 Brno, Czech Republic

Yi Hu
Advanced Analytical Centre, James Cook University, Townsville, Australia

Lei Huang
State Key Laboratory of Pollution Control and Resource Reuse, School of the Environment, Nanjing University, Nanjing, China

Oscar Hundley
Plant Physiology/Microbiology Laboratory, Department of Biology, P.O. Box 18540, College of Science, Engineering and Technology, Jackson State University, 1000 Lynch Street, Jackson, Mississippi 39217, USA

Yoshinori Ikenaka
Laboratory of Toxicology, Department of Environmental Veterinary Sciences, Graduate School of Veterinary Medicine, Hokkaido University, Kita 18, Nishi 9, Kita-ku, Sapporo 060-0818, Japan

Mayumi Ishizuka
Laboratory of Toxicology, Department of Environmental Veterinary Sciences, Graduate School of Veterinary Medicine, Hokkaido University, Kita 18, Nishi 9, Kita-ku, Sapporo 060-0818, Japan

Mohammad Shahidul Islam
Analytical Research Division, Bangladesh Council of Scientific and Industrial Research (BCSIR) Laboratories, Dhaka 1205, Bangladesh

Dilara Khanam
Department of Environmental Sciences, Jahangirnagar University, Dhaka 1342, Bangladesh

Rene Kizek
Department of Chemistry and Biochemistry, Mendel University of Agriculture and Forestry, Zemedelska 1, CZ-613 00 Brno, Czech Republic

Chenghua Li
School of Marine Sciences, Ningbo University, Ningbo, Zhejiang Province, People's Republic of China

Taiwu Li
Ningbo City College of Vocational Technology, Ningbo, People's Republic of China

Xiaoyu Li
State Key Laboratory of Forest and Soil Ecology, Institute of Applied Ecology, Chinese Academy of Sciences, Liaoning, China and State Key Laboratory of Desert and Oasis Ecology, Xinjiang Institute of Ecology and Geography, Chinese Academy of Sciences, Xinjiang, China

Ye Li
School of Marine Sciences, Ningbo University, Ningbo, Zhejiang Province, People's Republic of China

Zhen Li
School of Marine Sciences, Ningbo University, Ningbo, Zhejiang Province, People's Republic of China

Lijuan Liu
State Key Laboratory of Desert and Oasis Ecology, Xinjiang Institute of Ecology and Geography, Chinese Academy of Sciences, Xinjiang, China

Yang Liu
Department of Environmental Health, Rollins School of Public Health, Emory University, Atlanta, Georgia, United States of America

Soledad Lobos
Laboratorio de Espectroscopía, Facultad de Farmacia, Universidad de Valparaíso, Playa Ancha, Valparaíso, Chile

Geping Luo
State Key Laboratory of Desert and Oasis Ecology, Xinjiang Institute of Ecology and Geography, Chinese Academy of Sciences, Xinjiang, China

Zong-Wei Ma
State Key Laboratory of Pollution Control and Resource Reuse, School of the Environment, Nanjing University, Nanjing, China

T. M. M. Mahmud
Institute of Tropical Agriculture, Universiti Putra Malaysia, 43400 UPM Serdang, Selangor, Malaysia

Gloria Miller
Plant Physiology/Microbiology Laboratory, Department of Biology, P.O. Box 18540, College of Science, Engineering and Technology, Jackson State University, 1000 Lynch Street, Jackson, Mississippi 39217, USA

Naoharu Mizuno
Department of Pharmacology, School of Veterinary Medicine Rakuno Gakuen University, Ebetsu 069-8501, Japan

Kaampwe Muzandu
Department of Biomedical Studies, School of Veterinary Medicine, University of Zambia, P.O. Box 32379, Lusaka, Zambia

Shouta M. M. Nakayama
Laboratory of Toxicology, Department of Environmental Veterinary Sciences, Graduate School of Veterinary Medicine, Hokkaido University, Kita 18, Nishi 9, Kita-ku, Sapporo 060-0818, Japan

Jennifer Ntoni
Plant Physiology/Microbiology Laboratory, Department of Biology, P.O. Box 18540, College of Science, Engineering and Technology, Jackson State University, 1000 Lynch Street, Jackson, Mississippi 39217, USA

Felix E. Okieimen
Research Laboratory, GeoEnvironmental & Climate Change Adaptation Research Centre, University of Benin, Benin City 300283, Nigeria

Abdolhossein Parizanganeh
Environmental Science Research Laboratory, Department of Environmental Science, Faculty of Science, University of Zanjan, Zanjan, Iran

Nicholas A. Paul
School of Marine and Tropical Biology & Centre for Sustainable Tropical Fisheries and Aquaculture, James Cook University, Townsville, Australia

Chang-Sheng Qu
State Key Laboratory of Pollution Control and Resource Reuse, School of the Environment, Nanjing University, Nanjing, China and Jiangsu Provincial Academy of Environmental Science, Nanjing, China

Syed Hafizur Rahman
Department of Environmental Sciences, Jahangirnagar University, Dhaka 1342, Bangladesh

Francis Raoul
Department of Chrono-Environment, UMR UFC/CNRS 6249 USC INRA, University of Franche-Comté, Besançon, France

Dominique Rieffel
Department of Chrono-Environment, UMR UFC/CNRS 6249 USC INRA, University of Franche-Comté, Besançon, France

Luis A. Rojas
Laboratorio de Microbiología Molecular y Biotecnología Ambiental, Departamento de Química and Center for Nanotechnology and Systems Biology, Universidad Técnica Federico Santa María, Valparaíso, Chile and Laboratorio de Espectroscopía, Facultad de Farmacia, Universidad de Valparaíso, Playa Ancha, Valparaíso, Chile

Zuzana Ruferova
Research Centre for Environmental Chemistry and Ecotoxicology, Faculty of Science, Masaryk University, Kotlarska 2, CZ-611 37 Brno, Czech Republic

Richard J. Saunders
School of Marine and Tropical Biology & Centre for Sustainable Tropical Fisheries and Aquaculture, James Cook University, Townsville, Australia

Renaud Scheifler
Department of Chrono-Environment, UMR UFC/CNRS 6249 USC INRA, University of Franche-Comté, Besançon, France

Michael Seeger
Laboratorio de Microbiología Molecular y Biotecnología Ambiental, Departamento de Química and Center for Nanotechnology and Systems Biology, Universidad Técnica Federico Santa María, Valparaíso, Chile

Kornelia Smalla
Julius Kühn-Institut, Federal Research Centre for Cultivated Plants (JKI), Institute for Epidemiology and Pathogen Diagnostics, Braunschweig, Germany

M. Soleimani
Department of Environmental Science, Faculty of Natural Resources, Isfahan University of Technology, Isfahan, 84156-83111, Iran

Xiurong Su
School of Marine Sciences, Ningbo University, Ningbo, Zhejiang Province, People's Republic of China

F. Hosseini Tayefeh
Department of Wild Life Management, Faculty of Forestry, Universiti Putra Malaysia, 43400 UPM Serdang, Selangor, Malaysia

Hiroki Teraoka
Department of Pharmacology, School of Veterinary Medicine Rakuno Gakuen University, Ebetsu 069-8501, Japan

Libuse Trnkova
Department of Chemistry, Faculty of Science, Masaryk University, Kotlarska 2, CZ-611 37 Brno, Czech Republic

Yugang Wang
State Key Laboratory of Desert and Oasis Ecology, Xinjiang Institute of Ecology and Geography, Chinese Academy of Sciences, Xinjiang, China

Raymond A. Wuana
Analytical Environmental Chemistry Research Group, Department of Chemistry, Benue State University, Makurdi 970001, Nigeria

Mohammad Reza Yaftian
Phase Equilibria Research Laboratory, Department of Chemistry, Faculty of Science, University of Zanjan, Zanjan, Iran

Carolina Yáñez
Laboratorio de Microbiología Molecular y Biotecnología Ambiental, Departamento de Química and Center for Nanotechnology and Systems Biology, Universidad Técnica Federico Santa María, Valparaíso, Chile

Jin Yang
Department of Environmental Science & Engineering, Fudan University, Shanghai, China

Xiaoliang Yang
College of Environmental Science and Forestry, State University of New York, Syracuse, New York, United States of America

Abbas Ali Zamani
Phase Equilibria Research Laboratory, Department of Chemistry, Faculty of Science, University of Zanjan, Zanjan, Iran

Chundan Zhang
School of Marine Sciences, Ningbo University, Ningbo, Zhejiang Province, People's Republic of China

Shun-an Zheng
Agro-Environmental Protection Institute, Ministry of Agriculture, Tianjin, People's Republic of China, Key Laboratory of Production Environment and Agro-Product Safety, Ministry of Agriculture, Tianjin, People's Republic of China, and Tianjin Key Laboratory of Agro-Environment and Agro-Product Safety, Tianjin, People's Republic of China

Xiangqun Zheng
Agro-Environmental Protection Institute, Ministry of Agriculture, Tianjin, People's Republic of China, Key Laboratory of Production Environment and Agro-Product Safety, Ministry of Agriculture, Tianjin, People's Republic of China, and Tianjin Key Laboratory of Agro-Environment and Agro-Product Safety, Tianjin, People's Republic of China

Jun Zhou
School of Marine Sciences, Ningbo University, Ningbo, Zhejiang Province, People's Republic of China

INTRODUCTION

Heavy metals have been used by humans for thousands of years. Although adverse health effects of heavy metals have been known for a long time, exposure to heavy metals continues and is even increasing in some areas. The adequate protection and restoration of soil and water ecosystems contaminated by heavy metals require their characterization and remediation. Remediating heavy metal contaminated soils and water is necessary to reduce the associated health and ecological risks, make the land resource available for agricultural production, enhance food security and scale down land tenure problems. The chapters in this book discuss both the causes and the environmental impact of heavy metal contamination; the articles highlighted also discuss many exciting new methods of analysis and decontamination currently studied and applied in the field today.

Chapter 1 provides an overall introduction to the scholarship on heacy metals; Wuana and Okieimen compile the scattered literature to critically review the possible sources, chemistry, potential biohazards and best available remedial strategies for a number of heavy metals (lead, chromium, arsenic, zinc, cadmium, copper, mercury and nickel) commonly found in contaminated soils. The principles, advantages and disadvantages of immobilization, soil washing and phytoremediation techniques which are frequently listed among the best demonstrated available technologies for cleaning up heavy metal contaminated sites are presented. Remediation of heavy metal contaminated soils is necessary to reduce the associated risks, make the land resource available for agricultural production, enhance food security and scale down land tenure problems arising from changes in the land use pattern.

Chapter 2 focuses on the heavy metals that leach from contaminated soils under acid rain. In this study by Zheng and colleagues, simulated acid rain (SAR) was pumped through columns of artificially contaminated purple soil. Column leaching tests and sequential extraction were conducted for the heavy metals Cu, Pb, Cd, and Zn to determine the extent of their leaching as well as to examine the transformation of their speciation in the artificially contaminated soil columns. Results showed that the maximum leachate concentrations of Cu, Pb, Cd, and Zn were less than those specified in the Chinese Quality Standards for Groundwater (Grade IV), thereby suggesting that the heavy metals that leached from the polluted purple soil receiving acid rain may not pose as risks to water quality. Most of the Pb and Cd leachate concentrations were below their detection limits. By contrast, higher Cu and Zn leachate concentrations were found because they were released by the soil in larger amounts as compared with those of Pb and Cd. The differences in the Cu and Zn leachate concentrations between the controls (SAR at pH 5.6) and the treatments (SAR at pH 3.0 and 4.5) were significant. Similar trends were observed in the total leached amounts of Cu and Zn. The proportions of Cu, Pb, Cd, and Zn in the EXC and OX fractions were generally increased after the leaching experiment at three pH levels, whereas those of the RES, OM, and CAR fractions were slightly decreased. Acid rain favors the leaching of heavy metals from the contaminated purple soil and makes the heavy metal fractions become more labile. Moreover, a pH decrease from 5.6 to 3.0 significantly enhanced such effects.

Fritsch and colleagues discuss the field of "landscape ecotoxicology" in chapter 3. Concepts and developments for a new field in ecotoxicology, referred to as "landscape ecotoxicology," were proposed in the 1990s; however, to date, few studies have been developed in this emergent field. In fact, there is a strong interest in developing this area, both for renewing the concepts and tools used in ecotoxicology as well as for responding to practical issues, such as risk assessment. The aim of this study was to investigate the spatial heterogeneity of metal bioaccumulation in animals in order to identify the role of spatially explicit factors, such as landscape as well as total and extractable metal concentrations in soils. Over a smelter-impacted area, the authors studied the accumulation of trace metals (TMs: Cd, Pb and Zn) in invertebrates (the grove snail *Cepaea sp* and the glass snail *Oxychilus draparnaudi)* and vertebrates (the bank vole *Myodes glareolus* and the greater white-toothed shrew *Crocidura russula*). Total and CaCl$_2$-extractable concentrations of TMs were measured in soils from woody patches where the animals were captured. TM concentrations in animals exhibited a high spatial heterogeneity. They increased with soil pollution and were better explained by total rather than $CaCl_2$-extractable TM concentrations, except in *Cepaea sp*. TM levels in animals and their variations along the pollution gradient were modulated by the landscape, and this influence was species and metal specific. Median soil metal concentrations (predicted by universal kriging) were calculated in buffers of increasing size and were related to bioaccumulation. The spatial scale at which TM concentrations in animals and soils showed the strongest correlations varied between metals, species and landscapes. The potential underlying mechanisms of landscape influence (community functioning, behaviour, etc.) are discussed. Present results highlight the need for the further development of landscape ecotoxicology and multi-scale approaches, which would enhance our understanding of pollutant transfer and effects in ecosystems

Chapter 4, by Ikenake and colleagues, is a evaluation of heavy metal pollution in Zambia. Heavy metal pollution is one of the most important problems in Zambia and causes serious effects to humans and animals. The aim of the present study was to evaluate the spatial distribution of heavy metals in main areas of Zambia and understand the characteristics of the pollution in each area. River and lake sediments and soil samples were collected from a large area of Zambia and analyzed for ten heavy metals (Cr, Co, Ni, Cu, Zn, As, Cd, Pb, Sr and Hg). The results indicate that heavy metal pollution in Zambia has strong regional differences. Using cluster analysis, the patterns of heavy metal pollution were divided into three major clusters: (1) Kabwe, (2) Copperbelt and (3) Lusaka and other areas. Heavy metals in the Copperbelt area are transported to downstream areas by the Kafue River. Pollution was also detected in national parks, and Lake Itezhi-tezhi has been polluted with high concentrations of Cu, possibly from mining activities in the upper reaches of the river. However, areas geographically distant from mining beds had only moderate or low heavy metal concentrations, although the concentrations of Pb and Zn were highly correlated with the populations of each town. The findings of this study indicate that heavy metal pollution in Zambia is still increasing, due to human activities, especially mining.

Qu and colleagues also examine the effects of heavy metal pollution in chapter 5, this time in China. Heavy metal pollution is becoming a serious issue in developing countries such as China, and the public is increasingly aware of its adverse health impacts in recent years. The authors assessed the potential health risks in a lead-zinc mining area and attempted to identify the key exposure pathways. The study evaluated the spatial distributions of personal exposure using indigenous exposure factors and field monitoring results of water, soil, food, and indoor and outdoor air samples. The risks posed by ten metals and the contribution of inhalation, ingestion and dermal contact pathways to these risks were estimated. Human hair samples were also analyzed to indicate the exposure level in the human body. The results show that heavy metal pollution may pose high potential health risks to local residents, especially in the village closest to the mine (V1), mainly due to Pb, Cd and Hg. Correspondingly, the residents in V1 had higher Pb (8.14 mg/kg) levels in hair than those in the other two villages. Most of the estimated risks came from soil, the intake of self-produced vegetables and indoor air inhalation. This study highlights the importance of site-specific multipathway health risk assessments in studying heavy-metal exposures in China.

The purpose of the study in chapter 6, by Li and colleagues, is to investigate the current status of metal pollution of the sediments from urban-stream, estuary and Jinzhou Bay of the coastal industrial city, NE China. Forty surface sediment samples from river, estuary and bay and one sediment core from Jinzhou bay were collected and analyzed for heavy metal concentrations of Cu, Zn, Pb, Cd, Ni and Mn. The data reveals that there was a remarkable change in the contents of heavy metals among the sampling sediments, and all the mean values of heavy metal concentration were higher than the national guideline values of marine sediment quality of China (GB 18668-2002). This is one of the most polluted of the world's impacted coastal systems. Both the correlation analyses and geostatistical analyses showed that Cu, Zn, Pb and Cd have a very similar spatial pattern and come from the industrial activities, and the concentration of Mn mainly caused by natural factors. The estuary is the most polluted area with extremely high potential ecological risk; however the contamination decreased with distance seaward of the river estuary. This study clearly highlights the urgent need to make great efforts to control the industrial emission and the exceptionally severe heavy metal pollution in the coastal area, and the immediate measures should be carried out to minimize the rate of contamination, and extent of future pollution problems.

Fabrik and colleagues discuss a low molecular mass protein called metallothionein (MT), suitable for assessment of the heavy metal environmental pollution, in chapter 7. The aim of this work was to determine the metallothionein and total thiols content in larvae of freshwater midges (*Chironomus riparius*) sampled from laboratory exposure to cadmium(II) ions and from field studies using differential pulse voltammetry Brdicka reaction. Unique electrochemical instrument, stationary electrochemical analyser Autolab coupled with autosampler, was utilized for the analysis of the samples. The detection limit for MT was evaluated as 5 nM. The larvae exposed to two doses (50 ng/g or 50 μg/g) of cadmium(II) ions for fifteen days under laboratory controlled conditions were at the end of the exposure killed, homogenized and analysed. MT content in control samples was 1.2 μM, in larvae exposed to 50 ng Cd/g it was 2.0 μM and in larvae exposed to 50 μg Cd/g 2.9 μM. Moreover at field study chironomid larvae as well as sediment samples have been collected from eight field sites with different levels of pollution by heavy. The metals content (chromium, nickel, copper, zinc, arsenic, molybdenum, cadmium, tin and lead) in the sediment and or MT content in the chironomid larvae were determined by inductively coupled plasma mass spectrometry or Brdicka reaction, respectively.

Zamani and colleagues study the contamination of groundwater by heavy metal ions around a lead and zinc plant in chapter 8. As a case study groundwater contamination in Bonab Industrial Estate (Zanjan-Iran) for iron, cobalt, nickel, copper, zinc, cadmium and lead content was investigated using differential pulse polarography (DPP). Although, cobalt, copper and zinc were found correspondingly in 47.8%, 100.0%, and 100.0% of the samples, they did not contain these metals above their maximum contaminant levels (MCLs). Cadmium was detected in 65.2% of the samples and 17.4% of them were polluted by this metal. All samples contained detectable levels of lead and iron with 8.7% and 13.0% of the samples higher than their MCLs. Nickel was also found in 78.3% of the samples, out of which 8.7% were polluted. In general, the results revealed the contamination of groundwater sources in the studied zone. The higher health risks are related to lead, nickel, and cadmium ions. Multivariate statistical techniques were applied for interpreting the experimental data and giving a description for the sources. The data analysis showed correlations and similarities between investigated heavy metals and helps to classify these ion groups. Cluster analysis identified five clusters among the studied heavy metals. Cluster 1 consisted of Pb, Cu, and cluster 3 included Cd, Fe; also each of the elements Zn, Co and Ni was located in groups with single member. The same results were obtained by factor analysis. Statistical investigations revealed that anthropogenic factors and notably lead and zinc plant and pedo-geochemical pollution sources are influencing water quality in the studied area.

In chapter 9, Rahman and colleagues discuss how intense urbanization, large scale industrialization and unprecedented population growth in the last few decades have been responsible for lowering environmental quality. Soil contamination with metals is a serious concern due to their toxicity and ability to accumulate in the biota. Their work assessed the heavy metal contamination of agricultural soil in the close vicinity of the Dhaka Export Processing Zone (DEPZ) in both dry and wet seasons using different indices viz., index of geoaccumulation, contamination factor, degree of contamination, modified degree of contamination, and pollution load index. Samples were collected from the surface layer of soil and analyzed by Atomic Absorption Spectrophotometer (AAS). The trend of metals according to average concentration during the dry and wet seasons was As > Fe > Hg > Mn > Zn > Cu > Cr > Ni > Pb > Cd and As > Fe > Mn > Zn > Hg > Cu > Ni > Cr > Pb > Cd, respectively. Because of seasonal rainfall, dilution and other run-off during the wet season, metals from the upper layer of soil were flushed out to some extent and hence all the indices values were lower in this season compared to that of the dry season. Igeoresults revealed that the study area was strongly and moderately contaminated with As and Hg in the dry and wet seasons respectively. According to , soil was classified as moderately contaminated with Zn, Cr, Pb and Ni, considerably contaminated with Cu and highly contaminated with As and Hg. The general trend of the mean was Hg > As > Cu > Zn > Ni > Cr > Pb > Fe > Mn > Cd and As > Hg > Cu > Cd > Zn > Ni >Pb > Fe > Mn in dry and wet seasons, respectively. The mCf values in the dry and wet seasons were 575.13 and 244.44 respectively indicating an ultra high degree of contamination. The Cd values in both seasons were associated with a very high degree of contamination. PLI results indicated immediate intervention to ameliorate pollution in both seasons. The main sources of metals included effluents from wastewater treatment plants, treated and untreated wastewater from surrounding industrial establishments as well as agricultural activities. Protecting the agricultural soil is a formidable challenge in the study area, which requires modernization of industries, thereby improving the recovery and recycling of wastewater. Indices analysis presented in the present work could serve as a landmark for contemporary research in toxicology.

The environment has been contaminated with organic and inorganic pollutants. Organic pollutants are largely anthropogenic and are introduced to the environment in many ways. Soil contamination with toxic metals, such as Cd, Pb, Cr, Zn, Ni and Cu, as a result of worldwide industrialization has increased noticeably within the past few years. There are some conventional remediation technologies to clean polluted areas, specifically soils contaminated with metals. In spite of being efficient, these methods are expensive, time consuming, and environmentally devastating. Recently, phytoremediation as a cost effective and environmentally friendly technology has been developed by scientists and engineers in which biomass/microorganisms or live plants are used to remediate the polluted areas. It can be categorized into various applications, including phytofiltration, phytostabilization, phytoextraction, and phytodegradation. Chapter 10, by Ahmadour and colleagues, provides a brief review of phytoremediation of soils contaminated with heavy metals to provide an extensive applicability of this green technology.

Miller and colleagues, in chapter 11, explore how lead (Pb), depending upon the reactant surface, pH, redox potential and other factors can bind tightly to the soil with a retention time of many centuries. Soil-metal interactions by sorption, precipitation and complexation processes, and differences between plant species in metal uptake efficiency, transport, and susceptibility make a general prediction of soil metal bioavailability and risks of plant metal toxicity difficult. Moreover, the tight binding characteristic of Pb to soils and plant materials make a significant portion of Pb unavailable for uptake by plants. This experiment was conducted to determine whether the addition of ethylenediaminetetraacetic acid (EDTA), ethylene glycol tetraacetic acid (EGTA), or acetic acid (HAc) can enhance the phytoextraction of Pb by making the Pb soluble and more bioavailable for uptake by coffeeweed (*Sesbania exaltata* Raf.). The authors also wanted to assess the efficacy of chelates in facilitating translocation of the metal into the above-ground biomass of this plant. To test the effect of chelates on Pb solubility, 2 g of Pb-spiked soil (1000 mg Pb/kg dry soil) were added to each 15 mL centrifuge tube. Chelates (EDTA, EGTA, HAc) in a 1:1 ratio with the metal, or distilled deionized water were then added. Samples were shaken on a platform shaker then centrifuged at the end of several time periods. Supernatants were filtered with a 0.45 μm filter and quantified by inductively coupled plasma-optical emission spectrometry (ICP-OES) to determine soluble Pb concentrations. Results revealed that EDTA was the most effective in bringing Pb into solution, and that maximum solubility was reached 6 days after chelate amendment. Additionally, a greenhouse experiment was conducted by planting *Sesbania* seeds in plastic tubes containing top soil and peat (2:1, v:v) spiked with various levels (0, 1000, 2000 mg Pb/kg dry soil) of lead nitrate. At six weeks after emergence, aqueous solutions of EDTA and/or HAc (in a 1:1 ratio with the metal) or distilled deionized water were applied to the root zones. Plants were harvested at 6 days after chelate addition to coincide with the duration of maximum metal solubility previously determined in this study. Results of the greenhouse experiment showed that coffeeweed was relatively tolerant to moderate levels of Pb and chelates as shown by very slight reductions in root and no discernable effects on shoot biomass. Root Pb concentrations increased with increasing levels of soil-applied Pb. Further increases in root Pb concentrations were attributed to chelate amendments. In the absence of chelates, translocation of Pb from roots to shoots was minimal. However, translocation dramatically increased in treatments with EDTA alone or in combination with HAc. Overall, the results of this study indicated that depending on the nature and type of Pb-contaminated soil being remediated, the bioavailability and uptake of Pb by coffeeweed can be enhanced by amending the soil with chelates especially after the plants have reached maximum biomass.

Biosorption of heavy metals using dried algal biomass has been extensively described but rarely implemented. In chapter 12, Saunders and colleagues contend this is because available algal biomass is a valuable product with a ready market. Therefore, the authors considered an alternative and practical approach to algal bioremediation in which algae were cultured directly in the waste water stream. They cultured three species of algae with and without nutrient addition in water that was contaminated with heavy metals from an Ash Dam associated with coal-fired power generation and tested metal uptake and bioremediation potential. All species achieved high concentrations of heavy metals (to 8% dry mass). Two key elements, V and As, reached concentrations in the biomass of 1543 mg.kg^{-1} DW and 137 mg.kg^{-1} DW. Growth rates were reduced by more than half in neat Ash Dam water than when nutrients were supplied in excess. Growth rate and bioconcentration were positively correlated for most elements, but some elements (e.g. Cd, Zn) were concentrated more when growth rates were lower, indicating the potential to tailor bioremediation depending on the pollutant. The cosmopolitan nature of the macroalgae studied, and their ability to grow and concentrate a suite of heavy metals from industrial wastes, highlights a clear benefit in the practical application of waste water bioremediation.

Chapter 13, by Rojs and colleagues, states that mercury-polluted environments are often contaminated with other heavy metals. Therefore, bacteria with resistance to several heavy metals may be useful for bioremediation. Cupriavidus metallidurans CH34 is a model heavy metal-resistant bacterium, but possesses a low resistance to mercury compounds. To improve inorganic and organic mercury resistance of strain CH34, the IncP-1β plasmid pTP6 that provides novel merB, merG genes and additional other mer genes was introduced into the bacterium by biparental mating. The transconjugant *Cupriavidus metallidurans* strain MSR33 was genetically and biochemically characterized. Strain MSR33 maintained stably the plasmid pTP6 over 70 generations under non-selective conditions. The organomercurial lyase protein MerB and the mercuric reductase MerA of strain MSR33 were synthesized in presence of Hg^{2+}. The minimum inhibitory concentrations (mM) for strain MSR33 were: Hg^{2+}, 0.12 and CH$_3$Hg$^+$, 0.08. The addition of Hg^{2+} (0.04 mM) at exponential phase had not an effect on the growth rate of strain MSR33. In contrast, after Hg^{2+} addition at exponential phase the parental strain CH34 showed an immediate cessation of cell growth. During exposure to Hg^{2+} no effects in the morphology of MSR33 cells were observed, whereas CH34 cells exposed to Hg^{2+} showed a fuzzy outer membrane. Bioremediation with strain MSR33 of two mercury-contaminated aqueous solutions was evaluated. Hg^{2+} (0.10 and 0.15 mM) was completely volatilized by strain MSR33 from the polluted waters in presence of thioglycolate (5 mM) after 2 h. A broad-spectrum mercury-resistant strain MSR33 was generated by incorporation of plasmid pTP6 that was directly isolated from the environment into C. metallidurans CH34. Strain MSR33 is capable to remove mercury from polluted waters. This is the first study to use an IncP-1β plasmid directly isolated from the environment, to generate a novel and stable bacterial strain useful for mercury bioremediation.

Ferritin, an iron homeostasis protein, has important functions in transition and storage of toxic metal ions. In the final chapter, by Li and colleagues, the full-length cDNA of ferritin was isolated from *Dendrorhynchus zhejiangensis* by cDNA library and RACE approaches. The higher similarity and conserved motifs for ferritin were also identified in worm counterparts, indicating that it belonged to a new member of ferritin family. The temporal expression of worm ferritin in haemocytes was analyzed by RT-PCR, and revealed the ferritin could be induced by Cd^{2+}, Pb^{2+} and Fe^{2+}. The heavy metal binding activity of recombinant ferritin was further elucidated by atomic force microscopy (AFM). It was observed that the ferritin protein could form a chain of beads with different size against three metals exposure, and the largest one with 35~40 nm in height was identified in the Cd^{2+} challenge group. Our results indicated that worm ferritin was a promising candidate for heavy metals detoxification.

PART I

INTRODUCTION

CHAPTER 1

HEAVY METALS IN CONTAMINATED SOILS: A REVIEW OF SOURCES, CHEMISTRY, RISKS, AND BEST AVAILABLE STRATEGIES FOR REMEDIATION

RAYMOND A. WUANA AND FELIX E. OKIEIMEN

1.1 INTRODUCTION

Soils may become contaminated by the accumulation of heavy metals and metalloids through emissions from the rapidly expanding industrial areas, mine tailings, disposal of high metal wastes, leaded gasoline and paints, land application of fertilizers, animal manures, sewage sludge, pesticides, wastewater irrigation, coal combustion residues, spillage of petrochemicals, and atmospheric deposition [1, 2]. Heavy metals constitute an ill-defined group of inorganic chemical hazards, and those most commonly found at contaminated sites are lead (Pb), chromium (Cr), arsenic (As), zinc (Zn), cadmium (Cd), copper (Cu), mercury (Hg), and nickel (Ni) [3]. Soils are the major sink for heavy metals released into the environment by aforementioned anthropogenic activities and unlike organic contami-

This chapter was originally published under the Creative Commons Attribution License. Wuanal RA and Okieimen FE. Heavy Metals in Contaminated Soils: A Review of Sources, Chemistry, Risks and Best Available Strategies for Remediation. ISRN Ecology **2011** *(2011), doi: 10.5402/2011/402647.*

nants which are oxidized to carbon (IV) oxide by microbial action, most metals do not undergo microbial or chemical degradation [4], and their total concentration in soils persists for a long time after their introduction [5]. Changes in their chemical forms (speciation) and bioavailability are, however, possible. The presence of toxic metals in soil can severely inhibit the biodegradation of organic contaminants [6]. Heavy metal contamination of soil may pose risks and hazards to humans and the ecosystem through: direct ingestion or contact with contaminated soil, the food chain (soil-plant-human or soil-plant-animal-human), drinking of contaminated ground water, reduction in food quality (safety and marketability) via phytotoxicity, reduction in land usability for agricultural production causing food insecurity, and land tenure problems [7–9].

The adequate protection and restoration of soil ecosystems contaminated by heavy metals require their characterization and remediation. Contemporary legislation respecting environmental protection and public health, at both national and international levels, are based on data that characterize chemical properties of environmental phenomena, especially those that reside in our food chain [10]. While soil characterization would provide an insight into heavy metal speciation and bioavailability, attempt at remediation of heavy metal contaminated soils would entail knowledge of the source of contamination, basic chemistry, and environmental and associated health effects (risks) of these heavy metals. Risk assessment is an effective scientific tool which enables decision makers to manage sites so contaminated in a cost-effective manner while preserving public and ecosystem health [11].

Immobilization, soil washing, and phytoremediation techniques are frequently listed among the best demonstrated available technologies (BDATs) for remediation of heavy metal-contaminated sites [3]. In spite of their cost-effectiveness and environment friendliness, field applications of these technologies have only been reported in developed countries. In most developing countries, these are yet to become commercially available technologies possibly due to the inadequate awareness of their inherent advantages and principles of operation. With greater awareness by the governments and the public of the implications of contaminated soils on human and animal health, there has been increasing interest amongst the scientific community in the development of technologies to remediate contaminated sites [12]. In developing countries with great population density and scarce funds available for environmental restoration, low-cost and ecologically sustainable remedial options are required to restore contaminated lands so as to reduce the associated risks, make the land resource available for agricultural production, enhance food security, and scale down land tenure problems.

In this paper, scattered literature is utilized to review the possible sources of contamination, basic chemistry, and the associated environmental and health risks of priority heavy metals (Pb, Cr, As, Zn, Cd, Cu, Hg, and Ni) which can provide insight into heavy metal speciation, bioavailability, and hence selection of appropriate remedial options. The principles, advantages, and disadvantages of immobilization, soil washing, and phytoremediation techniques as options for soil cleanup are also presented.

1.2 SOURCES OF HEAVY METALS IN CONTAMINATED SOILS

Heavy metals occur naturally in the soil environment from the pedogenetic processes of weathering of parent materials at levels that are regarded as trace ($<1000 \, \text{mg kg}^{-1}$) and rarely toxic [10, 13]. Due to the disturbance and acceleration of nature's slowly occurring geochemical cycle of metals by man, most soils of rural and urban environments may accumulate one or more of the heavy metals above defined background values high enough to cause risks to human health, plants, animals, ecosystems, or other media [14]. The heavy metals essentially become contaminants in the soil environments because (i) their rates of generation via man-made cycles are more rapid relative to natural ones, (ii) they become transferred from mines to random environmental locations where higher potentials of direct exposure occur, (iii) the concentrations of the metals in discarded products are relatively high compared to those in the receiving environment, and (iv) the chemical form (species) in which a metal is found in the receiving environmental system may render it more bioavailable [14]. A simple mass balance of the heavy metals in the soil can be expressed as follows [15, 16]:

$$M_{total} = (M_p + M_a + M_f + M_{ag} + M_{ow} + M_{ip}) - (M_{cr} + M_l) \quad (1)$$

where "M" is the heavy metal, "p" is the parent material, "a" is the atmospheric deposition, "f" is the fertilizer sources, "ag" are the agrochemical sources, "ow" are the organic waste sources, "ip" are other inorganic pollutants, "cr" is crop removal, and "l" is the losses by leaching, volatilization, and so forth. It is projected that the anthropogenic emission into the atmosphere, for several heavy metals, is one-to-three orders of magnitude higher than natural fluxes [17]. Heavy metals in the soil from anthropogenic sources tend to be more mobile, hence bioavailable than pedogenic, or lithogenic ones [18, 19]. Metal-bearing solids at contaminated sites can originate from a wide variety of anthropogenic sources in the form of metal mine tailings, disposal of high metal wastes in improperly protected landfills, leaded gasoline and lead-based paints, land application of fertilizer, animal manures, biosolids (sewage sludge), compost, pesticides, coal combustion residues, petrochemicals, and atmospheric deposition [1, 2, 20] are discussed hereunder.

1.2.1 FERTILIZERS

Historically, agriculture was the first major human influence on the soil [21]. To grow and complete the lifecycle, plants must acquire not only macronutrients (N, P, K, S, Ca, and Mg), but also essential micronutrients. Some soils are deficient in the heavy metals (such as Co, Cu, Fe, Mn, Mo, Ni, and Zn) that are essential for healthy plant growth [22], and crops may be supplied with these as an addition to the soil or as a foliar spray. Cereal crops grown on Cu-deficient soils are occasionally treated with Cu as an addition to the soil, and Mn may similarly be supplied to cereal and root crops. Large quantities of fertilizers are regularly added to soils in intensive farming systems to provide adequate N, P, and K for crop growth. The compounds used to supply these elements contain trace amounts of heavy metals (e.g., Cd and Pb) as impurities, which, after continued fertilizer, application may significantly increase their content in the soil [23]. Metals, such as Cd and Pb, have no known physiological activity. Application of certain phosphatic fertilizers inadvertently adds Cd and other potentially toxic elements to the soil, including F, Hg, and Pb [24].

1.2.2 PESTICIDES

Several common pesticides used fairly extensively in agriculture and horticulture in the past contained substantial concentrations of metals. For instance in the recent past, about 10% of the chemicals have approved for use as insecticides and fungicides in UK were based on compounds which contain Cu, Hg, Mn, Pb, or Zn. Examples of such pesticides are copper-containing fungicidal sprays such as Bordeaux mixture (copper sulphate) and copper oxychloride [23]. Lead arsenate was used in fruit orchards for many years to control some parasitic insects. Arsenic-containing compounds were also used extensively to control cattle ticks and to control pests in banana in New Zealand and Australia, timbers have been preserved with formulations of Cu, Cr, and As (CCA), and there are now many derelict sites where soil concentrations of these elements greatly exceed background concentrations. Such contamination has the potential to cause problems, particularly if sites are redeveloped for other agricultural or nonagricultural purposes. Compared with fertilizers, the use of such materials has been more localized, being restricted to particular sites or crops [8].

1.2.3 BIOSOLIDS AND MANURES

The application of numerous biosolids (e.g., livestock manures, composts, and municipal sewage sludge) to land inadvertently leads to the accumulation of heavy metals such as As, Cd, Cr, Cu, Pb, Hg, Ni, Se, Mo, Zn, Tl, Sb, and so forth, in the soil [20]. Certain animal wastes such as poultry, cattle, and pig manures produced in agriculture are commonly applied to crops and pastures either as solids or slurries [25]. Although most manures are seen as valuable fertilizers, in the pig and poultry industry, the Cu and Zn added to diets as growth promoters and As contained in poultry health products may also have the potential to cause metal contamination of the soil [25, 26]. The manures produced from animals on such diets contain high concentrations of As, Cu, and Zn and, if repeatedly applied to restricted areas of land, can cause considerable buildup of these metals in the soil in the long run.

Biosolids (sewage sludge) are primarily organic solid products, produced by wastewater treatment processes that can be beneficially recycled [27]. Land application of biosolids materials is a common practice in many countries that allow the reuse of biosolids produced by urban populations [28]. The term sewage sludge is used in many references because of its wide recognition and its regulatory definition. However, the term biosolids is becoming more common as a replacement for sewage sludge because it is thought to reflect more accurately the beneficial characteristics inherent to sewage sludge [29]. It is estimated that in the United States, more than half of approximately 5.6 million dry tonnes of sewage sludge used or disposed of annually is land applied, and agricultural utilization of biosolids occurs in every region of the country. In the European community, over 30% of the sewage sludge is used as fertilizer in agriculture [29]. In Australia over 175 000 tonnes of dry biosolids are produced each year by the major metropolitan authorities, and currently most biosolids applied to agricultural land are used in arable cropping situations where they can be incorporated into the soil [8].

There is also considerable interest in the potential for composting biosolids with other organic materials such as sawdust, straw, or garden waste. If this trend continues, there will be implications for metal contamination of soils. The potential of biosolids for contaminating soils with heavy metals has caused great concern about their application in agricultural practices [30]. Heavy metals most commonly found in biosolids are Pb, Ni, Cd, Cr, Cu, and Zn, and the metal concentrations are governed by the nature and the intensity of the industrial activity, as well as the type of process employed during the biosolids treatment [31]. Under certain conditions, metals added to soils in applications of biosolids can be leached downwards through the soil profile and can have the potential to contaminate groundwater [32]. Recent studies on some New Zealand soils treated with biosolids have shown increased concentrations of Cd, Ni, and Zn in drainage leachates [33, 34].

1.2.4 WASTEWATER

The application of municipal and industrial wastewater and related effluents to land dates back 400 years and now is a common practice in many parts of the world [35]. Worldwide, it is estimated that 20 million hectares of arable land are irrigated with waste water. In several Asian and African cities, studies suggest that agriculture based on wastewater irrigation accounts for 50 percent of the vegetable supply to urban areas [36]. Farmers generally are not bothered about environmental benefits or hazards and are primarily interested in maximizing their yields and profits. Although the metal concentrations in wastewater effluents are usually relatively low, long-term irrigation of land with such can eventually result in heavy metal accumulation in the soil.

1.2.5 METAL MINING AND MILLING PROCESSES AND INDUSTRIAL WASTES

Mining and milling of metal ores coupled with industries have bequeathed many countries, the legacy of wide distribution of metal contaminants in soil. During mining, tailings (heavier and larger particles settled at the bottom of the flotation cell during mining) are directly discharged into natural depressions, including onsite wetlands resulting in elevated concentrations [37]. Extensive Pb and zinc Zn ore mining and smelting have resulted in contamination of soil that poses risk to human and ecological health. Many reclamation methods used for these sites are lengthy and expensive and may not restore soil productivity. Soil heavy metal environmental risk to humans is related to bioavailability. Assimilation pathways include the ingestion of plant material grown in (food chain), or the direct ingestion (oral bioavailability) of, contaminated soil [38].

Other materials are generated by a variety of industries such as textile, tanning, petrochemicals from accidental oil spills or utilization of petroleum-based products, pesticides, and pharmaceutical facilities and are highly variable in composition. Although some are disposed of on land, few have benefits to agriculture or forestry. In addition, many are potentially hazardous because of their contents of heavy metals (Cr, Pb, and Zn) or toxic organic compounds and are seldom, if ever, applied to land. Others are very low in plant nutrients or have no soil conditioning properties [25].

1.2.6 AIRBORNE SOURCES

Airborne sources of metals include stack or duct emissions of air, gas, or vapor streams, and fugitive emissions such as dust from storage areas or waste piles. Metals from airborne sources are generally released as particulates contained in the gas stream. Some metals such as As, Cd, and Pb can also volatilize during high-temperature processing. These metals will convert to oxides and condense as fine particulates unless a reducing atmosphere is maintained [39]. Stack emissions can be distributed over a wide area by natural air currents until dry and/or wet precipitation mechanisms remove them from the gas stream. Fugitive emissions are often distributed over a much smaller area because emissions are made near the ground. In general, contaminant concentrations are lower in fugitive emissions compared to stack emissions. The type and concentration of metals emitted from both types of sources will depend on site-specific conditions. All solid particles in smoke from fires and in other emissions from factory chimneys are eventually deposited on land or sea; most forms of fossil fuels contain some heavy metals and this is, therefore, a form of contamination which has been continuing on a large scale since the industrial revolution began. For example, very high concentration of Cd, Pb, and Zn has been found in plants and soils adjacent to smelting works. Another major source of soil contamination is the aerial emission of Pb from the combustion of petrol containing tetraethyl lead; this contributes substantially to the content of Pb in soils in urban areas and in those adjacent to major roads. Zn and Cd may also be added to soils adjacent to roads, the sources being tyres, and lubricant oils [40].

1.3 BASIC SOIL CHEMISTRY AND POTENTIAL RISKS OF HEAVY METALS

The most common heavy metals found at contaminated sites, in order of abundance are Pb, Cr, As, Zn, Cd, Cu, and Hg [40]. Those metals are important since they are capable of decreasing crop production due to the risk of bioaccumulation and biomagnification in the food chain. There's also the risk of superficial and groundwater contamination. Knowledge of the basic chemistry, environmental, and associated health effects of these heavy metals is necessary in understanding their speciation, bioavailability, and remedial options. The fate and transport of a heavy metal in soil depends significantly on the chemical form and speciation of the metal. Once in the soil, heavy metals are adsorbed by initial fast reactions (minutes, hours), followed by slow adsorption reactions (days, years) and are, therefore, redistributed into different chemical forms with varying bioavailability, mobility, and toxicity [41, 42]. This distribution is believed to be controlled by reactions of heavy metals in soils such as (i) mineral precipitation and dissolution, (ii) ion exchange, adsorption, and desorption, (iii) aqueous complexation, (iv) biological immobilization and mobilization, and (v) plant uptake [43].

1.3.1 LEAD

Lead is a metal belonging to group IV and period 6 of the periodic table with atomic number 82, atomic mass 207.2, density $11.4 \, g \, cm^{-3}$, melting point 327.4°C, and boiling point 1725°C. It is a naturally occurring, bluish-gray metal usually found as a mineral combined with other elements, such as sulphur (i.e., PbS, $PbSO_4$), or oxygen ($PbCO_3$), and ranges from 10 to $30 \, mg \, kg^{-1}$ in the earth's crust [44]. Typical mean Pb concentration for surface soils worldwide averages $32 \, mg \, kg^{-1}$ and ranges from 10 to $67 \, mg \, kg^{-1}$ [10]. Lead ranks fifth behind Fe, Cu, Al, and Zn in industrial production of metals. About half of the Pb used in the U.S. goes for the manufacture of Pb storage batteries. Other uses include solders, bearings, cable covers, ammunition, plumbing, pigments, and caulking. Metals commonly alloyed with Pb are antimony (in storage batteries), calcium (Ca) and tin (Sn) (in maintenance-free storage batteries), silver (Ag) (for solder and anodes), strontium (Sr) and Sn (as anodes in electrowinning processes), tellurium (Te) (pipe and sheet in chemical installations and nuclear shielding), Sn (solders), and antimony (Sb), and Sn (sleeve bearings, printing, and high-detail castings) [45].

Ionic lead, Pb(II), lead oxides and hydroxides, and lead-metal oxyanion complexes are the general forms of Pb that are released into the soil, groundwater, and surface waters. The most stable forms of lead are Pb(II) and lead-hydroxy complexes. Lead(II) is the most common and reactive form of Pb, forming mononuclear and polynuclear oxides and hydroxides [3]. The predominant insoluble Pb compounds are lead phosphates, lead carbonates (form when the pH is above 6), and lead (hydr)oxides [46]. Lead sulfide (PbS) is the most stable solid form within the soil matrix and forms under reducing conditions, when increased concentrations of sulfide are present. Under anaerobic conditions a volatile organolead (tetramethyl lead) can be formed due to microbial alkylation [3].

Lead(II) compounds are predominantly ionic (e.g., $Pb^{2+} SO_4^{2-}$), whereas Pb(IV) compounds tend to be covalent (e.g., tetraethyl lead, $Pb(C_2H_5)_4$). Some Pb (IV) compounds, such as PbO_2, are strong oxidants. Lead forms several basic salts, such as $Pb(OH)_2 \cdot 2PbCO_3$, which was once the most widely used white paint pigment and the source of considerable chronic lead poisoning to children who ate peeling white paint. Many compounds of Pb(II) and a few Pb(IV) compounds are useful. The two most common of these are lead dioxide and lead sulphate, which are participants in the reversible reaction that occurs during the charge and discharge of lead storage battery.

In addition to the inorganic compounds of lead, there are a number of organolead compounds such as tetraethyl lead. The toxicities and environmental effects of organolead compounds are particularly noteworthy because of the former widespread use and distribution of tetraethyllead as a gasoline additive. Although more than 1000 organolead compounds have been synthesized, those of commercial and toxicological importance are largely limited to the alkyl (methyl and ethyl) lead compounds and their salts (e.g., dimethyldiethyllead, trimethyllead chloride, and diethyllead dichloride).

Inhalation and ingestion are the two routes of exposure, and the effects from both are the same. Pb accumulates in the body organs (i.e., brain), which may lead to poisoning (plumbism) or even death. The gastrointestinal tract, kidneys, and central nervous system are also affected by the presence of lead. Children exposed to lead are at risk for impaired development, lower IQ, shortened attention span, hyperactivity, and mental deterioration, with children under the age of six being at a more substantial risk. Adults usually experience decreased reaction time, loss of memory, nausea, insomnia, anorexia, and weakness of the joints when exposed to lead [47]. Lead is not an essential element. It is well known to be toxic and its effects have been more extensively reviewed than the effects of other trace metals. Lead can cause serious injury to the brain, nervous system, red blood cells, and kidneys [48]. Exposure to lead can result in a wide range of biological effects depending on the level and duration of exposure. Various effects occur over a broad range of doses, with the developing young and infants being more sensitive than adults. Lead poisoning, which is so severe as to cause evident illness, is now very rare. Lead performs no known essential function in the human body, it can merely do harm after uptake from food, air, or water. Lead is a particularly dangerous chemical, as it can accumulate in individual organisms, but also in entire food chains.

The most serious source of exposure to soil lead is through direct ingestion (eating) of contaminated soil or dust. In general, plants do not absorb or accumulate lead. However, in soils testing high in lead, it is possible for some lead to be taken up. Studies have shown that lead does not readily accumulate in the fruiting parts of vegetable and fruit crops (e.g., corn, beans, squash, tomatoes, strawberries, and apples). Higher concentrations are more likely to be found in leafy vegetables (e.g., lettuce) and on the surface of root crops (e.g., carrots). Since plants do not take up large quantities of soil lead, the lead levels in soil considered safe for plants will be much higher than soil lead levels where eating of soil is a concern (pica). Generally, it has been considered safe to use garden produce grown in soils with total lead levels less than 300 ppm. The risk of lead poisoning through the food chain increases as the soil lead level rises above this concentration. Even at soil levels above 300 ppm, most of the risk is from lead contaminated soil or dust deposits on the plants rather than from uptake of lead by the plant [49].

1.3.2 CHROMIUM

Chromium is a first-row d-block transition metal of group VIB in the periodic table with the following properties: atomic number 24, atomic mass 52, density 7.19 g cm^{-3}, melting point 1875°C, and boiling point 2665°C. It is one of the less common elements and does not occur naturally in elemental form, but only in compounds. Chromium is mined as a primary ore product in the form of the mineral chromite, $FeCr_2O_4$. Major sources of Cr-contamination include releases from electroplating processes and the disposal of Cr containing wastes [39]. Chromium(VI) is the form of Cr commonly found at contaminated sites. Chromium can also occur in the +III oxidation state, depending on pH and redox conditions. Chromium(VI) is the dominant form of Cr in shallow aquifers where aerobic conditions exist. Chromium(VI) can be reduced to Cr(III) by soil organic matter, S^{2-} and Fe^{2+} ions under anaerobic conditions often encountered in deeper groundwater. Major Cr(VI) species include chromate (CrO_4^{2-}) and dichromate ($Cr_2O_7^{2-}$) which precipitate readily in the presence of metal cations (especially Ba^{2+}, Pb^{2+}, and Ag^+). Chromate and dichromate also adsorb on soil surfaces, especially iron and aluminum oxides. Chromium(III) is the dominant form of Cr at low pH (<4). Cr^{3+} forms solution complexes with NH_3, OH^-, Cl^-, F^-, CN^-, SO_4^{2-}, and soluble organic ligands. Chromium(VI) is the more toxic form of chromium and is also more mobile. Chromium(III) mobility is decreased by adsorption to clays and oxide minerals below pH 5 and low solubility above pH 5 due to the formation of $Cr(OH)_3(s)$ [50]. Chromium mobility depends on sorption characteristics of the soil, including clay content, iron oxide content, and the amount of organic matter present. Chromium can be transported by surface runoff to surface waters in its soluble or precipitated form. Soluble and un-adsorbed chromium complexes can leach from soil into groundwater. The leachability of Cr(VI) increases as soil pH increases. Most of Cr released into natural waters is particle associated, however, and is ultimately deposited into the sediment [39]. Chromium is associated with allergic dermatitis in humans [21].

1.3.3 ARSENIC

Arsenic is a metalloid in group VA and period 4 of the periodic table that occurs in a wide variety of minerals, mainly as As_2O_3, and can be recovered from processing of ores containing mostly Cu, Pb, Zn, Ag and Au. It is also present in ashes from coal combustion. Arsenic has the following properties: atomic number 33, atomic mass 75, density 5.72 g cm^{-3}, melting point 817°C, and boiling point 613°C, and exhibits fairly complex chemistry and can be present in several oxidation states (−III, 0, III, V) [39]. In aerobic environments, As (V) is dominant, usually in the form of arsenate (AsO_4^{3-}) in various protonation states: H_3AsO_4, $H_2AsO_4^-$, $HAsO_4^{2-}$, and AsO_4^{3-}. Arsenate and other anionic forms of arsenic behave as chelates and can precipitate when metal cations are present [51]. Metal arsenate complexes are stable only under certain conditions. Arsenic (V) can also coprecipitate with or adsorb onto iron oxyhydroxides under acidic and moderately reducing conditions. Coprecipitates are immobile under these conditions, but arsenic mobility increases as pH increases [39]. Under reducing conditions As(III) dominates, existing as arsenite ($AsO3\ 3-$), and its protonated forms H_3AsO_3, $H_2AsO_3^-$, and $HAsO_3^{2-}$. Arsenite can adsorb or coprecipitate with metal sulfides and has a high affinity for other sulfur compounds. Elemental arsenic and arsine, AsH_3, may be present under extreme reducing conditions. Biotransformation (via methylation) of arsenic creates methylated derivatives of arsine, such as dimethyl arsine $HAs(CH_3)_2$ and trimethylarsine $As(CH_3)_3$ which are highly volatile. Since arsenic is often present in anionic form, it does not form complexes with simple anions such as Cl^- and SO_4^{2-}. Arsenic speciation also includes organometallic forms such as methylarsinic acid $(CH_3)AsO_2H_2$ and dimethylarsinic acid $(CH_3)_2AsO_2H$. Many As compounds adsorb strongly to soils and are therefore transported only over short distances in groundwater and surface water. Arsenic is associated with skin damage, increased risk of cancer, and problems with circulatory system [21].

1.3.4 ZINC

Zinc is a transition metal with the following characteristics: period 4, group IIB, atomic number 30, atomic mass 65.4, density $7.14\,g\,cm^{-3}$, melting point 419.5°C, and boiling point 906°C. Zinc occurs naturally in soil (about $70\,mg\,kg^{-1}$ in crustal rocks) [52], but Zn concentrations are rising unnaturally, due to anthropogenic additions. Most Zn is added during industrial activities, such as mining, coal, and waste combustion and steel processing. Many foodstuffs contain certain concentrations of Zn. Drinking water also contains certain amounts of Zn, which may be higher when it is stored in metal tanks. Industrial sources or toxic waste sites may cause the concentrations of Zn in drinking water to reach levels that can cause health problems. Zinc is a trace element that is essential for human health. Zinc shortages can cause birth defects. The world's Zn production is still on the rise which means that more and more Zn ends up in the environment. Water is polluted with Zn, due to the presence of large quantities present in the wastewater of industrial plants. A consequence is that Zn-polluted sludge is continually being deposited by rivers on their banks. Zinc may also increase the acidity of waters. Some fish can accumulate Zn in their bodies, when they live in Zn-contaminated waterways. When Zn enters the bodies of these fish, it is able to biomagnify up the food chain. Water-soluble zinc that is located in soils can contaminate groundwater. Plants often have a Zn uptake that their systems cannot handle, due to the accumulation of Zn in soils. Finally, Zn can interrupt the activity in soils, as it negatively influences the activity of microorganisms and earthworms, thus retarding the breakdown of organic matter [53].

1.3.5 CADMIUM

Cadmium is located at the end of the second row of transition elements with atomic number 48, atomic weight 112.4, density $8.65\,g\,cm^{-3}$, melting point 320.9°C, and boiling point 765°C. Together with Hg and Pb, Cd is one of the big three heavy metal poisons and is not known for any essential biological function. In its compounds, Cd occurs as the divalent Cd(II) ion. Cadmium is directly below Zn in the periodic table and has a chemical similarity to that of Zn, an essential micronutrient for plants and animals. This may account in part for Cd's toxicity; because Zn being an essential trace element, its substitution by Cd may cause the malfunctioning of metabolic processes [54].

The most significant use of Cd is in Ni/Cd batteries, as rechargeable or secondary power sources exhibiting high output, long life, low maintenance, and high tolerance to physical and electrical stress. Cadmium coatings provide good corrosion resistance coating to vessels and other vehicles, particularly in high-stress environments such as marine and aerospace. Other uses of cadmium are as pigments, stabilizers for polyvinyl chloride (PVC), in alloys and electronic compounds. Cadmium is also present as an impurity in several products, including phosphate fertilizers, detergents and refined petroleum products. In addition, acid rain and the resulting acidification of soils and surface waters have increased the geochemical mobility of Cd, and as a result its surface-water concentrations tend to increase as lake water pH decreases [54]. Cadmium is produced as an inevitable byproduct of Zn and occasionally lead refining. The application of agricultural inputs such as fertilizers, pesticides, and biosolids (sewage sludge), the disposal of industrial wastes or the deposition of atmospheric contaminants increases the total concentration of Cd in soils, and the bioavailability of this Cd determines whether plant Cd uptake occurs to a significant degree [28]. Cadmium is very biopersistent but has few toxicological properties and, once absorbed by an organism, remains resident for many years.

Since the 1970s, there has been sustained interest in possible exposure of humans to Cd through their food chain, for example, through the consumption of certain species of shellfish or vegetables. Concern regarding this latter route (agricultural crops) led to research on the possible consequences of applying sewage sludge (Cd-rich biosolids) to soils used for crops meant for human consumption, or of using cadmium-enriched phosphate fertilizer [54]. This research has led to the stipulation of highest permissible concentrations for a number of food crops [8].

Cadmium in the body is known to affect several enzymes. It is believed that the renal damage that results in proteinuria is the result of Cd adversely affecting enzymes responsible for reabsorption of proteins in kidney tubules. Cadmium also reduces the activity of delta-aminolevulinic acid synthetase, arylsulfatase, alcohol dehydrogenase, and lipoamide dehydrogenase, whereas it enhances the activity of delta-aminolevulinic acid dehydratase, pyruvate dehydrogenase, and pyruvate decarboxylase [45]. The most spectacular and publicized occurrence of cadmium poisoning resulted from dietary intake of cadmium by people in the Jintsu River Valley, near Fuchu, Japan. The victims were afflicted by itai itai disease, which means ouch, ouch in Japanese. The symptoms are the result of painful osteomalacia (bone disease) combined with kidney malfunction. Cadmium poisoning in the Jintsu River Valley was attributed to irrigated rice contaminated from an upstream mine producing Pb, Zn, and Cd. The major threat to human health is chronic accumulation in the kidneys leading to kidney dysfunction. Food intake and tobacco smoking are the main routes by which Cd enters the body [45].

1.3.6 COPPER

Copper is a transition metal which belongs to period 4 and group IB of the periodic table with atomic number 29, atomic weight 63.5, density $8.96 \, \text{g cm}^{-3}$, melting point 1083°C and boiling point 2595°C. The metal's average density and concentrations in crustal rocks are $8.1 \times 103 \, \text{kg m}^{-3}$ and $55 \, \text{mg kg}^{-1}$, respectively [52].

Copper is the third most used metal in the world [55]. Copper is an essential micronutrient required in the growth of both plants and animals. In humans, it helps in the production of blood haemoglobin. In plants, Cu is especially important in seed production, disease resistance, and regulation of water. Copper is indeed essential, but in high doses it can cause anaemia, liver and kidney damage, and stomach and intestinal irritation. Copper normally occurs in drinking water from Cu pipes, as well as from additives designed to control algal growth. While Cu's interaction with the environment is complex, research shows that most Cu introduced into the environment is, or rapidly becomes, stable and results in a form which does not pose a risk to the environment. In fact, unlike some man-made materials, Cu is not magnified in the body or bioaccumulated in the food chain. In the soil, Cu strongly complexes to the organic implying that only a small fraction of copper will be found in solution as ionic copper, Cu(II). The solubility of Cu is drastically increased at pH 5.5 [56], which is rather close to the ideal farmland pH of 6.0–6.5 [57].

Copper and Zn are two important essential elements for plants, microorganisms, animals, and humans. The connection between soil and water contamination and metal uptake by plants is determined by many chemical and physical soil factors as well as the physiological properties of the crops. Soils contaminated with trace metals may pose both direct and indirect threats: direct, through negative effects of metals on crop growth and yield, and indirect, by entering the human food chain with a potentially negative impact on human health. Even a reduction of crop yield by a few percent could lead to a significant long-term loss in production and income. Some food importers are now specifying acceptable maximum contents of metals in food, which might limit the possibility for the farmers to export their contaminated crops [36].

1.3.7 MERCURY

Mercury belongs to same group of the periodic table with Zn and Cd. It is the only liquid metal at stp. It has atomic number 80, atomic weight 200.6, density $13.6\,g\,cm^{-3}$, melting point $-13.6°C$, and boiling point $357°C$ and is usually recovered as a byproduct of ore processing [39]. Release of Hg from coal combustion is a major source of Hg contamination. Releases from manometers at pressure-measuring stations along gas/oil pipelines also contribute to Hg contamination. After release to the environment, Hg usually exists in mercuric (Hg^{2+}), mercurous (Hg_2^{2+}), elemental (Hgo), or alkylated form (methyl/ethyl mercury). The redox potential and pH of the system determine the stable forms of Hg that will be present. Mercurous and mercuric mercury are more stable under oxidizing conditions. When mildly reducing conditions exist, organic or inorganic Hg may be reduced to elemental Hg, which may then be converted to alkylated forms by biotic or abiotic processes. Mercury is most toxic in its alkylated forms which are soluble in water and volatile in air [39]. Mercury(II) forms strong complexes with a variety of both inorganic and organic ligands, making it very soluble in oxidized aquatic systems [51]. Sorption to soils, sediments, and humic materials is an important mechanism for the removal of Hg from solution. Sorption is pH dependent and increases as pH increases. Mercury may also be removed from solution by coprecipitation with sulphides. Under anaerobic conditions, both organic and inorganic forms of Hg may be converted to alkylated forms by microbial activity, such as by sulfur-reducing bacteria. Elemental mercury may also be formed under anaerobic conditions by demethylation of methyl mercury, or by reduction of Hg(II). Acidic conditions (pH < 4) also favor the formation of methyl mercury, whereas higher pH values favor precipitation of HgS(s) [39]. Mercury is associated with kidney damage [21].

1.3.8 NICKEL

Nickel is a transition element with atomic number 28 and atomic weight 58.69. In low pH regions, the metal exists in the form of the nickelous ion, Ni(II). In neutral to slightly alkaline solutions, it precipitates as nickelous hydroxide, $Ni(OH)_2$, which is a stable compound. This precipitate readily dissolves in acid solutions forming Ni(III) and in very alkaline conditions; it forms nickelite ion, $HNiO_2$, that is soluble in water. In very oxidizing and alkaline conditions, nickel exists in form of the stable nickelo-nickelic oxide, Ni_3O_4, that is soluble in acid solutions. Other nickel oxides such as nickelic oxide, Ni_2O_3, and nickel peroxide, NiO_2, are unstable in alkaline solutions and decompose by giving off oxygen. In acidic regions, however, these solids dissolve producing Ni^{2+} [58].

Nickel is an element that occurs in the environment only at very low levels and is essential in small doses, but it can be dangerous when the maximum tolerable amounts are exceeded. This can cause various kinds of cancer on different sites within the bodies of animals, mainly of those that live near refineries. The most common application of Ni is an ingredient of steel and other metal products. The major sources of nickel contamination in the soil are metal plating industries, combustion of fossil fuels, and nickel mining and electroplating [59]. It is released into the air by power plants and trash incinerators and settles to the ground after undergoing precipitation reactions. It usually takes a long time for nickel to be removed from air. Nickel can also end up in surface water when it is a part of wastewater streams. The larger part of all Ni compounds that are released to the environment will adsorb to sediment or soil particles and become immobile as a result. In acidic soils, however, Ni becomes more mobile and often leaches down to the adjacent groundwater. Microorganisms can also suffer from growth decline due to the presence of Ni, but they usually develop resistance to Ni after a while. Nickel is not known to accumulate in plants or animals and as a result Ni has not been found to biomagnify up the food chain. For animals Ni is an essential foodstuff in small amounts. The primary source of mercury is the sulphide ore cinnabar.

1.4 SOIL CONCENTRATION RANGES AND REGULATORY GUIDELINES FOR SOME HEAVY METALS

The specific type of metal contamination found in a contaminated soil is directly related to the operation that occurred at the site. The range of contaminant concentrations and the physical and chemical forms of contaminants will also depend on activities and disposal patterns for contaminated wastes on the site. Other factors that may influence the form, concentration, and distribution of metal contaminants include soil and ground-water chemistry and local transport mechanisms [3].

Soils may contain metals in the solid, gaseous, or liquid phases, and this may complicate analysis and interpretation of reported results. For example, the most common method for determining the concentration of metals contaminants in soil is via total elemental analysis (USEPA Method 3050). The level of metal contamination determined by this method is expressed as mg metal kg^{-1} soil. This analysis does not specify requirements for the moisture content of the soil and may therefore include soil water. This measurement may also be reported on a dry soil basis. The level of contamination may also be reported as leachable metals as determined by leach tests, such as the toxicity characteristic leaching procedure (TCLP) (USEPA Method 1311) or the synthetic precipitation-leaching procedure, or SPLP test (USEPA Method 1312). These procedures measure the concentration of metals in leachate from soil contacted with an acetic acid solution (TCLP) [60] or a dilute solution of sulfuric and nitric acid (SPLP). In this case, metal contamination is expressed in mgL^{-1} of the leachable metal. Other types of leaching tests have been proposed including sequential extraction procedures [61, 62] and extraction of acid volatile sulfide [63]. Sequential procedures contact the solid with a series of extractant solutions that are designed to dissolve different fractions of the associated metal. These tests may provide insight into the different forms of metal contamination present. Contaminant concentrations can be measured directly in metals-contaminated water. These concentrations are most commonly expressed as total dissolved metals in mass concentrations ($mg L^{-1}$ or gL^{-1}) or in molar concentrations ($mol L^{-1}$). In dilute solutions, a $mg L^{-1}$

is equivalent to one part per million (ppm), and a gL^{-1} is equivalent to one part per billion (ppb).

Riley et al. [64] and NJDEP [65] have reported soil concentration ranges and regulatory guidelines for some heavy metals (Table 1). In Nigeria, in the interim period, whilst suitable parameters are being developed, the Department of Petroleum Resources [60] has recommended guidelines on remediation of contaminated land based on two parameters intervention values and target values (Table 2).

TABLE 1: Soil concentration ranges and regulatory guidelines for some heavy metals.

Metal ($mg\,kg^{-1}$)	Soil concentration range† ($mg\,kg^{-1}$)	Regulatory limits‡
Pb	1.00–69 000	600
Cd	0.10–345	100
Cr	0.05–3 950	100
Hg	<0.01–1 800	270
Zn	150–5 000	1 500

†[64]; ‡Nonresidential direct contact soil clean-up criteria [65].

TABLE 2: Target and intervention values for some metals for a standard soil [60].

Metal ($mg\,kg^{-1}$)	Target value ($mg\,kg^{-1}$)	Intervention value
Ni	140.00	720.00
Cu	0.30	10.00
Zn	—	—
Cd	100.00	380.00
Pb	35.00	210.00
As	200	625
Cr	20	240
Hg	85	530

The intervention values indicate the quality for which the functionality of soil for human, animal, and plant life are, or threatened with being seriously impaired. Concentrations in excess of the intervention values correspond to serious contamination. Target values indicate the soil quality required for sustainability or expressed in terms of remedial policy, the soil quality required for the full restoration of the soil's functionality for human, animal, and plant life. The target values therefore indicate the soil quality levels ultimately aimed at.

1.5 REMEDIATION OF HEAVY METAL-CONTAMINATED SOILS

The overall objective of any soil remediation approach is to create a final solution that is protective of human health and the environment [66]. Remediation is generally subject to an array of regulatory requirements and can also be based on assessments of human health and ecological risks where no legislated standards exist or where standards are advisory. The regulatory authorities will normally accept remediation strategies that centre on reducing metal bioavailability only if reduced bioavailability is equated with reduced risk, and if the bioavailability reductions are demonstrated to be long term [66]. For heavy metal-contaminated soils, the physical and chemical form of the heavy metal contaminant in soil strongly influences the selection of the appropriate remediation treatment approach. Information about the physical characteristics of the site and the type and level of contamination at the site must be obtained to enable accurate assessment of site contamination and remedial alternatives. The contamination in the soil should be characterized to establish the type, amount, and distribution of heavy metals in the soil. Once the site has been characterized, the desired level of each metal in soil must be determined. This is done by comparison of observed heavy metal concentrations with soil quality standards for a particular regulatory domain, or by performance of a site-specific risk assessment. Remediation goals for heavy metals may be set as total metal concentration or as leachable metal in soil, or as some combination of these.

Several technologies exist for the remediation of metal-contaminated soil. Gupta et al. [67] have classified remediation technologies of contaminated soils into three categories of hazard-alleviating measures: (i) gentle in situ remediation, (ii) in situ harsh soil restrictive measures, and (iii) in situ or ex situ harsh soil destructive measures. The goal of the last two harsh alleviating measures is to avert hazards either to man, plant, or animal while the main goal of gentle in situ remediation is to restore the malfunctionality of soil (soil fertility), which allows a safe use of the soil. At present, a variety of approaches have been suggested for remediating contaminated soils. USEPA [68] has broadly classified remediation technologies for contaminated soils into (i) source control and (ii) containment remedies. Source control involves in situ and ex situ treatment technologies for sources of contamination. In situ or in place means that the contaminated soil is treated in its original place; unmoved, unexcavated; remaining at the site or in the subsurface. In situ treatment technologies treat or remove the contaminant from soil without excavation or removal of the soil. Ex situ means that the contaminated soil is moved, excavated, or removed from the site or subsurface. Implementation of ex situ remedies requires excavation or removal of the contaminated soil. Containment remedies involve the construction of vertical engineered barriers (VEB), caps, and liners used to prevent the migration of contaminants.

Another classification places remediation technologies for heavy metal-contaminated soils under five categories of general approaches to remediation (Table 3): isolation, immobilization, toxicity reduction, physical separation, and extraction [3]. In practice, it may be more convenient to employ a hybrid of two or more of these approaches for more cost effectiveness. The key factors that may influence the applicability and selection of any of the available remediation technologies are: (i) cost, (ii) long-term effectiveness/permanence, (iii) commercial availability, (iv) general acceptance, (v) applicability to high metal concentrations, (vi) applicability to mixed wastes (heavy metals and organics), (vii) toxicity reduction, (viii) mobility reduction, and (ix) volume reduction. The present paper focuses on soil washing, phytoremediation, and immobilization techniques since they are among the best demonstrated available technologies (BDATs) for heavy metal-contaminated sites.

TABLE 3: Technologies for remediation of heavy metal-contaminated soils.

Category	Remediation technologies
Isolation	(i) Capping (ii) subsurface barriers.
Immobilization	(i) Solidification/stabilization (ii) vitrification (iii) chemical treatment.
Toxicity and/or mobility reduction	(i) Chemical treatment (ii) permeable treatment walls (iii) biological treatment bioaccumulation, phytoremediation (phytoextraction, phytostabilization, and rhizofiltration), bioleaching, biochemical processes.
Physical separation	
Extraction	(i) Soil washing, pyrometallurgical extraction, in situ soil flushing, and electrokinetic treatment.

1.5.1 IMMOBILIZATION TECHNIQUES

Ex situ and in situ immobilization techniques are practical approaches to remediation of metal-contaminated soils. The ex situ technique is applied in areas where highly contaminated soil must be removed from its place of origin, and its storage is connected with a high ecological risk (e.g., in the case of radio nuclides). The method's advantages are: (i) fast and easy applicability and (ii) relatively low costs of investment and operation. The method's disadvantages include (i) high invasivity to the environment, (ii) generation of a significant amount of solid wastes (twice as large as volume after processing), (iii) the byproduct must be stored on a special landfill site, (iv) in the case of changing of the physicochemical condition in the side product or its surroundings, there is serious danger of the release of additional contaminants to the environment, and (v) permanent control of the stored wastes is required. In the in situ technique, the fixing agents amendments are applied on the unexcavated soil. The technique's advantages are (i) its low invasivity, (ii) simplicity and rapidity, (iii) relatively inexpensive, and (iv) small amount of wastes are produced, (v) high public acceptability, (vi) covers a broad spectrum of inorganic pollutants. The disadvantages of in situ immobilization are (i) its only a temporary solution (contaminants are still in the environment), (ii) the activation of pollutants may occur when soil physicochemical properties change, (iii) the reclamation process is applied only to the surface layer of soil (30–50 cm), and (iv) permanent monitoring is necessary [66, 69].

Immobilization technology often uses organic and inorganic amendment to accelerate the attenuation of metal mobility and toxicity in soils. The primary role of immobilizing amendments is to alter the original soil metals to more geochemically stable phases via sorption, precipitation, and complexation processes [70]. The mostly applied amendments include clay, cement, zeolites, minerals, phosphates, organic composts, and microbes [3, 71]. Recent studies have indicated the potential of low-cost industrial residues such as red mud [72, 73] and termitaria [74] in immobilization of heavy metals in contaminated soils. Due to the complexity of soil matrix and the limitations of current analytical techniques, the exact immobilization mechanisms have not been clarified, which could include precipitation, chemical adsorption and ion exchange, surface precipitation, formation of stable complexes with organic ligands, and redox reaction [75].

Most immobilization technologies can be performed ex situ or in situ. In situ processes are preferred due to the lower labour and energy requirements, but implementation of in situ will depend on specific site conditions.

1.5.1.1. SOLIDIFICATION/STABILIZATION (S/S)

Solidification involves the addition of binding agents to a contaminated material to impart physical/dimensional stability to contain contaminants in a solid product and reduce access by external agents through a combination of chemical reaction, encapsulation, and reduced permeability/surface area. Stabilization (also referred to as fixation) involves the addition of reagents to the contaminated soil to produce more chemically stable constituents. Conventional S/S is an established remediation technology for contaminated soils and treatment technology for hazardous wastes in many countries in the world [76].

The general approach for solidification/stabilization treatment processes involves mixing or injecting treatment agents to the contaminated soils. Inorganic binders (Table 4), such as clay (bentonite and kaolinite), cement, fly ash, blast furnace slag, calcium carbonate, Fe/Mn oxides, charcoal, zeolite [9, 77], and organic stabilizers (Table 5) such as bitumen, composts, and manures [78], or a combination of organic-inorganic amendments may be used. The dominant mechanism by which metals are immobilized is by precipitation of hydroxides within the solid matrix [79, 80]. Solidification/ stabilization technologies are not useful for some forms of metal contamination, such as species that exist as oxyanions (e.g., $Cr_2O_7^{2-}$, AsO_3^-) or metals that do not have low-solubility hydroxides (e.g., Hg). Solidification/stabilization may not be applicable at sites containing wastes that include organic forms of contamination, especially if volatile organics are present. Mixing and heating associated with binder hydration may release organic vapors. Pretreatment, such as air stripping or incineration, may be used to remove the organics and prepare the waste for metal stabilization/solidification [39]. The application of S/S technologies will also be affected by the chemical composition of the contaminated matrix, the amount of water present, and the ambient temperature. These factors can interfere with the solidification/stabilization process by inhibiting bonding of the waste to the binding material, retarding the setting of the mixtures, decreasing the stability of the matrix, or reducing the strength of the solidified area [81].

TABLE 4: Organic amendments for heavy metal immobilization [82].

Material	Heavy metal immobilized
Bark saw dust (from timber industry)	Cd, Pb, Hg, Cu
Xylogen (from paper mill wastewater)	Zn, Pb, Hg
Chitosan (from crab meat canning industry)	Cd, Cr, Hg
Bagasse (from sugar cane)	Pb
Poultry manure (from poultry farm)	Cu, Pb, Zn, Cd
Cattle manure (from cattle farm)	Cd
Rice hulls (from rice processing)	Cd, Cr, Pb
Sewage sludge	Cd
Leaves	Cr, Cd
Straw	Cd, Cr, Pb

Cement-based binders and stabilizers are common materials used for implementation of S/S technologies [83]. Portland cement, a mixture of Ca silicates, aluminates, aluminoferrites, and sulfates, is an important cement-based material. Pozzolanic materials, which consist of small spherical particles formed by coal combustion (such as fly ash) and in lime and cement kilns, are also commonly used for S/S. Pozzolans exhibit cement-like properties, especially if the silica content is high. Portland cement and pozzolans can be used alone or together to obtain optimal properties for a particular site [84]. Organic binders may also be used to treat metals through polymer microencapsulation. This process uses organic materials such as bitumen, polyethylene, paraffins, waxes, and other polyolefins as thermoplastic or thermosetting resins. For polymer encapsulation, the organic materials are heated and mixed with the contaminated matrix at elevated temperatures (120° to 200°C). The organic materials polymerize and agglomerate the waste, and the waste matrix is encapsulated [84]. Organics are volatilized and collected, and the treated material is extruded for disposal or possible reuse (e.g., as paving material) [39]. The contaminated material may require pretreatment to separate rocks and debris and dry the feed material. Polymer encapsulation requires more energy and more complex equipment than cement-based S/S operations. Bitumen (asphalt) is the cheapest and most common thermoplastic binder [84]. Solidification/stabilization is achieved by mixing the contaminated material with appropriate amounts of binder/stabilizer and water. The mixture sets and cures to form a solidified matrix and contain the waste. The cure time and pour characteristics of the mixture and the final properties of the hardened cement depend upon the composition (amount of cement, pozzolan, and water) of the binder/stabilizer.

Ex situ S/S can be easily applied to excavated soils because methods are available to provide the vigorous mixing needed to combine the binder/stabilizer with the contaminated material. Pretreatment of the waste may be necessary to screen and crush large rocks and debris. Mixing can be performed via in-drum, in-plant, or area-mixing processes. In-drum mixing may be preferred for treatment of small volumes of waste or for toxic wastes. In-plant processes utilize rotary drum mixers for batch processes or pug mill mixers for continuous treatment. Larger volumes of waste may be excavated and moved to a contained area for area mixing. This process involves layering the contaminated material with the stabilizer/binder, and subsequent mixing with a backhoe or similar equipment. Mobile and fixed treatment plants are available for ex situ S/S treatment. Smaller pilot-scale plants can treat up to 100 tons of contaminated soil per day while larger portable plants typically process 500 to over 1000 tons per day [39]. Stabilization/stabilization techniques are available to provide mixing of the binder/stabilizer with the contaminated soil in situ. In situ S/S is less labor and energy intensive than ex situ process that require excavation, transport, and disposal of the treated material. In situ S/S is also preferred if volatile or semivolatile organics are present because excavation would expose these contaminants to the air [85]. However, the presence of bedrock, large boulders cohesive soils, oily sands, and clays may preclude the application of in situ S/S at some sites. It is also more difficult to provide uniform and complete mixing through in situ processes. Mixing of the binder and contaminated matrix may be achieved using in-place mixing, vertical auger mixing, or injection grouting. In-place mixing is similar to ex situ area mixing except that the soil is not excavated prior to treatment. The in situ process is useful for treating surface or shallow contamination and involves spreading and mixing the binders with the waste using conventional excavation equipment such as draglines, backhoes, or clamshell buckets. Vertical auger mixing uses a system of augers to inject and mix the binding reagents with the waste. Larger (6–12 ft diameter) augers are used for shallow (10–40 ft) drilling and can treat 500–1000 cubic yards per day [86, 87]. Deep stabilization/solidification (up to 150 ft) can be achieved by using ganged augers (up to 3 ft in diameter each) that can treat 150–400 cubic yards per day. Finally injection grouting may be performed to inject the binder containing suspended or dissolved reagents into the treatment area under pressure. The binder permeates the surrounding soil and cures in place [39].

1.5.1.2 VITRIFICATION

The mobility of metal contaminants can be decreased by high-temperature treatment of the contaminated area that results in the formation of vitreous material, usually an oxide solid. During this process, the increased temperature may also volatilize and/or destroy organic contaminants or volatile metal species (such as Hg) that must be collected for treatment or disposal. Most soils can be treated by vitrification, and a wide variety of inorganic and organic contaminants can be targeted. Vitrification may be performed ex situ or in situ although in situ processes are preferred due to the lower energy requirements and cost [88]. Typical stages in ex situ vitrification processes may include excavation, pretreatment, mixing, feeding, melting and vitrification, off-gas collection and treatment, and forming or casting of the melted product. The energy requirement for melting is the primary factor influencing the cost of ex situ vitrification. Different sources of energy can be used for this purpose, depending on local energy costs. Process heat losses and water content of the feed should be controlled in order to minimize energy requirements. Vitrified material with certain characteristics may be obtained by using additives such as sand, clay, and/or native soil. The vitrified waste may be recycled and used as clean fill, aggregate, or other reusable materials [39].

In situ vitrification (ISV) involves passing electric current through the soil using an array of electrodes inserted vertically into the contaminated region. Each setting of four electrodes is referred to as a melt. If the soil is too dry, it may not provide sufficient conductance, and a trench containing flaked graphite and glass frit (ground glass particles) must be placed between the electrodes to provide an initial flow path for the current. Resistance heating in the starter path melts the soil. The melt grows outward and down as the molten soil usually provides additional conductance for the current. A single melt can treat up to 1000 tons of contaminated soil to depths of 20 feet, at a typical treatment rate of 3 to 6 tons per hour. Larger areas are treated by fusing together multiple individual vitrification zones. The main requirement for in situ vitrification is the ability of the soil melt to carry current and solidify as it cools. If the alkali content (as Na_2O and K_2O) of the soil is too high (1.4 wt%), the molten soil may not provide enough conductance to carry the current [89].

Vitrification is not a classical immobilization technique. The advantages include (i) easily applied for reclamation of heavily contaminated soils (Pb, Cd, Cr, asbestos, and materials containing asbestos), (ii) in the course of applying this method qualification of wastes (from hazardous to neutral) could be changed.

1.5.1.3 ASSESSMENT OF EFFICIENCY AND CAPACITY OF IMMOBILIZATION

The efficiency (E) and capacity (P) of different additives for immobilization and field applications can be evaluated using the expressions

$$E(\%) = (M_o - M_e)/M_o \times 100$$

$$P = ((M_o - M_e)V)/m \tag{2}$$

where E = efficiency of immobilization agent; P = capacity of immobilization agent; M_e = equilibrium extractable concentration of single metal in the immobilized soil ($mg\,L^{-1}$); M_o = initial extractable concentration of single metal in preimmobilized soil (mg L^{-1}); V = volume of metal salt solution (mg L^{-1}); m = weight of immobilization agent (g) [90]. High values of E and P represent the perfect efficiency and capacity of an additive that can be used in field studies of metal immobilization. After screening out the best efficient additive, another experiment could be conducted to determine the best ratio (soil/additive) for the field-fixing treatment. After the fixing treatment of contaminated soils, a lot of methods including biological and physiochemical experiments could be used to assess the remediation efficiency. Environmental risk could also be estimated after confirming the immobilized efficiency and possible release [89].

1.5.2 SOIL WASHING

Soil washing is essentially a volume reduction/waste minimization treatment process. It is done on the excavated (physically removed) soil (ex situ) or on-site (in situ). Soil washing as discussed in this review refers to ex situ techniques that employ physical and/or chemical procedures to extract metal contaminants from soils. During soil washing, (i) those soil particles which host the majority of the contamination are separated from the bulk soil fractions (physical separation), (ii) contaminants are removed from the soil by aqueous chemicals and recovered from solution on a solid substrate (chemical extraction), or (iii) a combination of both [91]. In all cases, the separated contaminants then go to hazardous waste landfill (or occasionally are further treated by chemical, thermal, or biological processes). By removing the majority of the contamination from the soil, the bulk fraction that remains can be (i) recycled on the site being remediated as relatively inert backfill, (ii) used on another site as fill, or (iii) disposed of relatively cheaply as nonhazardous material.

Ex situ soil washing is particularly frequently used in soil remediation because it (i) completely removes the contaminants and hence ensures the rapid cleanup of a contaminated site [92], (ii) meets specific criteria, (iii) reduces or eliminates long-term liability, (iv) may be the most cost-effective solution, and (v) may produce recyclable material or energy [93]. The disadvantages include the fact that the contaminants are simply moved to a different place, where they must be monitored, the risk of spreading contaminated soil and dust particles during removal and transport of contaminated soil, and the relatively high cost. Excavation can be the most expensive option when large amounts of soil must be removed, or disposal as hazardous or toxic waste is required.

Acid and chelator soil washing are the two most prevalent removal methods [94]. Soil washing currently involves soil flushing an in situ process in which the washing solution is forced through the in-place soil matrix, ex situ extraction of heavy metals from the soil slurry in reactors, and soil heap leaching. Another heavy metal removal technology is electro-remediation, which mostly involves electrokinetic movement of charged particles suspended in the soil solution, initiated by an electric gradient [35]. The metals can be removed by precipitation at the electrodes. Removal of the majority of the contaminants from the soil does not mean that the contaminant-depleted bulk is totally contaminant free. Thus, for soil washing to be successful, the level of contamination in the treated bulk must be below a site-specific action limit (e.g., based on risk assessment). Cost effectiveness with soil washing is achieved by offsetting processing costs against the ability to significantly reduce the amount of material requiring costly disposal at a hazardous waste landfill [95].

Typically the cleaned fractions from the soil washing process should be >70–80% of the original mass of the soil, but, where the contaminants have a very high associated disposal cost, and/or where transport distances to the nearest hazardous waste landfill are substantial, a 50% reduction might still be cost effective. There is also a generally held opinion that soil washing based on physical separation processes is only cost effective for sandy and granular soils where the clay and silt content (particles less than 0.063 mm) is less than 30–35% of the soil. Soil washing by chemical dissolution of the contaminants is not constrained by the proportion of clay as this fraction can also be leached by the chemical agent. However, clay-rich soils pose other problems such as difficulties with materials handling and solid-liquid separation [96]. Full-scale soil washing plants exist as fixed centralized treatment centres, or as mobile/transportable units. With fixed centralized facilities, contaminated soil is brought to the plant, whereas with mobile/transportable facilities, the plant is transported to a contaminated site, and soil is processed on the site. Where mobile/transportable plant is used, the cost of mobilization and demobilization can be significant. However, where large volumes of soil are to be treated, this cost can be more than offset by reusing clean material on the site (therefore avoiding the cost of transport to an off-site centralized treatment facility, and avoiding the cost of importing clean fill).

1.5.2.1 PRINCIPLES OF SOIL WASHING

Soil washing is a volume reduction/waste minimization treatment technology based on physical and/or chemical processes. With physical soil washing, differences between particle grain size, settling velocity, specific gravity, surface chemical behaviour, and rarely magnetic properties are used to separate those particles which host the majority of the contamination from the bulk which are contaminant-depleted. The equipment used is standard mineral processing equipment, which is more generally used in the mining industry [91]. Mineral processing techniques as applied to soil remediation have been reviewed in literature [97].

With chemical soil washing, soil particles are cleaned by selectively transferring the contaminants on the soil into solution. Since heavy metals are sparingly soluble and occur predominantly in a sorbed state, washing the soils with water alone would be expected to remove too low an amount of cations in the leachates, chemical agents have to be added to the washing water [98]. This is achieved by mixing the soil with aqueous solutions of acids, alkalis, complexants, other solvents, and surfactants. The resulting cleaned particles are then separated from the resulting aqueous solution. This solution is then treated to remove the contaminants (e.g., by sorption on activated carbon or ion exchange) [91, 95].

The effectiveness of washing is closely related to the ability of the extracting solution to dissolve the metal contaminants in soils. However, the strong bonds between the soil and metals make the cleaning process difficult [99]. Therefore, only extractants capable of dissolving large quantities of metals would be suitable for cleaning purposes. The realization that the goal of soil remediation is to remove the metal and preserve the natural soil properties limits the choice of extractants that can be used in the cleaning process [100].

1.5.2.2 CHEMICAL EXTRACTANTS FOR SOIL WASHING

Owing to the different nature of heavy metals, extracting solutions that can optimally remove them must be carefully sought during soil washing. Several classes of chemicals used for soil washing include surfactants, cosolvents, cyclodextrins, chelating agents, and organic acids [101–106]. All these soil washing extractants have been developed on a case-by-case basis depending on the contaminant type at a particular site. A few studies have indicated that the solubilization/exchange/extraction of heavy metals by washing solutions differs considerably for different soil types. Strong acids attack and degrade the soil crystalline structure at extended contact times. For less damaging washes, organic acids and chelating agents are often suggested as alternatives to straight mineral acid use [107].

Natural, low-molecular-weight organic acids (LMWOAs) including oxalic, citric, formic, acetic, malic, succinic, malonic, maleic, lactic, aconitic, and fumaric acids are natural products of root exudates, microbial secretions, and plant and animal residue decomposition in soils [108]. Thus metal dissolution by organic acids is likely to be more representative of a mobile metal fraction that is available to biota [109]. The chelating organic acids are able to dislodge the exchangeable, carbonate, and reducible fractions of heavy metals by washing procedures [94]. Although many chelating compounds including citric acid [108], tartaric acid [110], and EDTA [94, 100, 111] for mobilizing heavy metals have been evaluated, there remain uncertainties as to the optimal choice for full-scale application. The identification and quantification of coexisting solid metal species in the soil before and after treatment are essential to design and assess the efficiency of soil-washing technology [4]. A recent study [112] showed that changes in Ni, Cu, Zn, Cd, and Pb speciation and uptake by maize in a sandy loam before and after washing with three chelating organic acids indicated that EDTA and citric acid appeared to offer greater potentials as chelating agents for remediating the permeable soil. Tartaric acid was, however, recommended in events of moderate contamination.

The use of soil washing to remediate contaminated fine-grained soils that contained more than 30% fines fraction has been reported by several workers [113–115]. Khodadoust et al. [59, 116] have also studied the removal of various metals (Pb, Ni, and Zn) from field and clay (kaolin) soil samples using a broad spectrum of extractants (chelating agents and organic acids). Chen and Hong [117] reported on the chelating extraction of Pb and Cu from an authentic contaminated soil using derivatives of iminodiacetic acid and L-cyestein. Wuana et al. [118] investigated the removal of Pb and Cu from kaolin and bulk clay soils using two mineral acids (HCl and H2SO4) and chelating agents (EDTA and oxalic acid). The use of chelating organic acids—citric acid, tartaric acid and EDTA in the simultaneous removal of Ni, Cu, Zn, Cd, and Pb from an experimentally contaminated sandy loam was carried out by Wuana et al. [112]. These studies furnished valuable information on the distribution of heavy metals in the soils and their removal using various extracting solutions.

1.5.3 PHYTOREMEDIATION

Phytoremediation, also called green remediation, botanoremediation, agroremediation, or vegetative remediation, can be defined as an in situ remediation strategy that uses vegetation and associated microbiota, soil amendments, and agronomic techniques to remove, contain, or render environmental contaminants harmless [119, 120]. The idea of using metal-accumulating plants to remove heavy metals and other compounds was first introduced in 1983, but the concept has actually been implemented for the past 300 years on wastewater discharges [121, 122]. Plants may break down or degrade organic pollutants or remove and stabilize metal contaminants. The methods used to phytoremediate metal contaminants are slightly different from those used to remediate sites polluted with organic contaminants. As it is a relatively new technology, phytoremediation is still mostly in its testing stages and as such has not been used in many places as a full-scale application. However, it has been tested successfully in many places around the world for many different contaminants. Phytoremediation is energy efficient, aesthetically pleasing method of remediating sites with low-to-moderate levels of contamination, and it can be used in conjunction with other more traditional remedial methods as a finishing step to the remedial process.

The advantages of phytoremediation compared with classical remediation are that (i) it is more economically viable using the same tools and supplies as agriculture, (ii) it is less disruptive to the environment and does not involve waiting for new plant communities to recolonize the site, (iii) disposal sites are not needed, (iv) it is more likely to be accepted by the public as it is more aesthetically pleasing then traditional methods, (v) it avoids excavation and transport of polluted media thus reducing the risk of spreading the contamination, and (vi) it has the potential to treat sites polluted with more than one type of pollutant. The disadvantages are as follow (i) it is dependant on the growing conditions required by the plant (i.e., climate, geology, altitude, and temperature), (ii) large-scale operations require access to agricultural equipment and knowledge, (iii) success is dependant on the tolerance of the plant to the pollutant, (iv) contaminants collected in senescing tissues may be released back into the environment in autumn, (v) contaminants may be collected in woody tissues used as fuel, (vi) time taken to remediate sites far exceeds that of other technologies, (vii) contaminant solubility may be increased leading to greater environmental damage and the possibility of leaching. Potentially useful phytoremediation technologies for remediation of heavy metal-contaminated soils include phytoextraction (phytoaccumulation), phytostabilization, and phytofiltration [123].

1.5.3.1 PHYTOEXTRACTION (PHYTOACCUMULATION)

Phytoextraction is the name given to the process where plant roots uptake metal contaminants from the soil and translocate them to their above soil tissues. A plant used for phytoremediation needs to be heavy-metal tolerant, grow rapidly with a high biomass yield per hectare, have high metal-accumulating ability in the foliar parts, have a profuse root system, and a high bioaccumulation factor [21, 124]. Phytoextraction is, no doubt, a publicly appealing (green) remediation technology [125]. Two approaches have been proposed for phytoextraction of heavy metals, namely, continuous or natural phytoextraction and chemically enhanced phytoextraction [126, 127].

Continuous or Natural Phytoextraction

Continuous phytoextraction is based on the use of natural hyperaccumulator plants with exceptional metal-accumulating capacity. Hyperaccumulators are species capable of accumulating metals at levels 100-fold greater than those typically measured in shoots of the common nonaccumulator plants. Thus, a hyperaccumulator plant will concentrate more than $10 \, mg \, kg^{-1}$ Hg, $100 \, mg \, kg^{-1}$ Cd, $1000 \, mg \, kg^{-1}$ Co, Cr, Cu, and Pb; $10 \, 000 \, mg \, kg^{-1}$ Zn and Ni [128, 129]. Hyperaccumulator plant species are used on metalliferous sites due to their tolerance of relatively high levels of pollution. Approximately 400 plant species from at least 45 plant families have been so far, reported to hyperaccumulate metals [22, 127]; some of the families are *Brassicaceae, Fabaceae, Euphorbiaceae, Asterraceae, Lamiaceae,* and *Scrophulariaceae* [130, 131]. Crops like alpine pennycress (*Thlaspi caerulescens*), Ipomea alpine, *Haumaniastrum robertii, Astragalus racemosus, Sebertia acuminate* have very high bioaccumulation potential for Cd/Zn, Cu, Co, Se, and Ni, respectively [22]. Willow (*Salix viminalis L.*), Indian mustard (*Brassica juncea L.*), corn (*Zea mays L.*), and sunflower (*Helianthus annuus L.*) have reportedly shown high uptake and tolerance to heavy metals [132]. A list of some plant hyperaccumulators are given in Table 6. A number of processes are involved during phytoextraction of metals from soil: (i) a metal fraction is sorbed at root surface, (ii) bioavailable metal moves across cellular membrane into root cells, (iii) a fraction of the metal absorbed into roots is immobilized in the vacuole, (iv) intracellular mobile metal crosses cellular membranes into root vascular tissue (xylem), and (v) metal is translocated from the root to aerial tissues (stems and leaves) [22]. Once inside the plant, most metals are too insoluble to move freely in the vascular system so they usually form carbonate, sulphate, or phosphate precipitate immobilizing them in apoplastic (extracellular) and symplastic (intracellular) compartments [46]. Hyperaccumulators have several beneficial characteristics but may tend to be slow growing and produce low biomass, and years or decades are needed to clean up contaminated sites. To overcome these shortfalls, chemically enhanced phytoextraction has been developed. The approach makes use of high biomass crops that are induced to take up large amounts of metals when their mobility in soil is enhanced by chemical treatment with chelating organic acids [133].

TABLE 6: Some metal hyperaccumulating plants [21].

Plant	Metal	Concentration (mg kg^{-1})
Dicotyledons		
Cystus ladanifer	Cd	309
	Co	2 667
	Cr	2 667
	Ni	4 164
	Zn	7 695
Thlaspi caerulescens	Cd	10 000–15 000
	Zn	10 000–15 000
Arabidopsis halleri	Cd	5 900–31 000
Alyssum sp.	Ni	4 200–24 400
Brassica junica	Pb	10 000–15 000
	Zn	2 600
Betula	Zn	528
Grasses		
Vetiveria zizaniodes		
Paspalum notatum	Zn	0.03
Stenotaphrum secundatum		
Pennisetum glaucum		

Chelate-Assisted (Induced) Phytoextraction

For more than 10 years, chelate-enhanced phytoextraction of metals from contaminated soils have received much attention as a cost-effective alternative to conventional techniques of enhanced soil remediation [133, 134]. When the chelating agent is applied to the soil, metal-chelant complexes are formed and taken up by the plant, mostly through a passive apoplastic pathway [133]. Unless the metal ion is transported as a noncationic chelate, apoplastic transport is further limited by the high cation exchange capacity of cell walls [46]. Chelators have been isolated from plants that are strongly involved in the uptake of heavy metals and their detoxification. The chelating agent EDTA has become one of the most tested mobilizing amendments for less mobile/available metals such as Pb [135, 136]. Chelators have been isolated from plants that are strongly involved in the uptake of heavy metals and their detoxification. The addition of EDTA to a Pb-contaminated soil (total soil Pb 2500 mg kg^{-1}) increased shoot lead concentration of *Zea mays L.* (corn) and *Pisun sativum* (pea) from less than 500 mg kg^{-1} to more than 10,000 mg kg^{-1}. Enhanced accumulation of metals by plant species with EDTA treatment is attributed to many factors working either singly or in combination. These factors include (i) an increase in the concentration of available metals, (ii) enhanced metal-EDTA complex movement to roots, (iii) less binding of metal-EDTA complexes with the negatively charged cell wall constituents, (iv) damage to physiological barriers in roots either due to greater concentration of metals or EDTA or metal-EDTA complexes, and (v) increased mobility of metals within the plant body when complexed with EDTA compared to free-metal ions facilitating the translocation of metals from roots to shoots [134, 137]. For the chelates tested, the order of effectiveness in increasing Pb desorption from the soil was EDTA > hydroxyethylethylene-diaminetriacetic acid (HEDTA) > diethylenetriaminepentaacetic acid (DTPA) > ethylenediamine di(o-hyroxyphenylacetic acid) EDDHA [135]. Vassil et al. [138] reported that *Brassica juncea* exposed to Pb and EDTA in hydroponic solution was able to accumulate up to 55 mM kg^{-1} Pb in dry shoot tissue (1.1% w/w). This represents a 75-fold concentration of lead in shoot over that in solution. A 0.25 mM threshold concentration of EDTA was required to stimulate this dramatic accumulation of both lead and EDTA in shoots. Since EDTA has been associated with high toxicity and persistence in the environment, several other alternatives have been proposed. Of all those, EDDS ([S,S]-ethylenediamine disuccinate) has been introduced as a promising and environmentally friendlier mobilizing agent, especially for Cu and Zn [135, 139, 140]. Once the plants have grown and absorbed the metal pollutants, they are harvested and disposed of safely. This process is repeated several times to reduce contamination to acceptable levels.

Interestingly, in the last few years, the possibility of planting metal hyperaccumulator crops over a low-grade ore body or mineralized soil, and then harvesting and incinerating the biomass to produce a commercial bio-ore has been proposed [141] though this is usually reserved for use with precious metals. This process called phytomining offers the possibility of exploiting ore bodies that are otherwise uneconomic to mine, and its effect on the environment is minimal when compared with erosion caused by opencast mining [123, 141].

Assessing the Efficiency of Phytoextraction

Depending on heavy metal concentration in the contaminated soil and the target values sought for in the remediated soil, phytoextraction may involve repeated cropping of the plant until the metal concentration drops to acceptable levels. The ability of the plant to account for the decrease in soil metal concentrations as a function of metal uptake and biomass production plays an important role in achieving regulatory acceptance. Theoretically, metal removal can be accounted for by determining metal concentration in the plant, multiplied by the reduction in soil metal concentrations [127]. It should, however, be borne in mind that this approach may be challenged by a number of factors working together during field applications. Practically, the bioaccumulation factor, f, amount of metal extracted, M (mg/kg plant) and phytoremediation time, t_0 (year) [142] can be used to evaluate the plant's phytoextraction efficiency and calculated according to equation (3) [143] by assuming that the plant can be cropped n times each year and metal pollution occurs only in the active rooting zone, that is, top soil layer (0–20 cm) and still assuming a soil bulk density of $1.3 t/m^3$, giving a total soil mass of 2600 t/ha.

f = (Metal concentration in plant shoot)/(Metal concentration in soil)

M (mg/kg plant) = Metal concentration in plant tissue x Biomass

t_p (year) = (Metal concentration in soil needed to decrease x Soil mass) /
(Metal concentation in plant shoot x Plant shoot biomass x n)   (3)

Prospects of Phytoextraction

One of the key aspects of the acceptance of phytoextraction pertains to its performance, ultimate utilization of byproducts, and its overall economic viability. Commercialization of phytoextraction has been challenged by the expectation that site remediation should be achieved in a time comparable to other clean-up technologies [123]. Genetic engineering has a great role to play in supplementing the list of plants available for phytoremediation by the use of engineering tools to insert into plants those genes that will enable the plant to metabolize a particular pollutant [144]. A major goal of plant genetic engineering is to enhance the ability of plants to metabolize many of the compounds that are of environmental concern. Currently, some laboratories are using traditional breeding techniques, others are creating protoplast-fusion hybrids, and still others are looking at the direct insertion of novel genes to enhance the metabolic capabilities of plants [144]. On the whole, phytoextraction appears a very promising technology for the removal of metal pollutants from the environment and is at present approaching commercialization.

Possible Utilization of Biomass after Phytoextraction

A serious challenge for the commercialization of phytoextraction has been the disposal of contaminated plant biomass especially in the case of repeated cropping where large tonnages of biomass may be produced. The biomass has to be stored, disposed of or utilized in an appropriate manner so as not to pose any environmental risk. The major constituents of biomass material are lignin, hemicellulose, cellulose, minerals, and ash. It possesses high moisture and volatile matter, low bulk density, and calorific value [127]. Biomass is solar energy fixed in plants in form of carbon, hydrogen, and oxygen (oxygenated hydrocarbons) with a possible general chemical formula $CH_{1.44}O_{0.66}$. Controlled combustion and gasification of biomass can yield a mixture of producer gas and/or pyro-gas which leads to the generation of thermal and electrical energy [145]. Composting and compacting can be employed as volume reduction approaches to biomass reuse [146]. Ashing of biomass can produce bio-ores especially after the phytomining of precious metals. Heavy metals such as Co, Cu, Fe, Mn, Mo, Ni, and Zn are plant essential metals, and most plants have the ability to accumulate them [147]. The high concentrations of these metals in the harvested biomass can be "diluted" to acceptable concentrations by combining the biomass with clean biomass in formulations of fertilizer and fodder.

1.5.3.2 PHYTOSTABILIZATION

Phytostabilization, also referred to as in-place inactivation, is primarily concerned with the use of certain plants to immobilize soil sediment and sludges [148]. Contaminant are absorbed and accumulated by roots, adsorbed onto the roots, or precipitated in the rhizosphere. This reduces or even prevents the mobility of the contaminants preventing migration into the groundwater or air and also reduces the bioavailability of the contaminant thus preventing spread through the food chain. Plants for use in phytostabilization should be able to (i) decrease the amount of water percolating through the soil matrix, which may result in the formation of a hazardous leachate, (ii) act as barrier to prevent direct contact with the contaminated soil, and (iii) prevent soil erosion and the distribution of the toxic metal to other areas [46]. Phytostabilization can occur through the process of sorption, precipitation, complexation, or metal valence reduction. This technique is useful for the cleanup of Pb, As, Cd, Cr, Cu, and Zn [147]. It can also be used to reestablish a plant community on sites that have been denuded due to the high levels of metal contamination. Once a community of tolerant species has been established, the potential for wind erosion (and thus spread of the pollutant) is reduced, and leaching of the soil contaminants is also reduced. Phytostabilization is advantageous because disposal of hazardous material/biomass is not required, and it is very effective when rapid immobilization is needed to preserve ground and surface waters [147, 148].

1.5.3.3 PHYTOFILTRATION

Phytofiltration is the use of plant roots (rhizofiltration) or seedlings (blastofiltration), is similar in concept to phytoextraction, but is used to absorb or adsorb pollutants, mainly metals, from groundwater and aqueous-waste streams rather than the remediation of polluted soils [3, 123]. Rhizosphere is the soil area immediately surrounding the plant root surface, typically up to a few millimetres from the root surface. The contaminants are either adsorbed onto the root surface or are absorbed by the plant roots. Plants used for rhizofiltration are not planted directly in situ but are acclimated to the pollutant first. Plants are hydroponically grown in clean water rather than soil, until a large root system has developed. Once a large root system is in place, the water supply is substituted for a polluted water supply to acclimatize the plant. After the plants become acclimatized, they are planted in the polluted area where the roots uptake the polluted water and the contaminants along with it. As the roots become saturated, they are harvested and disposed of safely. Repeated treatments of the site can reduce pollution to suitable levels as was exemplified in Chernobyl where sunflowers were grown in radioactively contaminated pools [21].

1.6 CONCLUSION

Background knowledge of the sources, chemistry, and potential risks of toxic heavy metals in contaminated soils is necessary for the selection of appropriate remedial options. Remediation of soil contaminated by heavy metals is necessary in order to reduce the associated risks, make the land resource available for agricultural production, enhance food security, and scale down land tenure problems. Immobilization, soil washing, and phytoremediation are frequently listed among the best available technologies for cleaning up heavy metal contaminated soils but have been mostly demonstrated in developed countries. These technologies are recommended for field applicability and commercialization in the developing countries also where agriculture, urbanization, and industrialization are leaving a legacy of environmental degradation.

REFERENCES

1. S. Khan, Q. Cao, Y. M. Zheng, Y. Z. Huang, and Y. G. Zhu, "Health risks of heavy metals in contaminated soils and food crops irrigated with wastewater in Beijing, China," Environmental Pollution, vol. 152, no. 3, pp. 686–692, 2008.
2. M. K. Zhang, Z. Y. Liu, and H. Wang, "Use of single extraction methods to predict bioavailability of heavy metals in polluted soils to rice," Communications in Soil Science and Plant Analysis, vol. 41, no. 7, pp. 820–831, 2010.
3. GWRTAC, "Remediation of metals-contaminated soils and groundwater," Tech. Rep. TE-97-01,, GWRTAC, Pittsburgh, Pa, USA, 1997, GWRTAC-E Series.
4. T. A. Kirpichtchikova, A. Manceau, L. Spadini, F. Panfili, M. A. Marcus, and T. Jacquet, "Speciation and solubility of heavy metals in contaminated soil using X-ray microfluorescence, EXAFS spectroscopy, chemical extraction, and thermodynamic modeling," Geochimica et Cosmochimica Acta, vol. 70, no. 9, pp. 2163–2190, 2006.

5. D. C. Adriano, Trace Elements in Terrestrial Environments: Biogeochemistry, Bioavailability and Risks of Metals, Springer, New York, NY, USA, 2nd edition, 2003.

6. P. Maslin and R. M. Maier, "Rhamnolipid-enhanced mineralization of phenanthrene in organic-metal co-contaminated soils," Bioremediation Journal, vol. 4, no. 4, pp. 295–308, 2000.

7. M. J. McLaughlin, B. A. Zarcinas, D. P. Stevens, and N. Cook, "Soil testing for heavy metals," Communications in Soil Science and Plant Analysis, vol. 31, no. 11–14, pp. 1661–1700, 2000.

8. M. J. McLaughlin, R. E. Hamon, R. G. McLaren, T. W. Speir, and S. L. Rogers, "Review: a bioavailability-based rationale for controlling metal and metalloid contamination of agricultural land in Australia and New Zealand," Australian Journal of Soil Research, vol. 38, no. 6, pp. 1037–1086, 2000.

9. W. Ling, Q. Shen, Y. Gao, X. Gu, and Z. Yang, "Use of bentonite to control the release of copper from contaminated soils," Australian Journal of Soil Research, vol. 45, no. 8, pp. 618–623, 2007.

10. A. Kabata-Pendias and H. Pendias, Trace Metals in Soils and Plants, CRC Press, Boca Raton, Fla, USA, 2nd edition, 2001.

11. Q. Zhao and J. J. Kaluarachchi, "Risk assessment at hazardous waste-contaminated sites with variability of population characteristics," Environment International, vol. 28, no. 1-2, pp. 41–53, 2002.

12. N. S. Bolan, B.G. Ko, C.W.N. Anderson, and I. Vogeler, "Solute interactions in soils in relation to bioavailability and remediation of the environment," in Proceedings of the 5th International Symposium of Interactions of Soil Minerals with Organic Components and Microorganisms, Pucón, Chile, November 2008.

13. G. M. Pierzynski, J. T. Sims, and G. F. Vance, Soils and Environmental Quality, CRC Press, London, UK, 2nd edition, 2000.

14. J. J. D'Amore, S. R. Al-Abed, K. G. Scheckel, and J. A. Ryan, "Methods for speciation of metals in soils: a review," Journal of Environmental Quality, vol. 34, no. 5, pp. 1707–1745, 2005.

15. B. J. Alloway, Heavy Metals in Soils, Blackie Academic and Professional, London, UK, 2nd edition, 1995.

16. E. Lombi and M. H. Gerzabek, "Determination of mobile heavy metal fraction in soil: results of a pot experiment with sewage sludge," Communications in Soil Science and Plant Analysis, vol. 29, no. 17-18, pp. 2545–2556, 1998.

17. G. Sposito and A. L. Page, "Cycling of metal ions in the soil environment," in Metal Ions in Biological Systems, H. Sigel, Ed., vol. 18 of Circulation of Metals in the Environment, pp. 287–332, Marcel Dekker, Inc., New York, NY, USA, 1984.

18. S. Kuo, P. E. Heilman, and A. S. Baker, "Distribution and forms of copper, zinc, cadmium, iron, and manganese in soils near a copper smelter," Soil Science, vol. 135, no. 2, pp. 101–109, 1983.

19. M. Kaasalainen and M. Yli-Halla, "Use of sequential extraction to assess metal partitioning in soils," Environmental Pollution, vol. 126, no. 2, pp. 225–233, 2003.

20. N. T. Basta, J. A. Ryan, and R. L. Chaney, "Trace element chemistry in residual-treated soil: key concepts and metal bioavailability," Journal of Environmental Quality, vol. 34, no. 1, pp. 49–63, 2005.

21. A. Scragg, Environmental Biotechnology, Oxford University Press, Oxford, UK, 2nd edition, 2006.

22. M.M. Lasat, "Phytoextraction of metals from contaminated soil: a review of plant/soil/metal interaction and assessment of pertinent agronomic issues," Journal of Hazardous Substances Research, vol. 2, pp. 1–25, 2000.

23. L. H. P. Jones and S. C. Jarvis, "The fate of heavy metals," in The Chemistry of Soil Processes, D. J. Green and M. H. B. Hayes, Eds., p. 593, John Wiley & Sons, New York, NY, USA, 1981.

24. P. H. Raven, L. R. Berg, and G. B. Johnson, Environment, Saunders College Publishing, New York, NY, USA, 2nd edition, 1998.

25. M. E. Sumner, "Beneficial use of effluents, wastes, and biosolids," Communications in Soil Science and Plant Analysis, vol. 31, no. 11–14, pp. 1701–1715, 2000.

26. R. L. Chaney and D. P. Oliver, "Sources, potential adverse effects and remediation of agricultural soil contaminants," in Contaminants and the Soil Environments in the Australia-Pacific Region, R. Naidu, Ed., pp. 323–359, Kluwer Academic Publishers, Dordrecht, The Netherlands, 1996.

27. USEPA, "A plain english guide to the EPA part 503 biosolids rule," USEPA Rep. 832/R-93/003, USEPA, Washington, DC, USA, 1994.

28. K. Weggler, M. J. McLaughlin, and R. D. Graham, "Effect of Chloride in Soil Solution on the Plant Availability of Biosolid-Borne Cadmium," Journal of Environmental Quality, vol. 33, no. 2, pp. 496–504, 2004.

29. M. L. A. Silveira, L. R. F. Alleoni, and , and L. R. G. Guilherme, "Biosolids and heavy metals in soils," Scientia Agricola, vol. 60, no. 4, pp. 64–111, 2003.

30. R. Canet, F. Pomares, F. Tarazona, and M. Estela, "Sequential fractionation and plant availability of heavy metals as affected by sewage sludge applications to soil," Communications in Soil Science and Plant Analysis, vol. 29, no. 5-6, pp. 697–716, 1998.

31. S. V. Mattigod and A. L. Page, "Assessment of metal pollution in soil," in Applied Environmental Geochemistry, pp. 355–394, Academic Press, London, UK, 1983.

32. R. G. McLaren, L. M. Clucas, and M. D. Taylor, "Leaching of macronutrients and metals from undisturbed soils treated with metal-spiked sewage sludge. 3. Distribution of residual metals," Australian Journal of Soil Research, vol. 43, no. 2, pp. 159–170, 2005.

33. C. Keller, S. P. McGrath, and S. J. Dunham, "Trace metal leaching through a soil-grassland system after sewage sludge application," Journal of Environmental Quality, vol. 31, no. 5, pp. 1550–1560, 2002.

34. R. G. McLaren, L. M. Clucas, M. D. Taylor, and T. Hendry, "Leaching of macronutrients and metals from undisturbed soils treated with metal-spiked sewage sludge. 2. Leaching of metals," Australian Journal of Soil Research, vol. 42, no. 4, pp. 459–471, 2004.

35. S. C. Reed, R. W. Crites, and E. J. Middlebrooks, Natural Systems for Waste Management and Treatment, McGraw-Hill, New York, NY, USA, 2nd edition, 1995.

36. J. Bjuhr, Trace Metals in Soils Irrigated with Waste Water in a Periurban Area Downstream Hanoi City, Vietnam, Seminar Paper, Institutionen för markvetenskap, Sveriges lantbruksuniversitet (SLU), Uppsala, Sweden, 2007.

37. P. S. DeVolder, S. L. Brown, D. Hesterberg, and K. Pandya, "Metal bioavailability and speciation in a wetland tailings repository amended with biosolids compost,

wood ash, and sulfate," Journal of Environmental Quality, vol. 32, no. 3, pp. 851–864, 2003.

38. N. T. Basta and R. Gradwohl, "Remediation of heavy metal-contaminated soil using rock phosphate," Better Crops, vol. 82, no. 4, pp. 29–31, 1998.

39. L. A. Smith, J. L. Means, A. Chen, et al., Remedial Options for Metals-Contaminated Sites, Lewis Publishers, Boca Raton, Fla, USA,, 1995.

40. USEPA, Report: recent Developments for In Situ Treatment of Metals contaminated Soils, U.S. Environmental Protection Agency, Office of Solid Waste and Emergency Response, 1996.

41. J. Shiowatana, R. G. McLaren, N. Chanmekha, and A. Samphao, "Fractionation of arsenic in soil by a continuous-flow sequential extraction method," Journal of Environmental Quality, vol. 30, no. 6, pp. 1940–1949, 2001.

42. J. Buekers, Fixation of cadmium, copper, nickel and zinc in soil: kinetics, mechanisms and its effect on metal bioavailability, Ph.D. thesis, Katholieke Universiteit Lueven, 2007, Dissertationes De Agricultura, Doctoraatsprooefschrift nr.

43. D. B. Levy, K. A. Barbarick, E. G. Siemer, and L. E. Sommers, "Distribution and partitioning of trace metals in contaminated soils near Leadville, Colorado," Journal of Environmental Quality, vol. 21, no. 2, pp. 185–195, 1992.

44. USDHHS, Toxicological profile for lead, United States Department of Health and Human Services, Atlanta, Ga, USA, 1999.

45. S.E. Manahan, Toxicological Chemistry and Biochemistry, CRC Press, Limited Liability Company (LLC), 3rd edition, 2003.

46. I. Raskin and B. D. Ensley, Phytoremediation of Toxic Metals:Using Plants to Clean Up the Environment, John Wiley & Sons, New York, NY, USA, 2000.

47. NSC, Lead Poisoning, National Safety Council, 2009, http://www.nsc.org/news_resources/Resources/Documents/Lead_Poisoning.pdf.

48. D. R. Baldwin and W. J. Marshall, "Heavy metal poisoning and its laboratory investigation," Annals of Clinical Biochemistry, vol. 36, no. 3, pp. 267–300, 1999.

49. C.J. Rosen, Lead in the home garden and urban soil environment, Communication and Educational Technology Services, University of Minnesota Extension, 2002.

50. P. Chrostowski, J. L. Durda, and K. G. Edelmann, "The use of natural processes for the control of chromium migration," Remediation, vol. 2, no. 3, pp. 341–351, 1991.

51. I. Bodek, W. J. Lyman, W. F. Reehl, and D. H. Rosenblatt, in Environmental Inorganic Chemistry: Properties, Processes and Estimation Methods, Pergamon Press, Elmsford, NY, USA, 1988.

52. B. E. Davies and L. H. P. Jones, "Micronutrients and toxic elements," in Russell's Soil Conditions and Plant Growth, A. Wild, Ed., pp. 781–814, John Wiley & Sons; Interscience, New York, NY, USA, 11th edition, 1988.

53. K. M. Greany, An assessment of heavy metal contamination in the marine sediments of Las Perlas Archipelago, Gulf of Panama, M.S. thesis, School of Life Sciences Heriot-Watt University, Edinburgh, Scotland, 2005.

54. P. G. C. Campbell, "Cadmium-A priority pollutant," Environmental Chemistry, vol. 3, no. 6, pp. 387–388, 2006.

55. VCI, Copper history/Future, Van Commodities Inc., 2011, http://trademetalfutures.com/copperhistory.html.

56. C. E. Martínez and H. L. Motto, "Solubility of lead, zinc and copper added to mineral soils," Environmental Pollution, vol. 107, no. 1, pp. 153–158, 2000.

57. J. Eriksson, A. Andersson, and R. Andersson, "The state of Swedish farmlands," Tech. Rep. 4778, Swedish Environmental Protection Agency, Stockholm, Sweden, 1997.

58. M. Pourbaix, Atlas of Electrochemical Equilibria, Pergamon Press, New York, NY, USA, 1974, Translated from French by J.A. Franklin.

59. A. P. Khodadoust, K. R. Reddy, and K. Maturi, "Removal of nickel and phenanthrene from kaolin soil using different extractants," Environmental Engineering Science, vol. 21, no. 6, pp. 691–704, 2004.

60. DPR-EGASPIN, Environmental Guidelines and Standards for the Petroleum Industry in Nigeria (EGASPIN), Department of Petroleum Resources, Lagos, Nigeria, 2002.

61. A. Tessier, P. G. C. Campbell, and M. Blsson, "Sequential extraction procedure for the speciation of particulate trace metals," Analytical Chemistry, vol. 51, no. 7, pp. 844–851, 1979.

62. A. M. Ure, PH. Quevauviller, H. Muntau, and B. Griepink, "Speciation of heavy metals in soils and sediments. An account of the improvement and harmonization of extraction techniques undertaken under the auspices of the BCR of Commission of the European Communities," International Journal of Environmental Analytical Chemistry, vol. 51, no. 1, pp. 35–151, 1993.

63. D. M. DiToro, J. D. Mahony, D. J. Hansen, K. J. Scott, A. R. Carlson, and G. T. Ankley, "Acid volatile sulfide predicts the acute toxicity of cadmium and nickel in sediments," Environmental Science and Technology, vol. 26, no. 1, pp. 96–101, 1992.

64. R. G. Riley, J. M. Zachara, and F. J. Wobber, "Chemical contaminants on DOE lands and selection of contaminated mixtures for subsurface science research," US-DOE, Energy Resource Subsurface Science Program, Washington, DC, USA, 1992.

65. NJDEP, Soil Cleanup Criteria, New Jersey Department of Environmental Protection, Proposed Cleanup Standards for Contaminated Sites, NJAC 7:26D, 1996.

66. T. A. Martin and M. V. Ruby, "Review of in situ remediation technologies for lead, zinc and cadmium in soil," Remediation, vol. 14, no. 3, pp. 35–53, 2004.

67. S. K. Gupta, T. Herren, K. Wenger, R. Krebs, and T. Hari, "In situ gentle remediation measures for heavy metal-polluted soils," in Phytoremediation of Contaminated Soil and Water, N. Terry and G. Bañuelos, Eds., pp. 303–322, Lewis Publishers, Boca Raton, Fla, USA, 2000.

68. USEPA, "Treatment technologies for site cleanup: annual status report (12th Edition)," Tech. Rep. EPA-542-R-07-012, Solid Waste and Emergency Response (5203P), Washington, DC, USA, 2007.

69. USEPA, "Recent developments for in situ treatment of metal contaminated soils," Tech. Rep. EPA-542-R-97-004, USEPA, Washington, DC, USA, 1997.

70. Y. Hashimoto, H. Matsufuru, M. Takaoka, H. Tanida, and T. Sato, "Impacts of chemical amendment and plant growth on lead speciation and enzyme activities in a shooting range soil: an X-ray absorption fine structure investigation," Journal of Environmental Quality, vol. 38, no. 4, pp. 1420–1428, 2009.

71. N. Finžgar, B. Kos, and D. Leštan, "Bioavailability and mobility of Pb after soil treatment with different remediation methods," Plant, Soil and Environment, vol. 52, no. 1, pp. 25–34, 2006.

72. J. Boisson, M. Mench, J. Vangronsveld, A. Ruttens, P. Kopponen, and T. De Koe, "Immobilization of trace metals and arsenic by different soil additives: evaluation by means of chemical extractions," Communications in Soil Science and Plant Analysis, vol. 30, no. 3-4, pp. 365–387, 1999.

73. E. Lombi, F. J. Zhao, G. Zhang et al., "In situ fixation of metals in soils using bauxite residue: chemical assessment," Environmental Pollution, vol. 118, no. 3, pp. 435–443, 2002.

74. C. O. Anoduadi, L. B. Okenwa, F. E. Okieimen, A. T. Tyowua, and E.G. Uwumarongie-Ilori, "Metal immobilization in CCA contaminated soil using laterite and termite mound soil. Evaluation by chemical fractionation," Nigerian Journal of Applied Science, vol. 27, pp. 77–87, 2009.

75. L. Q. Wang, L. Luo, Y. B Ma, D. P. Wei, and L. Hua, "In situ immobilization remediation of heavy metals-contaminated soils: a review," Chinese Journal of Applied Ecology, vol. 20, no. 5, pp. 1214–1222, 2009.

76. F. R. Evanko and D. A. Dzombak, "Remediation of metals contaminated soils and groundwater," Tech. Rep. TE-97-01, Groundwater Remediation Technologies Analysis Centre, Pittsburg, Pa, USA, 1997.

77. E. M. Fawzy, "Soil remediation using in situ immobilisation techniques," Chemistry and Ecology, vol. 24, no. 2, pp. 147–156, 2008.

78. M. Farrell, W. T. Perkins, P. J. Hobbs, G. W. Griffith, and D. L. Jones, "Migration of heavy metals in soil as influenced by compost amendments," Environmental Pollution, vol. 158, no. 1, pp. 55–64, 2010.

79. P. Bishop, D. Gress, and J. Olafsson, "Cement stabilization of heavy metals:Leaching rate assessment," in Industrial Wastes- Proceedings of the 14th Mid-Atlantic Industrial Waste Conference, Technomics, Lancaster, Pa, USA, 1982.

80. W. Shively, P. Bishop, D. Gress, and T. Brown, "Leaching tests of heavy metals stabilized with Portland cement," Journal of the Water Pollution Control Federation, vol. 58, no. 3, pp. 234–241, 1986.

81. USEPA, "Interference mechanisms in waste stabilization/solidification processes," Tech. Rep. EPA/540/A5-89/004, United States Environmental Protection Agency, Office of Research and Development, Cincinnati, Ohio, USA, 1990.

82. G. Guo, Q. Zhou, and L. Q. Ma, "Availability and assessment of fixing additives for the in situ remediation of heavy metal contaminated soils: a review," Environmental Monitoring and Assessment, vol. 116, no. 1–3, pp. 513–528, 2006.

83. J. R. Conner, Chemical Fixation and Solidification of Hazardous Wastes, Van Nostrand Reinhold, New York, NY, USA, 1990.

84. USEPA, "Stabilization/solidification of CERCLA and RCRA wastes," Tech. Rep. EPA/625/6-89/022, United States Environmental Protection Agency, Center for Environmental Research Information, Cincinnati, Ohio, USA, 1989.

85. USEPA, "International waste technologies/geo-con in situ stabilization/solidification," Tech. Rep. EPA/540/A5-89/004, United States Environmental Protection Agency, Office of Research and Development, Cincinnati, Ohio, USA, 1990.

86. B. H. Jasperse and C. R. Ryan, "Stabilization and fixation using soil mixing," in Proceedings of the ASCE Specialty Conference on Grouting, Soil Improvement, and Geosynthetics, ASCE Publications, Reston, Va, USA, 1992.

87. C. R. Ryan and A. D. Walker, "Soil mixing for soil improvement," in Proceedings of the 23rd Conference on In situ Soil Modification, Geo-Con, Inc., Louisville, Ky, USA, 1992.

88. USEPA, "Vitrification technologies for treatment of Hazardous and radioactive waste handbook," Tech. Rep. EPA/625/R-92/002, United States Environmental Protection Agency, Office of Research and Development, Washington, DC, USA, 1992.

89. J. L. Buelt and L. E. Thompson, The In situ Vitrification Integrated Program: Focusing on an Innovative Solution on Environmental Restoration Needs, Battelle Pacific Northwest Laboratory, Richland, Wash, USA, 1992.

90. A. Jang, Y. S. Choi, and I. S. Kim, "Batch and column tests for the development of an immobilization technology for toxic heavy metals in contaminated soils of closed mines," Water Science and Technology, vol. 37, no. 8, pp. 81–88, 1998.

91. G. Dermont, M. Bergeron, G. Mercier, and M. Richer-Laflèche, "Soil washing for metal removal: a review of physical/chemical technologies and field applications," Journal of Hazardous Materials, vol. 152, no. 1, pp. 1–31, 2008.

92. P. Wood, "Remediation methods for contaminated sites," in Contaminated Land and Its Reclamation, R. Hester and R. Harrison, Eds., Royal Society of Chemistry, Cambridge, UK, 1997.

93. GOC, "Site Remediation Technologies: A Reference Manual," 2003, Contaminated Sites Working Group, Government of Canada, Ontario, Canada.

94. R. W. Peters, "Chelant extraction of heavy metals from contaminated soils," Journal of Hazardous Materials, vol. 66, no. 1-2, pp. 151–210, 1999.

95. CLAIRE, "Understanding soil washing, contaminated land: applications in real environments," Tech. Rep. TB13, 2007.

96. M. Pearl and P. Wood, "Review of pilot and full scale soil washing plants," Warren Spring Laboratory Report LR 1018, Department of the Environment, AEA Technology National Environmental Technology Centre, 1994, B551 Harwell, Oxfordshire, OX11 0RA.

97. A. Gosselin, M. Blackburn, and M. Bergeron, Assessment Protocol of the applicability of ore-processing technology to Treat Contaminated Soils, Sediments and Sludges, prepared for Eco-Technology innovation Section, Eco-Technology Innovation Section, Technology Development and Demonstration Program, Environment Canada, Canada, 1999.

98. A. P. Davis and I. Singh, "Washing of zinc(II) from contaminated soil column," Journal of Environmental Engineering, vol. 121, no. 2, pp. 174–185, 1995.

99. D. Gombert, "Soil washing and radioactive contamination," Environmental Progress, vol. 13, no. 2, pp. 138–142, 1994.

100. R. S. Tejowulan and W. H. Hendershot, "Removal of trace metals from contaminated soils using EDTA incorporating resin trapping techniques," Environmental Pollution, vol. 103, no. 1, pp. 135–142, 1998.

101. USEPA, "Engineering bulletin: soil washing treatment," Tech. Rep. EPA/540/2-90/017, Office of Emergency and Remedial Response, United States Environmental Protection Agency, Washington, DC, USA, 1990.

102. A. L. Wood, D. C. Bouchard, M. L. Brusseau, and P. S. C. Rao, "Cosolvent effects on sorption and mobility of organic contaminants in soils," Chemosphere, vol. 21, no. 4-5, pp. 575–587, 1990.

103. W. Chu and K. H. Chan, "The mechanism of the surfactant-aided soil washing system for hydrophobic and partial hydrophobic organics," Science of the Total Environment, vol. 307, no. 1–3, pp. 83–92, 2003.

104. Y. Gao, J. He, W. Ling, H. Hu, and F. Liu, "Effects of organic acids on copper and cadmium desorption from contaminated soils," Environment International, vol. 29, no. 5, pp. 613–618, 2003.

105. K. Maturi and K. R. Reddy, "Extractants for the removal of mixed contaminants from soils," Soil and Sediment Contamination, vol. 17, no. 6, pp. 586–608, 2008.

106. H. Zhang, Z. Dang, L. C. Zheng, and X. Y. Yi, "Remediation of soil co-contaminated with pyrene and cadmium by growing maize (Zea mays L.)," International Journal of Environmental Science and Technology, vol. 6, no. 2, pp. 249–258, 2009.

107. J. Yu and D. Klarup, "Extraction kinetics of copper, zinc, iron, and manganese from contaminated sediment using disodium ethylenediaminetetraacetate," Water, Air, and Soil Pollution, vol. 75, no. 3-4, pp. 205–225, 1994.

108. R. Naidu and R. D. Harter, "Effect of different organic ligands on cadmium sorption by and extractability from soils," Soil Science Society of America Journal, vol. 62, no. 3, pp. 644–650, 1998.

109. J. Labanowski, F. Monna, A. Bermond et al., "Kinetic extractions to assess mobilization of Zn, Pb, Cu, and Cd in a metal-contaminated soil: EDTA vs. citrate," Environmental Pollution, vol. 152, no. 3, pp. 693–701, 2008.

110. X. Ke, P. J. Li, Q. X. Zhou, Y. Zhang, and T. H. Sun, "Removal of heavy metals from a contaminated soil using tartaric acid," Journal of Environmental Sciences, vol. 18, no. 4, pp. 727–733, 2006.

111. B. Sun, F. J. Zhao, E. Lombi, and S. P. McGrath, "Leaching of heavy metals from contaminated soils using EDTA," Environmental Pollution, vol. 113, no. 2, pp. 111–120, 2001.

112. R. A. Wuana, F. E. Okieimen, and J. A. Imborvungu, "Removal of heavy metals from a contaminated soil using organic chelating acids," International Journal of Environmental Science and Technology, vol. 7, no. 3, pp. 485–496, 2010.

113. H. Farrah and W. F. Pickering, "Extraction of heavy metal ions sorbed on clays," Water, Air, and Soil Pollution, vol. 9, no. 4, pp. 491–498, 1978.

114. B. J. W. Tuin and M. Tels, "Removing heavy metals from contaminated clay soils by extraction with hydrochloric acid, edta or hypochlorite solutions," Environmental Technology, vol. 11, no. 11, pp. 1039–1052, 1990.

115. K. R. Reddy and S. Chinthamreddy, "Comparison of extractants for removing heavy metals from contaminated clayey soils," Soil and Sediment Contamination, vol. 9, no. 5, pp. 449–462, 2000.

116. A. P. Khodadoust, K. R. Reddy, and K. Maturi, "Effect of different extraction agents on metal and organic contaminant removal from a field soil," Journal of Hazardous Materials, vol. 117, no. 1, pp. 15–24, 2005.

117. T. C. Chen and A. Hong, "Chelating extraction of lead and copper from an authentic contaminated soil using N-(2-acetamido)iminodiacetic acid and S-carboxymethyl-L-cysteine," Journal of Hazardous Materials, vol. 41, no. 2-3, pp. 147–160, 1995.

118. R. A. Wuana, F. E. Okieimen, and R. E. Ikyereve, "Removal of lead and copper from contaminated kaolin and bulk clay soils using acids and chelating agents," Journal of Chemical Society of Nigeria, vol. 33, no. 1, pp. 213–219, 2008.

119. S. D. Cunningham and D. W. Ow, "Promises and prospects of phytoremediation," Plant Physiology, vol. 110, no. 3, pp. 715–719, 1996.

120. H. S. Helmisaari, M. Salemaa, J. Derome, O. Kiikkilä, C. Uhlig, and T. M. Nieminen, "Remediation of heavy metal-contaminated forest soil using recycled organic matter and native woody plants," Journal of Environmental Quality, vol. 36, no. 4, pp. 1145–1153, 2007.

121. R. L. Chaney, M. Malik, Y. M. Li et al., "Phytoremediation of soil metals," Current Opinion in Biotechnology, vol. 8, no. 3, pp. 279–284, 1997.

122. R. J. Henry, An Overview of the Phytoremediation of Lead and Mercury, United States Environmental Protection Agency Office of Solid Waste and Emergency Response Technology Innovation office, Washington, DC, USA, 2000.

123. C. Garbisu and I. Alkorta, "Phytoextraction: a cost-effective plant-based technology for the removal of metals from the environment," Bioresource Technology, vol. 77, no. 3, pp. 229–236, 2001.

124. C. D. Jadia and M. H. Fulekar, "Phytotoxicity and remediation of heavy metals by fibrous root grass (sorghum)," Journal of Applied Biosciences, vol. 10, no. 1, pp. 491–499, 2008.

125. M. Vysloužilová, P. Tlustoš, J. Száková, and D. Pavlíková, "As, Cd, Pb and Zn uptake by Salix spp. clones grown in soils enriched by high loads of these elements," Plant, Soil and Environment, vol. 49, no. 5, pp. 191–196, 2003.

126. E. Lombi, F. J. Zhao, S. J. Dunham, and S. P. McGrath, "Phytoremediation of heavy metal-contaminated soils: natural hyperaccumulation versus chemically enhanced phytoextraction," Journal of Environmental Quality, vol. 30, no. 6, pp. 1919–1926, 2001.

127. M. Ghosh and S. P. Singh, "A review on phytoremediation of heavy metals and utilization of its byproducts," Applied Ecology and Environmental Research, vol. 3, no. 1, pp. 1–18, 2005.

128. A. J. M. Baker and R. R. Brooks, "Terrestrial higher plants which hyperaccumulate metallic elements: a review of their distribution, ecology and phytochemistry," Biorecovery, vol. 1, pp. 81–126, 1989.

129. M. M. Lasat, "Phytoextraction of toxic metals: a review of biological mechanisms," Journal of Environmental Quality, vol. 31, no. 1, pp. 109–120, 2002.

130. D. E. Salt, R. D. Smith, and I. Raskin, "Phytoremediation," Annual Reviews in Plant Physiology & Plant Molecular Biology, vol. 49, pp. 643–668, 1998.

131. S. Dushenkov, "Trends in phytoremediation of radionuclides," Plant and Soil, vol. 249, no. 1, pp. 167–175, 2003.

132. U. Schmidt, "Enhancing phytoextraction: the effect of chemical soil manipulation on mobility, plant accumulation and leaching of heavy metals," Journal of Environmental Quality, vol. 32, no. 6, pp. 1939–1954, 2003.

133. B. Nowack, R. Schulin, and B. H. Robinson, "Critical assessment of chelant-enhanced metal phytoextraction," Environmental Science and Technology, vol. 40, no. 17, pp. 5225–5232, 2006.

134. M. W. H. Evangelou, M. Ebel, and A. Schaeffer, "Chelate assisted phytoextraction of heavy metals from soil. Effect, mechanism, toxicity, and fate of chelating agents," Chemosphere, vol. 68, no. 6, pp. 989–1003, 2007.

135. J. W. Huang, J. Chen, W. R. Berti, and S. D. Cunningham, "Phytoremediadon of lead-contaminated soils: role of synthetic chelates in lead phytoextraction," Environmental Science and Technology, vol. 31, no. 3, pp. 800–805, 1997.

136. Saifullah, E. Meers, M. Qadir, et al., "EDTA-assisted Pb phytoextraction," Chemosphere, vol. 74, no. 10, pp. 1279–1291, 2009.

137. Y. Xu, N. Yamaji, R. Shen, and J. F. Ma, "Sorghum roots are inefficient in uptake of EDTA-chelated lead," Annals of Botany, vol. 99, no. 5, pp. 869–875, 2007.

138. A. D. Vassil, Y. Kapulnik, I. Raskin, and D. E. Sait, "The role of EDTA in lead transport and accumulation by Indian mustard," Plant Physiology, vol. 117, no. 2, pp. 447–453, 1998.

139. B. Kos and D. Leštan, "Chelator induced phytoextraction and in situ soil washing of Cu," Environmental Pollution, vol. 132, no. 2, pp. 333–339, 2004.

140. S. Tandy, K. Bossart, R. Mueller et al., "Extraction of heavy metals from soils using biodegradable chelating agents," Environmental Science and Technology, vol. 38, no. 3, pp. 937–944, 2004.

141. R. R. Brooks, M. F. Chambers, L. J. Nicks, and B. H. Robinson, "Phytomining," Trends in Plant Science, vol. 3, no. 9, pp. 359–362, 1998.

142. P. Zhuang, Z. H. Ye, C. Y. Lan, Z. W. Xie, and W. S. Shu, "Chemically assisted phytoextraction of heavy metal contaminated soils using three plant species," Plant and Soil, vol. 276, no. 1-2, pp. 153–162, 2005.

143. X. Zhang, H. Xia, Z. Li, P. Zhuang, and B. Gao, "Potential of four forage grasses in remediation of Cd and Zn contaminated soils," Bioresource Technology, vol. 101, no. 6, pp. 2063–2066, 2010.

144. L. A. Newman, S. E. Strand, N. Choe et al., "Uptake and biotransformation of trichloroethylene by hybrid poplars," Environmental Science and Technology, vol. 31, no. 4, pp. 1062–1067, 1997.

145. P. V. R. Iyer, T. R. Rao, and P. D. Grover, Biomass Thermochemical Characterization Characterization, Indian Institute of Technology, Delhi, India, 3rd edition, 2002.

146. M.D. Hetland, J. R. Gallagher, D. J. Daly, D. J. Hassett, and L. V. Heebink, "Processing of plants used to phytoremediate lead-contaminated sites," A. Leeson, E. A. Forte, M. K. Banks, and V. S. Magar, Eds., pp. 129–136, Batelle Press.

147. C. D. Jadia and M. H. Fulekar, "Phytoremediation of heavy metals: recent techniques," African Journal of Biotechnology, vol. 8, no. 6, pp. 921–928, 2009.

148. USEPA, "Introduction to phytoremediation," Tech. Rep. EPA 600/R-99/107, United States Environmental Protection Agency, Office of Research and Development, Cincinnati, Ohio, USA, 2000.

PART II

HEAVY METAL CONTAMINATION

CHAPTER 2

LEACHING BEHAVIOR OF HEAVY METALS AND TRANSFORMATION OF THEIR SPECIATION IN POLLUTED SOIL RECEIVING SIMULATED ACID RAIN

SHUN-AN ZHENG, XIANGQUN ZHENG, AND CHUN CHEN

2.1 INTRODUCTION

Acid rain has been a well-known environmental problem for decades and can lead to acidification of surface waters and soils. Acid deposition is formed from SO_2 and NO_x emitted to the atmosphere, largely because of fossil-fuel combustion. The most important sources are energy production, especially coal- and oil-fired power plants, and transportation sources, such as vehicles and ships. The air pollutants are transformed in the atmosphere to H_2SO_4 and HNO_3, transported across distances potentially as far as hundreds of kilometers, and deposited as precipitation (wet deposition) and as gas and particles (dry deposition) [1], [2]. Acid deposition is also an environmental problem of increasing concern in China, where acid rain is mainly distributed in the areas of Yangtze River to the south,

This chapter was originally published under the Creative Commons Attribution License. Zheng S-A, Zheng X, Chen C. Leaching Behavior of Heavy Metals and Transformation of Their Speciation in Polluted Soil Receiving Simulated Acid Rain. PLoS ONE 7,11 (2012). doi:10.1371/journal.pone.0049664..

Qinghai-Tibet Plateau to the east, and in the Sichuan Basin. About 40% of the total territory of China is affected by the acid rain [3]. Sichuan Basin is one of the most severely hit-area of acid rain in southern China, where the rapid industrialization for last few decades has caused fast growth in sulfur emissions. Based on the monitoring data for 21 cities in Sichuan Province within the State-Controlled-Network of China, the number of cities with the annual average pH value of acid rain lower than 5.6 was 19. According to the environmental protection and monitoring agencies in Sichuan Province, the direct economic loss due to acid rain is estimated to be U. S. $3 billion for one year [4].

Sichuan basin is also known as the "Red Basin" because it is mainly covered by red or purple rock series of the Trias–Cretaceous system, from which the purple soils, one of the most important soils for agricultural production in subtropical areas of China, are developed and formed. Purple soil, classified as Eutric Regosols in FAO Taxonomy or Pup-Calric-Entisol in the Chinese soil taxonomy, is typically characterized by thin soil horizons and inherited many of the characteristics of parent materials or rocks, such as its color ranging from purple to red. But other changes in purple soil properties have taken place as a result of land use changes, agricultural practices, or eco-environment disturbances [5], [6]. Due to the soil background and human activities, the soil has been severely contaminated by heavy metals (commonly including Cu, Pb, Cd and Zn) in many areas, resulting in potential risk to local human health and environment [7], [8].

There is accumulating evidence that acid rain is able to enhance metal mobilization in soil ecosystems. Previous studies demonstrated that the H^+ ion in the acidic water displaces the cations from their binding sites, reduces the cation exchange capacity (CEC), and increases the concentrations of these cations in the soil-water system. The negatively charged sulfate and nitrate ions in the acid rain can act as "counter-ions", which allow cations to be leached from the soil. Through a series of chemical reactions, cations such as K^+, Na^+, Ca^{2+} and Mg^{2+} are leached out and become unavailable to plants as nutrients [9], [10]. Likewise, toxic ions, such as Cu, Pb and Cd, usually bound to the negatively charged surface of soil particles can be displaced by H^+ ion too [11].

Leaching is the process by which contaminants are transferred from a stabilized matrix to liquid medium, such as water or other solutions.

However, the influence of acid rain on the leaching behavior of heavy metals and transformation of their speciation in polluted purple soil is not investigated in detail. The objectives of this study were to evaluate the leaching of heavy metals Cu, Pb, Cd and Zn in a contaminated purple soil affected by simulated acid rain (SAR) over a range of pH, and to identify how simulated acid rain influences the chemical speciation of these metals in purple soil. Column experiments were used in this study to provide information about element release, transport in soil, and chemistry of soil and leachates.

2.2 MATERIALS AND METHODS

2.2.1 SOIL SAMPLE COLLECTION AND ANALYSIS

The analyzed soil sample was collected from a 0 cm to 20 cm layer of agricultural purple soil in the Pengzhou Agro-ecological Station of the Chinese Academy of Agricultural Sciences in Sichuan Province, China. The study area has a subtropical humid monsoon climate with an average annual precipitation of 850 mm to 1000 mm and an average annual temperature of 15°C to 16°C. The soil samples were air-dried, ground, and passed through a 2 mm sieve. Selected soil characteristics determined by standard methods [12] were 6.27 for pH, 0.64 $g \cdot kg^{-1}$ for $CaCO_3$, 26.51 $g \cdot kg^{-1}$ for organic matter, 20.81 $g \cdot kg^{-1}$ for CEC, 265.41 $g \cdot kg^{-1}$ for clay (<0.002 mm), 51.84 $mg \cdot kg^{-1}$ for total copper, 43.92 $mg \cdot kg^{-1}$ for total lead, 0.36 $mg \cdot kg^{-1}$ for total cadmium, 121.23 $mg \cdot kg^{-1}$ for total zinc.

2.2.2 CONTAMINATED SOIL PREPARATION

Artificially contaminated soils were composed of the collected purple soil. The load quantities of Cu [$Cu(NO_3)_2 \cdot 3H_2O$], Pb [$Pb(NO_3)_2$], Cd [$Cd(NO_3)_2 \cdot 4H_2O$], and Zn [$Zn(NO_3)_2 \cdot 6H_2O$] were as follows: Cu+Pb+Cd+Zn = 400 $mg \cdot kg^{-1}$+500 $mg \cdot kg^{-1}$+1 $mg \cdot kg^{-1}$+500 $mg \cdot kg^{-1}$

Based on the levels of polluted soil defined by the Chinese Environmental Protection Agency, the selected concentrations in this study represented moderately contaminated soils (Grade III of the National Soil Heavy Metals Standards GB15618-1995). The thoroughly mixed soil samples were stored and incubated at a 75% water holding capacity for 12 mon. The chemical speciation of aged soil was then determined by the sequential extraction procedure of Tessier et al. [13]. The chemical reagents, extraction conditions, and their corresponding fractions are defined as follows:

Exchangeable fraction (EXC): 2 g of the soil sample (oven-dry weight), 16 mL 1.0 $mol \cdot L^{-1}$ $MgCl_2$, pH 7, shake vigorously in a reciprocating shaker for 1 h, 20°C.

1. Carbonate-bound fraction (CAR): 16 mL of pH 5, 1.0 $mol \cdot L^{-1}$ sodium acetate, shake vigorously in the reciprocating shaker for 5 h, 20°C.
2. The Fe/Mn oxide-bound fraction (OX): 40 mL of 0.04 $mol \cdot L^{-1}$ $NH_4OH \cdot HCl$ in 25% (v/v) acetic acid at pH 3 for 5 h at 96°C with occasional agitation.
3. The OM-bound fraction: 6 mL of 0.02 $mol \cdot L^{-1}$ HNO_3 and 10 mL of 30% H_2O_2 (pH adjusted to 2 with HNO_3), water bath, 85°C for 5 h with occasional agitation. 10 mL of 3.2 $mol \cdot L^{-1}$ NH_4OAc in 20% (v/v) HNO_3, shake vigorously in the reciprocating shaker for 30 min.
4. Residual fraction (RES): Dried in a force-air oven at 40°C, 24 h. Subsamples after sieving with 0.149 mm openings were used for determining Cu, Zn, Pb and Cd contents.

Extractions were performed in 100 mL polypropylene centrifuge tubes. Between each successive extraction, the supernatant was centrifuged at 1500×g for 30 min and then filtered using a membrane filter (0.45-μm nominal pore size).

2.2.3 SIMULATED ACID RAIN PREPARATION

Simulated acid rain was designed according to the main ion composition and pH of the local rain water (the pH of rainfall varied from 3.0 to 5.6)

[4]. Synthetic acid rain with pH values of 3.0, 4.5, and 5.6 was prepared from a stock H_2SO_4–HNO_3 solution (4:1, v/v). The concentrations of Ca^{2+}, NH^{4+}, Mg^{2+}, SO_4^{2-}, CO_3^{2+}, Cl^-, and K^+ in SAR were 1.5, 2.62, 1.00, 10.00, 2.61, 11.17, and 1.78 mg·L^{-1}, respectively.

2.2.4 LEACHING EXPERIMENT

The column (Fig. 1A) was oriented vertically and slowly saturated from the bottom with deionized water until it reached the field-holding capacity. The soil column was allowed to stabilize for 24 h. The feed solutions were composed of SAR at pH 3.0, 4.5, and 5.6, with the last group as the control. The feed solutions were then introduced into the system using a peristaltic pump to percolate through the packed soil columns at a flow rate of (60±5) mL·h^{-1}, which corresponds to the field infiltration velocity of 3.0 cm·h–1 [14] (Fig. 1B). The redox potential was measured in one-third of the columns to check the aeration of columns and to avoid waterlogging conditions. Each column was flushed with 3000 mL (1530 mm) of the incoming solution, which corresponds to 1.5 yr of precipitation (rain) in the study area. Experiments were conducted in triplicate at each pH treatment.

The leachate from the soil column was filtered through a 0.45 μm membrane filter, and 200 mL of the filtered leachate (equivalent to 100 mm of added SAR) was sampled using glass collectors to determine the heavy metal concentrations, pH, and electrical conductivity. After the tests, the columns were separated and extruded. All of the soils from the entire depth range were thoroughly mixed, dried, and ground to analyze the chemical speciation of the heavy metal residues using the sequential extraction procedure [13].

2.2.5 METAL DETERMINATION AND QUALITY CONTROL

The Cu, Zn, Pb, and Cd concentrations in soil were determined by digesting 0.5 g of the soil samples (oven-dry weight) with a HNO_3–HF–$HClO_4$ mixture followed by elemental analysis. The concentrations of these metals in all the solutions were analyzed by graphite furnace atomic absorption spectrometry (AA220Z; Varian, USA). The detection limits for Cu,

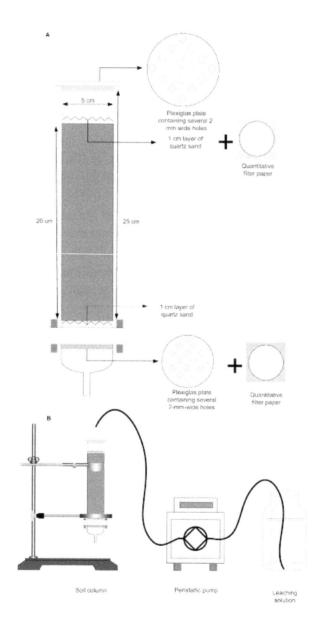

FIGURE 1: Leaching experimental design. (A) Schematic diagram of soil column. (B) Schematic analytical setup for the measurement of metal concentrations in the leaching experiment.

Zn, Pb, and Cd were 2, 2, 1, and 0.5 $\mu g \cdot L^{-1}$, respectively. All the reagents used for analysis were of analytical grade or higher. All the containers were soaked in 10% HCl, rinsed thoroughly in deionized water, and dried before use. The standard substances such as the geochemical standard reference sample soil in China (GSS-15) were used to examine the precision and accuracy of determination. The relative errors (REs) between the sum of the metal concentration in individual fractions and the measured total metal concentration in the soil samples, which ranged from −12.36% to 9.44%, were calculated to check the reliability of the sequential extraction procedure.

2.2.6 STATISTICAL ANALYSIS

Data were analyzed using the Origin 8.5.1 for Windows software at the 5% and 1% significance levels.

2.3 RESULTS AND DISCUSSION

2.3.1 HEAVY METAL CONCENTRATIONS IN THE LEACHATE

The Pb and Cd concentrations in the leachate from SAR treatments at pH 3.0, 4.5, and 5.6 were generally very low throughout the leaching period, with most of them below their detection limits (data not shown). The maximum Pb and Cd concentrations in the leachate were only (17.0±2.16) and (2.8±0.42) $\mu g \cdot L^{-1}$, respectively, which are both less than the limits of the Chinese drinking water quality standards (50 and 10 $\mu g \cdot L^{-1}$, respectively). No significant differences were observed among the treatments of SAR at different pH levels. This result suggested that the polluted purple soil, which received SAR at a given pH in our study, may not cause groundwater contamination by Pb and Cd leaching. Low leachate concentrations of Pb and Cd can probably be attributed to the low mobility of Pb in soil as well as the comparatively low levels of Cd in artificially contaminated purple soil [15], [16].

The different results were observed with Cu and Zn. The leaching concentrations of Cu and Zn under SAR conditions are shown in Fig. 2. Higher Cu and Zn leachate concentrations were observed throughout the entire leaching period (ranging from 0.021 $mg \cdot L^{-1}$ to 1.49 $mg \cdot L^{-1}$ for Cu; 0.019 $mg \cdot L^{-1}$ to 2.34 $mg \cdot L^{-1}$ for Zn) as compared with those of Pb and Cd. However, the maximum Cu and Zn concentrations were still below the Chinese Quality Standards for Groundwater (Grade IV), thereby suggesting that leaching from the polluted purple soil under acid rain at the pH used in this study is unlikely to cause Cu and Zn contamination in water systems.

Fig. 2A shows the Cu concentrations in the leachates when SAR was used as the leaching reagent. Similar trends were observed for the changes in the Cu concentration of each treatment. The application of SAR produced a pulse of Cu in the leachates during the first stage of the experiment, with peaks of 1.49, 1.11, and 0.88 $mg \cdot L^{-1}$ at a pH of 3.0, 4.5, and 5.6, respectively. The concentrations decreased thereafter and reached <0.2 $mg \cdot L^{-1}$ at the end of the experiment. A small amount of Cu (<0.18 $mg \cdot L^{-1}$) was leached from the soil column in the first 200 mm of rainfall, which may be attributed to the required time for Cu to move from the upper surface layer to the bottom of the soil column and for the Cu-bound compounds to functionalize with the soil components. The SAR treatments at pH 3.0 and 4.5 yielded higher Cu concentrations than those in the control (SAR at pH 5.6), particularly before 1000 mm of SAR was added. The Cu leachate concentration increased as the pH values of SAR decreased. However, the Cu leachate concentrations for the treatments at pH 3.0 and 4.5 were approximately equal to those in the control as the leaching amounts were increased, thereby suggesting that the leachable Cu in all of the treatments was decreased. After 1200 mm of SAR was added, no significant differences were observed between the control and the treatments at pH 3.0 and 4.5.

The changes in the Zn leachate concentrations against the leaching amounts in the SAR treatments at different pH levels are shown in Fig. 2B. The variation curves of Zn with the three treatments were generally similar. Two peak values of the Zn concentration were found during the leaching experiment. The first peak appeared after approximately 300 mm

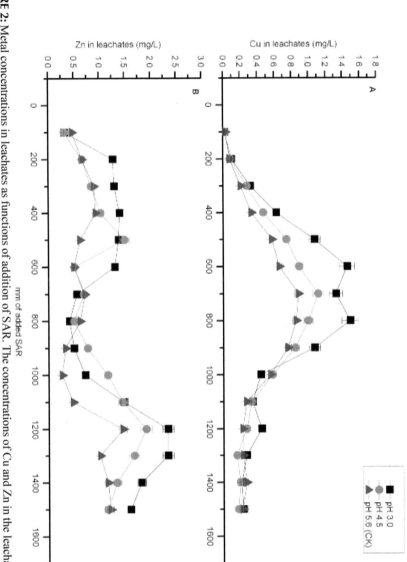

FIGURE 2: Metal concentrations in leachates as functions of addition of SAR at different pHs. (A) Copper. (B) Zinc.

of addition of SAR at different pHs. (A) Copper. (B) Zinc.

The concentrations of Cu and Zn in the leachates as functions

to 500 mm of continuous leaching, and the second one appeared after 1200 mm of SAR was added. The peak values of the Zn concentration in the leachates were 2.34, 1.92, and 1.46 mg·L^{-1} for the treatments at pH 3.0, 4.5, and 5.6, respectively. These results suggested that the release of Zn from the contaminated purple soil can be observed in two stages. In the first leaching stage, leached Zn mainly existed as water soluble and exchangeable fractions. After being leached out, these mobile Zn began to adsorb and desorb in the soil column with the downward movement of leachate, and after 300–500 mm rainfall Zn concentration in the leachate reached the first peak value. During the second stage, Zn was released from different fractions such as the CAR-, OX-, and OM-bound Cu. A comparable amount of Zn was shifted from those fractions, which revealed the second peak values after 1200 mm of rainfall was added. Many studies have demonstrated that Zn in EXC and the water soluble fraction usually account for only 0.5% to 7% of the total Zn concentration [17]–[19], which could explain why the second peak values were higher than the first ones.

2.3.2 TOTAL LEACHING AMOUNTS OF HEAVY METALS

An analysis of the total leaching amounts of heavy metals can directly reflect the leaching strength of the said heavy metals. The total leaching amounts of Pb and Cd were not calculated because most of their concentrations in the leachate were below the detection limits. The total leaching amounts of Cu and Zn in each treatment are shown in Table 1. Two-way ANOVA showed that the total amounts of the leached metals from the soil columns were significantly affected by the pH level and metal species as well as the interaction between these two factors (Table 2). Zn leached more easily from the soil column than Cu at each pH level (Tables 1 and 2). Similarly, Guo et al. [20] revealed that more Zn flowed out of the soil column than Cu in acidic soils under SAR treatments. The amount of leached metals increased at pH 3.0 and 4.5 as compared with that in the controls (pH 5.6), whereas the magnitude of this increase depended on the metal. Metallic elements become more soluble (leachable) under acidic conditions [21]. pH may likewise control the nature of the interactions between metals and soil surfaces [22]. Thus, higher Cu and Zn concentrations were induced by SAR and dissolved in columns at lower pH (Table 1). The H+ ion in acid rain displaces the cations from their binding sites, which causes the increased amount of heavy metal desorption, as shown in the following reactions [Eq. (1–2)]:

$$SOH \bullet \bullet \bullet X^2 + H^+ \leftrightarrow SOH_2 + X^2 \tag{1}$$

$$SOX^{2+} + H^+ \leftrightarrow SOH + X^{2+} \tag{2}$$

TABLE 1: Total amounts of the metals leached from the soil columns in mg of metal per kg of soil.

Treatment	Total leaching amounts (mgkg⁻¹)	
	Cu	Zn
pH 3.0	(3.93 ± 0.44) a	(7.98 ± 0.61) a
pH 4.5	(2.98 ± 0.36) b	(6.47 ± 0.47) b
pH 5.6	(2.27 ± 0.39) c	(4.72 ± 0.33) c

Note: The values are means ± standard deviation. Different lower case letters show significant differences in the same treatment (ANOVA/LSD, P<0.05).

TABLE 2: Two-way ANOVA of pH levels and metal species effects for total amounts of the metals leached from the soil columns.

Source of variance	SS	MS	F
pH	48.22	48.22	245.75**
Metal species	16.93	8.46	43.14**
pH x metal species	1.64	0.82	4.18*
Error	2.35	0.20	
Total	69.14		

*Note: **significant at 99% probability level, *significant at 95% probability level.*

The different affinities of the binding sites to metals may cause the different rates and amounts of the desorbed metal. For Zn, greater amounts were released (Table 1). H⁺ ions can only replace some metal cations, whereas the surface OH– groups attached to the Fe or Al atoms can accept protons from the solution, thereby increasing the positive surface charges via the protonation of mineral surfaces [23], [24]. Protonation is a very rapid reaction during the initial stage and depends on the amounts and properties of ferric oxides.

$$Fe\text{-}OH + H^+ \leftrightarrow FE\text{-}OH_2 \leftrightarrow dissolution(FE^{3+}) \tag{3}$$

The surface protonation in soil promotes the dissolution of ferric oxides in acidic solutions [Eq. (3)], which exhibit high adsorptive capacities for heavy metals and consequently increase the solubilization of these metals in their oxide fractions.

2.3.3 CHEMICAL SPECIATION OF HEAVY METALS IN PURPLE SOIL AFFECTED BY SAR

The chemical fractions of Cu, Pb, Cd, and Zn in soils before and after the column leaching tests are shown in Fig. 3. For the original contaminated purple soil, Cu and Cd were dominantly associated with RES (37% to 41%), followed by OM (31% to 33%). Generally, EXC- and CAR-bound Cu and Cd accounted for <9% of the total amount of the respective metals. A significant fraction of Pb in the soils was bound in RES (66.31%), and four fractions accounted for <35% of the total Pb. Some studies have shown [15], [25] that Pb is mostly present in the RES of soil at the surface or profile scale and is widely considered to exhibit very low geochemical mobility. The amount of Zn in soils was mainly associated with RES (42.69%), then with the OX (25.96%), followed by the CAR (14.44%), and OM (12.23%). The percent of exchangeable Zn in the soils was relatively low (4.69%).

Some variations in the proportions of the four metals in five fractions after the tests are shown by Fig. 3. The relative concentrations of the elements in the EXC and OX fractions were generally increased after the leaching experiment was conducted for the three treatments, whereas those of RES, OM, and CAR fractions were slightly decreased. The Pb and Cd concentrations in EXC ranged from 3.28% to 4.39% and 4.47% to

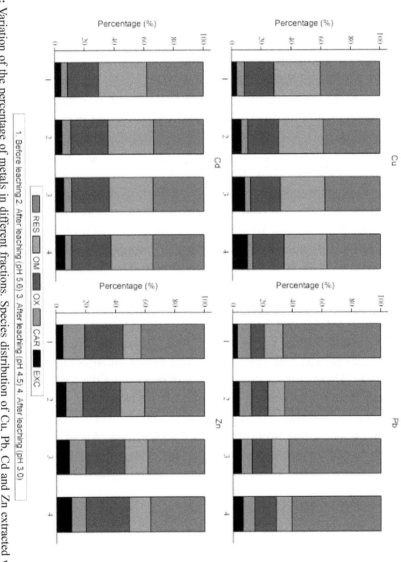

FIGURE 3: Variation of the percentage of metals in different fractions. Species distribution of Cu, Pb, Cd and Zn extracted with Tessier scheme before and after the column tests.

1. Before leaching 2. After leaching (pH 5.6) 3. After leaching (pH 4.5) 4. After leaching (pH 3.0)

6.79%, respectively. By contrast, these metals increased by more than 3% in OX. The increase in the Cu concentrations in EXC was greater than that in the OX (EXC 3% to 7% vs. OX <2%). The relative concentrations of Zn in OM and OX significantly increased, particularly in the EXC fraction (average 4.32%).

The EXC fraction was usually the first to be brought into the solution and is readily available for plant uptake. The four trace elements in EXC, with high bioavailability and mobility, were increased from ~1.2% to 5.03%, which indicated the increased direct risk of Cu, Pb, Cd, and Zn contamination in the soil/groundwater system as caused by acid rain. The amounts of the non-residual fractions represent the amounts of potentially active trace elements [25]. Generally, the high proportion of trace elements in the non-residual fractions of soils may suggest the large contributions of anthropogenic elements. The non-residual fractions of Cu, Pb, Cd, and Zn in the soil zones after the leaching tests averaged at 62.55%, 37.43%, 65.95%, and 61.61%, respectively. The non-residual fractions of Cu, Pb, Cd, and Zn in the original soil samples averaged at 59.75%, 33.69%, 62.04%, and 57.31%, respectively. The increased non-residual fractions represented the increased potential risk of Cu, Pb, Cd, and Zn contamination in the soil/groundwater system. Simple correlation analysis indicated that the pH was significantly correlated with the exchangeable heavy metal content ($r = 0.974$, $P<0.01$). Likewise, the pH was significantly correlated with the non-residual heavy metal content ($r = 0.968$, $P<0.01$). These data suggested that acid rain favors the leaching of heavy metals from the contaminated purple soil. Consequently, acid rain can affect the mobilization potential of heavy metals. The decrease in the pH from 5.6 to 3.0 significantly enhanced such effects of acid rain, which are likely to cause more serious harmful effects on soil-vegetation systems.

2.4 CONCLUSIONS

The maximum leachate concentrations of Cu, Pb, Cd, and Zn from the soil that received SAR were less than those in the Chinese Quality Standard for Groundwater (Grade IV). Moreover, most of the Pb and Cd leachate concentrations were below their detection limits. By contrast, higher Cu and Zn leachate concentrations were observed because these metals were released in higher amounts as compared with Pb and Cd. The differences in the Cu and Zn leachate concentrations between the control (SAR at pH 5.6) and the treatments (SAR at pH 3.0 and 4.5) were significant. Similar trends were observed in the total leaching amounts of Cu and Zn. The proportions of Cu, Pb, Cd, and Zn in the EXC and OX fractions were generally increased after the leaching experiment at three pH levels, whereas those of the RES, OM, and CAR fractions were slightly decreased. Acid rain favors the leaching of heavy metals from the contaminated purple soil and causes the heavy metal fractions to become more labile. A decrease in the pH from 5.6 to 3.0 significantly enhanced such effects. However, the experiment is only an indoor simulation that used artificially contaminated purple soil. Thus, long-term field experiments on soils contaminated with acid rain should be conducted to study the cycling effect of heavy metal residues in the soil or those that leach into the groundwater.

REFERENCES

1. Larssen T, Lydersen E, Tang D, He Y, Gao J, et al. (2006) Acid rain in China. Environ Sci Technol. 40: 418–425. doi: 10.1021/es0626133.
2. Singh A, Agrawal M (2008) Acid rain and its ecological consequences. J Environ. Biol. 29: 15–24.
3. Xie Z, Du Y, Zeng Y, Li Y, Yan M, et al. (2009) Effects of precipitation variation on severe acid rain in southern China. J Geogr Sci. 19: 489–501. doi: 10.1007/s11442-009-0489-y.
4. Ma L, Wang B, Yang J (2008) Spatial- Temporal Distribution of Acid Rain in Sichuan Province. Envion Sci Manage. 33: 26–29.
5. Li ZB, Li P, Han JG, Li M (2009) Sediment flow behavior in agro-watersheds of the purple soil region in China under different storm types and spatial scales. Soil Till Res. 105: 285–291. doi: 10.1016/j.still.2009.04.002.
6. Gao Y, Zhu B, Zhou P, Tang JL, Wang T, et al. (2009) Effects of vegetation cover on phosphorus loss from a hillslope cropland of purple soil under simulated rainfall: a case study in China. Nutr Cycl Agroecosys. 85: 263–273. doi: 10.1007/s10705-009-9265-8.
7. Wu F, Yang W, Zhang J, Zhou L (2010) Cadmium accumulation and growth responses of a poplar (Populus deltoids×Populus nigra) in cadmium contaminated purple soil and alluvial soil. J Hazard Mater. 177: 268–273. doi: 10.1016/j.jhazmat.2009.12.028.
8. Li Q, Wei C, Huang Y, Wang L, Wang D (2010) A study on the characteristics of heavy metals in orange ecosystem. Chin J Geochem. 29: 100–106. doi: 10.1007/s11631-010-0100-3.

9. Zhang JE, Ouyang Y, Ling DJ (2007) Impacts of simulated acid rain on cation leaching from the Latosol in south China. Chemosphere 67: 2131–2137. doi: 10.1016/j.chemosphere.2006.12.095.

10. Ling DJ, Huang QC, Ouyang Y (2010) Impacts of simulated acid rain on soil enzyme activities in a latosol. Ecotoxicol Environ Saf. 73: 1914–1918. doi: 10.1016/j.ecoenv.2010.07.024.

11. Wang M, Gu B, Ge Y, Liu Z, Jiang D, et al. (2009) Different responses of two Mosla species to potassium limitation in relation to acid rain deposition. J Zhejiang Univ-Sci B. 10: 563–571. doi: 10.1631/jzus.B0920037.

12. Lu RK (2000) Methods of soil and agrochemical analysis (in Chinese). Beijing: China Agricultural Science and Technology Press.

13. Tessier A, Campbell P, Bisson M (1979) Sequential extraction procedure for the speciation of particulate trace metals. Anal Chem. 51: 844–851. doi: 10.1021/ac50043a017.

14. Grolimund D, Borkovec M, Barmettler K, Sticher H (1996) Colloid-facilitated transport of strongly sorbing contaminants in natural porous media: a laboratory column study. Environ Sci Technol. 30: 3118–3123. doi: 10.1021/es960246x.

15. Martinez CE, Motto HL (2000) Solubility of lead, zinc and copper added to mineral soils. Environ Pollut. 107: 153–158. doi: 10.1016/S0269-7491(99)00111-6.

16. Lee SZ, Chang L, Yang HH, Chen CM, Liu MC (1998) Adsorption characteristics of lead onto soils. J Hazard Mater. 63: 37–49.

17. Kabala C, Singh BR (2001) Fractionation and mobility of copper, lead, and zinc in soil profiles in the vicinity of a copper smelter. J Environ Qual. 30: 485–492. doi: 10.2134/jeq2001.302485x.

18. Ma LQ, Rao GN (1997) Chemical fractionation of cadmium, copper, nickel, and zinc in contaminated soils. J Environ Qual. 26: 259–264. doi: 10.2134/jeq1997.00472425002600010036x.

19. Xiang HF, Tang HA, Ying QH (1995) Transformation and distribution of forms of zinc in acid, neutral and calcareous soils of China. Geoderma 66: 121–135. doi: 10.1016/0016-7061(94)00067-K.

20. Guo ZH, Liao BH, Huang CY (2005) Mobility and speciation of Cd, Cu, and Zn in two acidic soils affected by simulated acid rain. J Environ Sci. 17: 332–334.

21. Wilson MJ, Bell N (1996) Acid deposition and heavy metal mobilization. Appl Geochem. 11: 133–137. doi: 10.1016/0883-2927(95)00088-7.

22. Zhang H, Davison W, Tye AM, Crout NMJ, Young SD (2006) Kinetics of zinc and cadmium release in freshly contaminated soils. Environ Toxicol Chem. 25: 664–670. doi: 10.1897/04-664R.1.

23. Zhu MX, Jiang X, Ji GL (2005) Investigation of time-dependent reactions of H+ ions with variable and constant charge soils: a comparative study. Appl Geochem. 20: 169–178. doi: 10.1016/j.apgeochem.2004.06.003.

24. Zhang FS, Zhang XN, Yu TR (1991) Reactions of hydrogen ions with variable charge soils: I. Mechanisms of reaction. Soil Sci. 151: 436–442. doi: 10.1097/00010694-199106000-00005.

25. Tack F, Verloo MG (1995) Chemical speciation and fractionation in soil and sediment heavy metal analysis: a review. Int J Environ An Ch. 59: 225–238. doi: 10.1080/03067319508041330.

CHAPTER 3

SPATIALLY EXPLICIT ANALYSIS OF METAL TRANSFER TO BIOTA: INFLUENCE OF SOIL CONTAMINATION AND LANDSCAPE

CLÉMENTINE FRITSCH, MICHAËL COEURDASSIER, PATRICK GIRAUDOUX, FRANCIS RAOUL, FRANCIS DOUAY, DOMINIQUE RIEFFEL, ANNETTE DE VAUFLEURY, AND RENAUD SCHEIFLER

3.1 INTRODUCTION

Trace metals (TMs) are naturally present in the environment, however, soils can exhibit high levels of these persistent pollutants due to anthropogenic activities, and such contamination is recognised as a subject of concern for both organism and ecosystem health [1], [2]. Assessment of the environmental and ecological factors that may influence the transfer of pollutants in ecosystems is a key issue in ecotoxicology. However, the understanding of this phenomenon is hampered by the frequently high spatial and temporal variability of soil and landscape factors, which dramatically affects exposure pathways of receptors [3]–[6]. It is recognised that the exposure of organisms to contaminants varies spatially due to heterogeneity in the level of soil contamination, the environmental availability of the pollutant, the habitat and landscape characteristics, and some "host

This chapter was originally published under the Creative Commons Attribution License. Fritsch C, Cœurdassier M, Giraudoux P, Raoul F, Douay F, Rieffel D, de Vaufleury A, and Schiefler R. Spatially Explicit Analysis of Metal Transfer to Biota: Influence of Soil Contamination and Landscape. PLoS ONE 6,5 (2011). doi:10.1371/journal.pone.0020682.

factors" related to ecological and behavioural characteristics of the organism (habitat preferences, home range size, feeding behaviour, migratory behaviour, etc.) [6]–[10]. In heterogeneous landscapes, both the duration and the intensity of exposure may vary spatially because the time spent by the animal (due to foraging behaviour for instance) in the different patches constituting the landscape is likely to change according to land use [11]. Moreover, levels of TMs in soil can also vary between habitats or landscapes because interception and retention characteristics of certain habitats increase TM levels in soil. Additionally, differences in environmental TM availability between land uses may occur due to changes in soil characteristics, soil biodiversity and speciation of metals in soil [6], [12]–[20]. Finally, landscape patterns modulate the structure and the functioning of populations and communities [21], [22] and could therefore be an important factor governing the spatial heterogeneity of organism exposure.

Several authors have emphasised the necessity of studying factors that may affect pollutant bioavailability on the same spatial scale at which harmful effects on receptors occur (i.e., at the habitat or landscape level) [3], [4], [23]–[25]. Some authors have provided evidence for the influence of soil contamination, land use heterogeneity and home range size on organism exposure using modelling methods [10], [26]–[29]. In most cases, these studies developed individual-based random walk models and calculated the cumulative exposure of animals. An effort has also been made to introduce spatially explicit data, such as home range size, contamination heterogeneity and foraging behaviour, in the risk assessment [23], [30]–[32]. Concepts and developments for a new field in ecotoxicology referred to as "landscape ecotoxicology" were proposed in the 1990s [33], but not much research has been developed in this new field. While the influence of spatially explicit ecological variables on exposure has been explored with modelling, the effective differences in the exposure of organisms between land uses or landscapes of a specific site have rarely been assessed in the field [19]. To our knowledge, only one recent study has investigated the influence of habitat type on TM bioaccumulation, and it was found that the transfer of pollutants from soil to biota varied among habitats [19].

Land snails and small mammals have been extensively studied for use as biomonitors of TM environmental contamination due to their bioaccumulation abilities, which have shown that TM levels in tissues (referred to as "internal" levels hereafter) reflected the contamination level of sites [34]–[37]. The relationships between internal TM levels in animals and in soils have mainly been studied by comparing polluted sites to reference areas, which does not allow investigating the influence of the spatial heterogeneity of total or available soil TMs on bioaccumulation. By studying the accumulation of metals in small mammals from diffusely polluted flood plains, Wijnhoven et al. [38] investigated the relationship between TM concentrations in animals and soils sampled at the same locations. These authors found weak correlations between body and soil TM concentrations and suggested that dispersal of animals and changes in foraging and feeding behaviour due to periodic flooding influenced their exposure to contaminants. Conversely, in a recent study on TM accumulation in small mammals along a pollution gradient in a smelter-impacted area, we observed that internal TM levels were highly dependent upon local soil contamination [7]. While studying TM accumulation in wood mice, van den Brink et al. [39] showed intersite differences in soil-kidney Cd ratios, which they attributed to soil parameters that drive soil TM availability to plants and invertebrates. Such discrepancies highlight the potential influence of site-specific characteristics (for instance landscape, environmental particularities such as flooding or barriers, and soil properties) on TM transfer in food webs, and highlight the need for data on the influence of environmental characteristics on pollutant transfer in ecosystems, considering both the intensity and the spatial pattern of transfer.

Metal transfer from soil to biota is controlled in large part by TM bioavailability, which may be defined as the fraction of the total concentration of a metal that is available or can be made available for uptake and, as a consequence, for causing effects in organisms [8], [40]–[42]. TM availability is often estimated using chemical extractants; among these, calcium chloride ($CaCl_2$) is considered to be relevant for the assessment of the fraction of cadmium (Cd), lead (Pb) and zinc (Zn) that is bioavailable to some plants and invertebrates [40], [42]–[45]. Because plants and soil invertebrates may constitute the dietary items for a large variety of invertebrate and vertebrate consumers, it could be hypothesized that $CaCl_2$-extracts of TMs reflect their potential to be transferred in food webs. It is commonly admitted that organisms respond (accumulation and/or effects) to bioavailable rather than total metal concentrations in soils, and from a risk assessment perspective, it has been recommended that studies be developed based on bioavailable rather than total TM concentrations for risk estimates [45]. In the present work, we studied four sympatric species in a large polluted site surrounding a former Pb and Zn smelter. These species differed based on their phylogeny (invertebrates versus vertebrates), physiology (TM storage abilities), diet (herbivorous, omnivorous, and carnivorous), spatial mobility and habitat preferences. Small mammals have a larger spatial range of mobility than snails [10], [46]–[50]. The grove snail *Cepaea sp* is an herbivorous species [51], [52] while the glass snail *Oxychilus draparnaudi* is carnivorous [53]–[55]. The bank vole *Myodes* (ex-*Clethrionomys*) *glareolus* is herbivorous/granivorous and intermediary between strictly herbivorous voles and omnivorous mice [47], [56], [57], while the greater white-toothed shrew *Crocidura russula* is typically carnivorous [48]. We addressed three questions: (i) Are TM concentrations in animals better related to total or chemically-extractable soil TM concentrations? (ii) Does the landscape influence the relationship between TM concentrations in animals and in soils? (iii) At which spatial scale do TM concentrations in animals correlate best with soil TM concentrations? The latter question relies upon the hypothesis that the correlation will be strongest when the spatial range considered for soil contamination will approximate the surfaces exploited by the organism of concern.

3.2 MATERIALS AND METHODS

3.2.1 STUDY SITE AND SAMPLING STRATEGY

This study was conducted in the surrounding area of the former "Metaleurop-Nord" smelter in Northern France (Noyelles-Godault, Nord – Pas-de-Calais, 50°25′42 N, 3°00′55 E). The contamination of agricultural and urban soils around Metaleurop has been well documented [58]–[61], and soils of woody habitats were also recently studied [20], [62], thus showing that the area is highly polluted by Cd, Pb and Zn for both levels of contamination and surfaces of concern.

We defined a 40 km² (8×5 km) study area, which was divided into 160 squares (500×500 m), that constituted our sampling units. In each square, soils and animals were sampled in woody habitats. Habitats classified as "woody" consisted of natural forests, tree plantations (e.g., poplar groves), woodlots or copses and hedgerows in natural or cultivated lands and urban parks. Of the 160 squares, nine were excluded from the sampling; one was excluded because it corresponds to the location of the former smelter, which was undergoing rehabilitation process and the others because they were occupied by ploughed fields only and thus lacked any woody habitats. The number of squares actually sampled was therefore 151.

3.2.2 LANDSCAPE ANALYSIS

A land use analysis was performed to determine the landscape composition of each square. For this purpose, the study area was extended to 9×6 km by adding a line of squares around the initial 8×5 km grid with the aim of avoiding an edge effect in further statistical analyses. Land use mapping was accomplished using ArcGIS 8 (ESRI Co., USA) on the basis of the CORINE Land Cover (CLC) database (European Commission, 2000). The resolution of CLC is 25 ha. Based on aerial photographs (BD ORTHO® database from the Institut Géographique National, resolution of 0.5 m) and on field reconnaissance when necessary, the limits of the different units and the units smaller than 25 ha (such as hedgerows or copses) were manually digitalised in order to obtain a resolution of four meters. The resulting vector map was converted into a raster map (1 pixel = 4×4 m), which was composed of eight categories of land use: urban and industrial, ploughed field, short grass, shrub and tall grass, hedgerow and copse, forest, river and pond, and former Metaleurop Nord smelter. For each square, the number of pixels of each category of land use was computed. Seven groups of squares were identified using correspondence analysis and clustering, and these groups corresponded to seven landscape types. These types (hereafter referred to as "landscape") were named according to their land use matrix: agricultural lands, urban areas, woodlands, shrublands, mixed urban areas and agricultural lands, and mixed woodlands and grasslands. The last type was constituted of the former Metaleurop smelter itself. The four prevailing landscape types were agricultural lands, urban areas, woodlands and shrublands. These four landscape types were used to answer the second question of this work, which addressed the influence of landscape on the relationship between TM concentrations in soils and in individuals of the different species studied.

3.2.3 SOIL USE ANALYSIS

To answer the third question of this study, which addressed the scale at which TM concentrations in soils and animals correlate best, we needed to obtain soil contamination data in buffers of increasing radius size. To achieve this aim with the best accuracy, our data on "woody" soils were gathered along with data on agricultural and urban soils that were obtained from the soil database of the Equipe Sols et Environnement (LGCgE, Groupe ISA, Lille, France).

The land use map was used to build a soil use map with four types of soil use: woody soils (soils of woody patches), agricultural soils (soils of ploughed fields and grasslands) and urban soils (soils of urban areas such as gardens and parks). Dredged sediment deposits, which are extremely contaminated and do not technically

3.2.4 SOIL, SMALL MAMMAL AND SNAIL SAMPLING

In each of the 151 squares, one to 10 composite soil samples were obtained from woody patches during the autumn of 2006 [20], [62]; this sample number varied depending on logistical issues, such as the number of available woody patches and the accessibility of the patches. Each composite sample in a woody patch comprised 15 randomly placed elementary samplings, and these 15 sub-samples represented the same weight part in the composite sample. The first 25 cm of soil were sampled, and the litter layer (OL layer, which is a layer on the soil surface that accumulates little to no decomposed leaves and woody fragments) was removed. However, the humus layer (OF layer, which consisted of fragmented residues) was sampled with the top mineral soil material, consistent with the most frequently recommended protocol in Europe. A total of 262 soil samples were analysed over the 40 km² study area. Detailed data regarding soil physico-chemical parameters and contamination are published elsewhere [20], [62].

The sampling of animals, which could not have been performed on all 160 squares for logistical reasons, was conducted on 30 squares. These 30 squares were chosen in order to obtain three replicates from three levels of pollution (low, medium and high) in the four prevailing landscape types defined above (agricultural lands, urban areas, woodlands and shrublands). Because the landscape type "shrublands" was present only in the vicinity of the former smelter, we obtained only one level of pollution (high) for this type of landscape. Small mammals were captured during the autumn of 2006 using break-back traps baited with a mix of water, flour and peanut butter. Sampling authorisation was obtained from the DIrection Régionale de l'ENvironnement (DIREN) of Nord – Pas-de-Calais. In each square, 10 lines of 10 traps, each spaced three meters apart, were placed throughout the woody patches where the soils were sampled. In three squares, the available surfaces of woody patches were insufficient to place 10 lines of traps; the number of lines was therefore reduced to six or seven. The 289 trap lines were checked every morning for three consecutive days and were re-set/rebaited as necessary. The sampling effort consisted of 2820, 600, 2310 and 2940 trap-nights in agricultural lands, shrublands, urban areas and woodlands, respectively (one trap set for one night corresponds to a "trap-night"). The percentage of captures was calculated as the number of individuals trapped per 100 trap-nights. Snails were collected by hand searching in the same woody patches and at the same time as small mammals. Snails and small mammals were stored at −20°C until dissection. Each capture location (trap line for small mammals, woody patch for snails) was georeferenced using a GPS (Garmin eTrex®).

3.2.5 ANIMAL PREPARATION

Animals were identified at the species level using morphometric criteria [47], [54], [63], [64] and were dissected. Crystalline lens weighing is a standard method for estimating the relative age of small mammals and is relevant for the bank vole [65]. However, it was not used here because a significant proportion (10%) of the crystalline lenses of bank voles was not usable because they were broken by trapping and/or freezing. Moreover, to our knowledge, no validation of this age determination method has been published for the greater white-toothed shrew. Therefore, we decided to classify the small mammals (shrews as well as voles) into three classes of relative age (juveniles, non-reproductive adults and reproductive adults) on the basis of body size, body weight and reproductive status [7]. We found significant differences in crystalline weight between these three classes, suggesting that our age classification was acceptable.

Small mammals were dissected to sample the liver, which was dried in a 60°C oven to achieve a constant dry weight before acid digestion and TM analyses (see next paragraph). Snails of the *Cepaea* genus (*C. hortensis* and *C. nemoralis*) were grouped and will be referred to hereafter as *Cepaea sp* because specific determination is not possible for juveniles. Furthermore, preliminary analyses of our data showed that TM accumulation did not differ between adult *C. hortensis* and *C. nemoralis. Cepaea sp* snails were classified according to two classes of relative age (juveniles and adults) based on the presence of a clear lip at the mouth of their shell. The presence of this lip indicates that the snail has attained adulthood [66]. The age of *Oxychilus draparnaudi* specimens could not be determined because no published method was available. The soft bodies of the snails were separated from the shells and dried in a 60°C oven to achieve a constant dry weight before TM analysis. Three to five individuals of *Oxychilus draparnaudi* were combined to achieve a sufficient sample weight for TM analysis.

3.2.6 ANALYSES OF TM CONCENTRATIONS IN SOILS AND ANIMALS

Woody soils were analysed for total and $CaCl_2$-extractable metal concentrations. Samples were dried, disaggregated and homogenised before being sifted using a 250 μm mesh. Cd and Pb concentrations were measured with inductively-coupled argon plasma mass spectrometry (ICP-MS), and Zn concentrations were measured with inductively-coupled argon plasma atomic emission spectrometry (ICP-AES) after acid digestion according to the NF X31-147 standard procedure [67]. Measurements were performed by the Laboratoire d'Analyse des Sols of the Institut National de la Recherche Agronomique (INRA) of Arras (France), which benefits from the COFRAC accreditation n°1–1380 for the analytical quality of its TM measurements in soils. All precautions were taken with respect to protocol application and calibration; quality control was achieved using procedural blanks, Certified Reference Materials (CRM, namely: BCR 141 and 142; GBW 07401, 07402, 07404, 07405 and 07406), samples from inter-laboratory comparisons, internal control samples and duplicates of the analysis. Selective extractions using $CaCl_2$ (0.01 M) were conducted in triplicate on 3 g sub-samples of soil. Extractable metal concentrations were quantified using atomic absorption spectrometry (AAS, AA-6800, Shimadzu) by the Equipe Sols et Environnement (LGCgE, Groupe ISA, Lille, France). Quality control was performed on these measurements, using procedural blanks and CRM (BCR 483). The average recoveries of the CRM varied between 90% and 110%.

Metal concentrations in snail soft bodies and small mammal livers were measured using furnace (Cd, Pb) or flame (Zn) atomic absorption spectrometry (VARIAN 220Z and 220FS, respectively) at the Chrono-Environment department (UMR 6249 UFC/CNRS, Besançon, France). Digestion of samples was performed by dissolution in nitric acid (HNO_3, 65%, Carlo Erba analytical quality) in a dry oven (65°C) for 72 h. After digestion, samples were diluted by the addition of ultra-pure water (18.2 $M\Omega/cm^2$). Blanks (acid + ultra-pure water) and CRM (TORT-2 and DOLT-3, National Research Council, Canada) were prepared and analysed using the same methods as the samples. Average recoveries of the CRM were calculated at 95%±10% (n = 38) for Cd, 101%±17% (n = 42) for Pb, 80%±2% (n = 22) for Zn. Detection limits of the spectrometers (median±3 SD of blanks) were 0.17, 1.2 and 2.8 $\mu g.l^{-1}$ in the acid digests for Cd, Pb and Zn, respectively. Detection limits in snail soft bodies were 0.02, 0.12 and 0.29 $\mu g.g^{-1}$ in the grove snail and 0.09, 0.65 and 1.5 $\mu g.g^{-1}$ in the glass snail for Cd, Pb and Zn, respectively. Detection limits in small mammal livers were 0.02, 0.13 and 0.30 $\mu g.g^{-1}$ in the bank vole and 0.03, 0.24 and 0.56 $\mu g.g^{-1}$ in the greater white-toothed shrew for Cd, Pb and Zn, respectively. When measured values were under the detection limits, half of the detection limit value was used for statistical analyses. For both soil and animal samples, metal concentrations were expressed as microgram per gram dry weight ($\mu g.g^{-1}$ DW). Hereafter, body and hepatic TM concentrations of snails and small mammals will be referred to as "internal TM concentrations".

3.2.7 PREDICTION OF TM CONCENTRATIONS IN SOILS

To determine at which spatial scale the correlation between TM concentrations in animals and in soils was the best, a circular buffer technique was used in order to calculate soil metal concentration in the area surrounding the animal capture location. Several buffer radiuses were used: 50, 75, 100, 175, 250, 350, 500, 750 and 1000 meters around the sampling point. For each snail or small mammal, at each buffer size, median predicted Cd, Pb and Zn concentrations in the soil were computed. For this purpose, metal concentrations in soils were predicted (universal kriging) throughout the whole 54 km^2 study area at each node of a regular (100 m) grid. .

3.2.8 STATISTICAL ANALYSES OF TM CONCENTRATIONS IN ANIMALS

For each sampled individual, the following data were available: internal TM concentration (i.e., TM concentration in the liver for the mammals and in the whole soft body for the snails), total and $CaCl_2$-extractable TM concentration in the soil of the patch where the individual as was caught (referred to hereafter as TM concentration "at sampling point"), species, age (except for *Oxychilus draparnaudi*), landscape where the animal was found (i.e., landscape type of the square, as defined in the section "Landscape analysis") and median predicted total Cd, Pb and Zn concentrations in soils at several buffer sizes (as defined in the previous section). The normality of the data distribution was checked with the test of Shapiro-Wilk. Because metal concentrations in soils and animals were skewed, variables were log-transformed for statistical analyses using $\log_{10}(x+1)$. The analysis of the data was divided into three steps to address the three aims of this paper. Statistical analyses were performed using general linear models (LMs, [68]).

The purpose of the first step was to test whether animal and soil TM concentrations at the sampling point were related and whether total or $CaCl_2$-extractable soil TM concentrations best explained TM concentrations in animals. As age can influence TM accumulation in small mammals and snails [7], [34], [69], [70], we modelled the relationship between internal and soil TM concentrations controlling for age; residuals of the internal concentration/age relationship (which will be referred to hereafter as "internal concentration normalized to age") were modelled against total or $CaCl_2$-extractable TM concentrations in soils. On the basis of the R^2 of these models, we determined which soil TM concentration (total or $CaCl_2$-extractable) better explained the internal TM concentrations. Subsequently, the best variable was kept for further analyses. We analysed the differences in accumulation between species using LMs with internal metal as the dependent variable and total soil TM concentrations and species with interactions as explanatory variables. Comparisons of accumulation level between species were therefore performed conditionally to total TM concentrations in soils (i.e., placing the variable "total soil TM concentration" first in the model).

In the second step, we considered the influence of landscape on internal TM concentration by adding the variable "landscape" with interaction into the models. Comparisons of accumulation level between landscape types were therefore performed conditionally to age and total soil TM concentrations (i.e., placing these variables first in the model).

Finally, we determined at which spatial scale the animal and soil TM concentrations were most strongly correlated. We hypothesized that this correlation will be strongest when the buffer size approximates the surface exploited by the organism of concern. For this purpose, LMs were used with internal TM concentrations normalized to age as the dependent variable and median predicted total soil TM concentrations in buffers around sampled animals as independent variables. For each species, the R^2 of the models and the values of the coefficients (slopes) were computed for each buffer size, both on the whole dataset and by landscape. Because the sample size was small for the shrew in woodlands (n = 4), the grove snail in urban areas (n = 3) and null for snails in shrublands (no specimens), these landscape types were not analysed individually for these species. For each buffer radius size, we recorded the highest and lowest of the calculated partial R^2 values for soil concentrations and computed the differences between the values, which will be referred to hereafter as "ΔR^2". If the value of ΔR^2 was lower than 0.05, we assumed that the R^2 did not show straight-forward variations with buffer size increase.

In all cases, the significance of the variables was checked via permutation test (Monte-Carlo, 1000 permutations), the partial R^2 values were calculated using an analysis of variance (ANOVA). Pairwise differences were determined using the Tukey's post-hoc multiple comparison test (Tukey's "Honest Significant Difference" method).

All statistical analyses were performed using the software R 2.7.1 [71] with the following additional packages: ade4, geoR, gstat, maptools, pgirmess, pvclust, spdep, splancs, and vegan.

TABLE 1: Total and CaCl$_2$-extractable trace metal concentrations in soils of woody habitats from Metaleurop-impacted area.

Metal	Min.	Med.	Max.	DSD
Cd	0.10	5.0	236	2,402
Pb	16	303	7,331	41,960
Zn	44	460	7,264	38,760
Cd-CaCl$_2$	0.02	0.20	5.8	59
Pb-CaCl$_2$	0.27	0.66	14	1.7
Zn-CaCl$_2$	0.02	4.1	143	112

Min.: minimum, Med.: median, Max.: maximum. Concentrations in woody soils (n = 261) and dredged sediment deposit (DSD, n =1). Concentrations expressed as μg.g−1 DW.

3.3 RESULTS

Total soil TM concentrations varied from background levels to very high values, particularly in the dredged material deposit (Table 1). CaCl$_2$-extractable TM concentrations also varied widely within woody soils (Table 1). Spatial distribution and predictions of TM concentrations in soils showed a similar spatial pattern for the three TMs. Concentric circles were found around the former smelter with an enhancement of the contamination in downwind areas (Figure 1). Hot spots corresponding to dredged sediment deposits were also found within the study area (Figure 1).

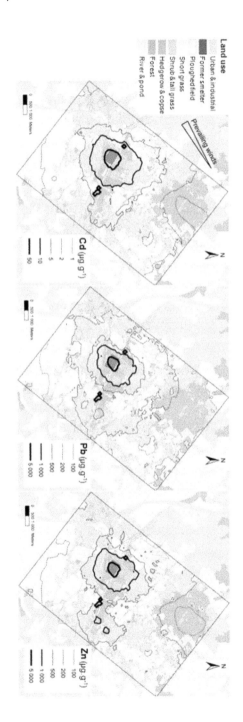

FIGURE 1: Iso-concentration lines of predicted total Cd, Pb and Zn concentrations in topsoils from Metaleurop-impacted area. Concentrations are expressed as µg.g⁻¹ DW.

A total of 131 grove snails (*Cepaea sp.*), 78 glass snails (*Oxychilus draparnaudi*), 248 bank voles (*Myodes* ex-*Clethrionomys glareolus*) and 163 greater white-toothed shrews (*Crocidura russula*) were captured (Table 2). Even if these four species were present over the whole area, the species were not evenly distributed among landscapes and pollution levels (Table 2). Neither of the two snail species selected were found in the shrubland landscape. Grove snails were abundant in both agricultural and woodland landscapes but rare in urban areas. The glass snail was often present in urban areas but less often in woodlands, and it was rarely found in agricultural areas. The bank vole was clearly more abundant in woodlands than in the three other landscapes, where it was roughly similar in abundance. Greater white-toothed shrews were equally abundant in agricultural and urban areas and were rare in both shrubland and woodland landscapes.

TABLE 2: Distribution of specimens by species, age and landscape and percentage of capture for small mammals.

Species	Age	Landscape				Total
		Agricultural lands	Shrublands	Urban areas	Woodlands	
Cepaea sp	Juv	9	0	2	19	30
	Ad	43	0	1	57	101
	Total	52	0	3	76	131
Oxychilus draparnaudi	Total	6	0	50	22	78
Myodes glareolus	Juv	7	10	1	54	72
	NR ad	2	7	9	44	62
	Ad	8	4	4	98	114
	Total	17	21	14	196	248
	PC	0.6	3.5	0.6	6.7	2.9
Crocidura russula	Juv	10	0	13	0	23
	NR ad	50	9	56	4	119
	Ad	15	0	6	0	21
	Total	75	9	75	4	163
	PC	2.7	1.5	3.2	0.1	1.9

Juv: juveniles, NR ad: non-reproductive adults and Ad: reproductive adults, PC: number of individuals trapped per 100 trap-nights.

3.3.1 INFLUENCE OF SPECIES, AGE AND TMS IN SOILS AT THE SAMPLING POINT ON INTERNAL TM CONCENTRATIONS

Considering all the species together, internal concentrations showed a large range of variation; they varied from under detection limits for Cd and Pb and 0.88 µg.g^{-1} DW for Zn to 741, 443 and 6619 for Cd in the greater white-toothed shrew, Pb in the grove snail and Zn in the glass snail, respectively (Table 3).

TABLE 3: Trace metal concentrations measured in snails (soft body) and small mammals (liver) from Metaleurop-impacted area.

Species	n	Cd			Pb			Zn		
		Min	Med	Max	Min	Med	Max	Min	Med	Max
Cepaea sp	131	UDL	42	170	1.0	23	443	0.88	447	1,927
Oxychilus draparnaudi	78	UDL	55	258	10	55	313	206	91	6,619
Myodes glareolus	243	0.13	11	116	0.08	1.8	200	19	89	173
Crocidura russula	163	3.0	72	741	UDL	8.8	159	65	139	271

Min.: minimum, Med.: median, Max.: maximum, UDL: under detection limit. Concentrations expressed as µg.g−1 DW.

Internal Cd concentrations, conditionally to soil total concentrations (i.e., placing the variable "total soil TM concentration" first in the model), ranked in the following order: greater white-toothed shrew ~ glass snail > bank vole > grove snail (p<0.001, Figure 2). Regarding Pb, internal concentrations, conditionally to total soil concentrations, were higher in small mammals than in snails and ranked in the following order: greater white-toothed shrew > bank vole ~ glass snail > grove snail (p<0.001), while for Zn the rankings were as follows: glass snail > grove snail > greater white-toothed shrew ~ bank vole (p<0.001).

Hepatic Cd concentrations slightly increased with age in small mammals (p<0.001, R^2 = 0.07 for the vole; p = 0.002, R^2 = 0.06 for the shrew, Table 4). Similarly, adult grove snails exhibited higher levels than juveniles (p<0.001, R^2 = 0.22). Pb concentrations, however, did not increase

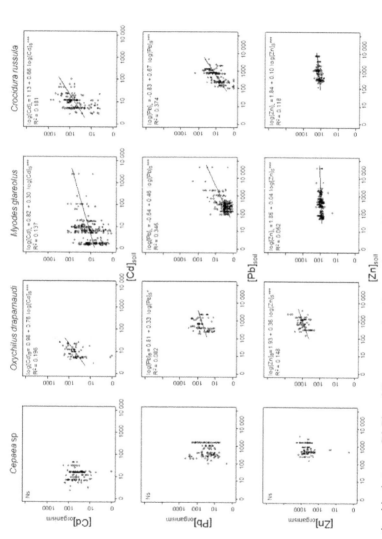

FIGURE 2: Relationships between Cd, Pb and Zn concentrations in organisms and in soils. TM concentrations in the soft body for snails ([C] B) and in the liver for small mammals ([C]L), total TM concentrations measured in soil at the sampling point ([C]s) (μg.g^{-1} DW). Statistical significance: general linear model, Ns for $p>0.05$, * for $p<0.05$, ** for $p<0.01$, *** for $p<0.001$, R^2: R-squared value of the model.

with age in small mammals although they differed with age in the grove snail, with juveniles exhibiting slightly higher concentrations than adults (p = 0.045, R2 = 0.03). Zn concentrations weakly increased with age in the bank vole (p<0.001, R^2 = 0.03) but did not vary for the shrew or for the grove snail.

TABLE 4: Multiple and partial R-squared for significant parameters of models linking internal TMs with studied variables.

Species	$[TM]_{internal}$	Partial R^2				R^2 Model
		Age	$log[TM]_{soil}$	Landscape	$[TM]_{soil}$*Landscape	
Cepae sp	log[Cd]	0.22			0.08	0.30
	log[Pb]	0.03		0.16		0.21
	log[Zn]					ns*
Oxychilus draparnaudi	log[Cd]		0.20		0.09	0.29
	log[Pb]		0.08	0.12	0.12	0.33
	log[Zn]		0.15	0.06	0.17	0.38
Myodes glareolus	log[Cd]	0.07	0.19	0.11	0.02	0.39
	log[Pb]		0.35	0.08	0.02	0.46
	log[Zn]	0.03	0.08	0.07		0.18
Crocidura russula	log[Cd]	0.06	0.17	0.07		0.31
	log[Pb]		0.37	0.09	0.03	0.50
	log[Zn]		0.12	0.05		0.19

*R^2 model: multiple R-squared, Partial R^2: partial R-squared, $[TM]_{internal}$: TM concentrations in animals, $[TM]_{soil}$: TM concentrations in soils at sampling location, $[TM]_{soil}$*Landscape: interaction between TM concentrations in soils and landscape, significance of paramters: p-value<0.05. *ns: no significant p-value>0.05*

Internal TM concentrations increased with total soil concentrations at the sampling points in the glass snail and the small mammals, although the grove snail did not show any increase in internal concentrations along the pollution gradient, regardless of the metal considered (Figure 2). Total TM concentrations explained 7 to 37% of the variation in internal concentration. Internal concentrations were always better correlated with total rather than $CaCl_2$-extractable TMs, except for the grove snail, for which internal concentrations were not related to either total or extractable TM concentrations (except for Cd-$CaCl_2$). Internal concentrations were not correlated with extractable TMs in the glass snail. Similarly, Zn in the bank vole and Pb in the greater white-toothed shrew were not significantly related to Zn-$CaCl_2$ and Pb-$CaCl_2$, respectively.

3.3.2 INFLUENCE OF LANDSCAPE ON INTERNAL TM CONCENTRATIONS

Accumulation of metals differed between landscapes for both the levels of internal TM (internal concentration normalized to age and conditionally to total soil TM concentration at the sampling point) and the evolution of internal TM concentrations along the soil pollution gradient, i.e., the slopes of the regressions between internal TM concentrations (normalized to age) and total soil TM concentrations at the sampling points (Figure 3). The variable "landscape" accounted for 5 to 16% of the variation in internal TM concentrations, and the interaction between soil contamination and landscape was sometimes significant, particularly for the glass snail (Table 4).

For *Cepaea* snails in urban areas, the sample size was insufficient and did not allow for reliable analysis. Internal Pb levels were higher in woodlands than in agricultural lands ($p<0.001$), while internal Cd and Zn levels did not differ among landscapes (Figure 3). Internal TM concentrations did not vary with soil TM concentrations, except in the case of Cd in woodlands where a negative correlation was revealed (Figure 3).

For sample size reasons, the influence of landscape on TM accumulation in *O. draparnaudi* could only be studied by comparing urban areas and woodlands. Internal Cd and Zn levels did not differ between landscapes, while internal Pb levels were higher in woodlands than in urban areas ($p = 0.003$, Figure 3). Internal Cd concentrations increased with soil TM concentrations in both urban areas and woodlands (without significant

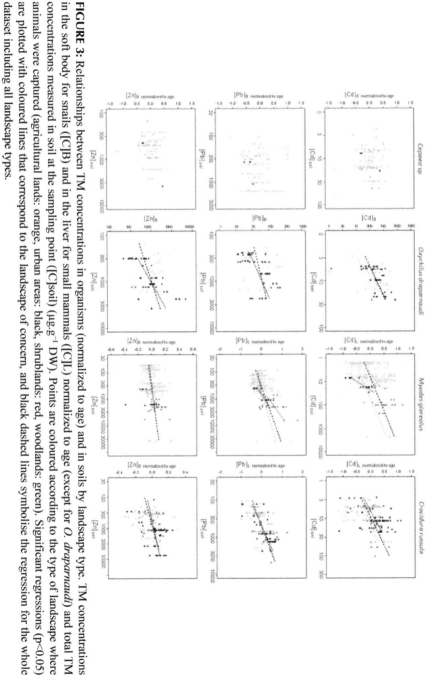

FIGURE 3: Relationships between TM concentrations in organisms (normalized to age) and in soils by landscape type. TM concentrations in the soft body for snails ([C]B) and in the liver for small mammals ([C]L) normalized to age (except for *O. draparnaudi*) and total TM concentrations measured in soil at the sampling point ([C]soil) (µg·g⁻¹ DW). Points are coloured according to the type of landscape where animals were captured (agricultural lands: orange, urban areas: black, shrublands: red, woodlands: green). Significant regressions (p<0.05) are plotted with coloured lines that correspond to the landscape of concern, and black dashed lines symbolise the regression for the whole dataset including all landscape types.

differences in slopes between landscapes), whereas internal Pb and Zn concentrations significantly increased along the pollution gradient in urban areas only (Figure 3).

For the greater white-toothed shrew, too few animals were collected on woodland and shrubland landscapes and thus no interpretation could be made. However, data from agricultural lands and urban areas showed that the level of internal Cd was higher in urban than in agricultural landscapes (p = 0.014, Figure 3), while the slopes of accumulation along the pollution gradient did not differ significantly. There were no differences for Pb and Zn in either the level of accumulation or in the increase of internal concentrations along the gradient (Figure 3).

Concerning the bank vole, we found higher levels of internal TMs in shrublands and woodlands compared to agricultural lands and urban areas for Cd (p<0.004) and urban areas for Zn (p<0.003, Figure 3). Internal Pb levels were higher in shrublands than in other landscape types (p<0.01). We did not detect a correlation between internal and soil Cd concentrations in shrublands, while regressions were significant in other landscapes (Figure 3). Although the increase in internal Cd concentrations with soil contamination was highest in urban areas and lowest in agricultural lands, the coefficients did not differ significantly between landscape. In contrast, we found a higher increase in internal Pb along the soil pollution gradient in shrublands compared to woodlands and agricultural areas (p<0.040, Figure 3). The relationships between internal and soil Zn concentrations were rarely significant and did not exhibit differences in slopes between landscapes (Figure 3).

Different patterns could be noticed between non-essential (Cd, Pb) and essential (Zn) metals in small mammals. First, most of the regressions were significant for non-essential metals, in contrast to what was observed for Zn. Second, the increase in internal concentrations along the pollution gradient was higher for non-essential metals (the lowest regression coefficient was 0.22 for Cd and Pb whereas the highest for Zn was 0.10). Finally, inter-individual variability was lower for Zn.

Several patterns linking internal and soil TM concentrations existed, showing various combinations of relatively high/low internal TM levels with sharp/slight increases in internal TMs along the pollution gradient. The relationships between internal and soil TM concentrations varied among landscapes, but these differences were metal-specific and species-specific, thus hampering any generalization (Figure 3, Table 4).

The partial R^2 for the landscape variable was found to be somewhat less important than soil contamination but as important as age for explaining internal TM concentrations (Table 4). The models gathering all the variables (total soil TM concentration, age (except for glass snail) and landscape) explained up to 39% of the variance of the dataset for Cd, 50% for Pb and 38% for Zn (Table 4). Therefore, if the studied variables significantly accounted for internal TM concentrations in snails and small mammals, a part of the inter-individual variability remained unexplained.

3.3.3 SPATIAL RANGE OF CORRELATION BETWEEN ANIMAL AND SOIL TM CONCENTRATIONS

Globally, TM concentrations in animals were related to soil TM concentrations in buffers, except for the grove snail (for which few relations were significant) and for Zn, which also showed few significant regressions. The number of significant relationships slightly increased from 27 when considering concentrations in soil at the sampling point to 31 (data not shown) when considering concentrations in soil at different buffer sizes. Figures 4 and 5 show the R^2 values of the regressions between internal concentrations (normalized to age) and concentrations in soils at sampling points, as well as the evolution of the R^2 values of the models, taking into account the TM concentration in soils at different buffer sizes.

We found four major patterns in the relationships between internal TM concentrations and soil TM concentrations in buffers (Figures 4 and 5). The first pattern is represented by better correlations between internal TM and soil TM concentrations at the sampling point or in buffers of small sizes (<100 m) compared to buffers of larger sizes. This pattern is illustrated by the case of Cd and Pb in bank voles from agricultural lands. A second pattern showed better correlations at the lowest buffer size (50 m) than at the sampling point, followed by a decrease in R^2 with increasing buffer size. Notably, this pattern can be seen in urban areas, for Cd and Pb in the bank vole, the shrew and the glass snail. However, in this case, the R^2 values for Cd in bank voles and glass snails were almost equal using

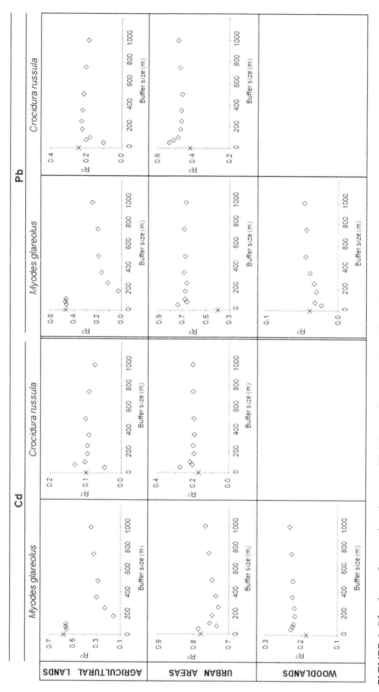

FIGURE 4: R^2 values of regressions between TMs in small mammals and in soils at several buffer sizes. Partial R^2 values for soil Cd and Pb concentrations in significant regressions between TM concentrations (normalized to age) in small mammals and total soil TM concentrations using measured values at the sampling point ($\times$) and predicted values at several buffer sizes in agricultural lands, urban areas and woodlands.

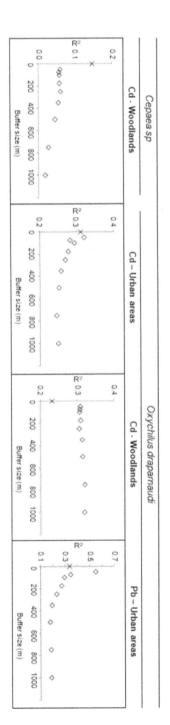

FIGURE 5: R^2 values of regressions between TMs in snails and in soils at several buffer sizes. Partial R^2 values for soil Cd and Pb concentrations in significant regressions between TM concentrations in snails (normalized to age for *Cepaea*) and soil TM concentrations using measured values at the sampling point (×) and predicted values at several buffer sizes in urban areas and woodlands.

TMs at the sampling point and in the 50 m buffer. A third pattern exhibited a better correlation using TM concentrations in buffers rather than at the sampling point, with the R^2 value remaining stable regardless of the buffer size. This is the case for Cd in the bank vole and in the glass snail from woodlands. Finally, a fourth pattern showed the lowest R^2 at a 50 m buffer size, but exhibited roughly similar R^2 values for TMs at the sampling point and in buffers from 75 to 1000 m. This pattern can be seen for Cd and Pb in shrews from agricultural lands. For the grove snail, calculations could be performed for Cd in woodlands only, and showed that internal concentrations were better correlated using Cd concentrations in soils at the sampling point rather than in buffers, albeit with a negative correlation.

As for the landscape effect, spatial patterns and correlations between animal and soil TM concentrations were found to be metal-dependent, species-dependent and landscape-dependent.

3.4 DISCUSSION

3.4.1 RELATIONSHIPS BETWEEN INTERNAL AND SOIL TM CONCENTRATIONS

The extensive, high-level TM contamination in soils surrounding the former Metaleurop smelter was reflected by the high metal levels measured in sampled organisms, which appeared among the highest reported in the literature for the various species studied [34], [36], [38], [69], [70], [72]–[75].

In contrast to the study of Notten et al. [74] on *Cepaea* snails in the Biesbosch floodplain, we did not observe an increase of TMs in soft bodies with total soil TMs, which could be due to a seasonal effect [70]. Differences in bioaccumulation between seasons may be due to several factors, most notably to variations in snail physiology, snail behaviour and/or contamination of their diet. Our study was conducted in autumn while the study of Notten et al. [74] was performed during the summer, and indeed, *Cepaea* snails sampled in the field or exposed in microcosms in the surroundings of Metaleurop during springtime exhibited an increase of internal TM concentrations along the pollution gradient (unpublished data and [76]). Previous field and laboratory studies have also shown that internal TM concentrations in *Cepaea* snails were better related to diet than to soil TM concentrations [74], [77]. The only significant relationship between TM concentrations in *Cepaea* snails and in soils was observed for $CaCl_2$-extractable Cd. This could be due to the fact that $CaCl_2$, which is usually a good extractant for the estimation of phytoavailable Cd concentrations [43], [44], may have estimated the concentration of their vegetal diet relatively well. This was not the case for the glass snail, which has a carnivorous diet and had internal TM concentrations that were only related to total soil TMs. *Cepaea* snails have been shown to excrete both Pb and Zn [70], [76]. Their ability to excrete these metals may explain the lack of a relationship between internal and soil concentrations. Oxychilus snails exhibited a stronger increase of internal Zn along the pollution gradient compared to *Cepaea* snails, suggesting the existence of different Zn regulation strategies among these taxa.

Recent research has been devoted to the study of the accumulation of TMs in small mammals in polluted sites, taking into account both total and extractable TM concentrations in soils In contrast to our results, in a study in Dutch floodplains, no correlation between internal Cd, Pb or Zn in and total concentrations in soils were found for both the bank vole or the greater white-toothed shrew [38], while correlations between internal Cd and soil Cd-$CaCl_2$ for the greater white-toothed shrew and between internal Zn and soil Zn-$CaCl_2$ for the bank vole were observed. In the same environmental context of the Dutch floodplains, van Gestel [41] found that soil invertebrates and small mammals showed increases in levels of Cd (and sometimes Cu and Pb) while soil pore water and $CaCl_2$ extracts of metals were low, highlighting that available soil metal concentrations were not suitable indicators of metal accumulation in the food chain. In a smelter-impacted site, Rogival et al. [78] used ammonium nitrate as chemical extractant and showed significant relationships between Cd and Pb concentrations in organs of the wood mouse and in its diet (acorns and earthworms). They found stronger relationships between the diet and soil TMs using total rather than extractable concentrations for several metals including Cd, Pb and Zn. Thus, if total concentrations in soils are better estimates of the contamination of the diet than extractable ones, this may explain why we found here better correlations between internal TMs and soil TMs using total rather than extractable concentrations. The slight increase in hepatic Zn concentrations in small mammals along the pollution gradient is consistent with previous studies, which have provided evidence for slight differences between polluted and control field sites and the ability of small mammals to regulate internal levels of this essential element [7], [34], [69], [79], [80].

The use of chemical extractants to estimate bioavailability of metals is strongly dependent on the site, the metal, and the species under consideration, rendering any generalization risky. Based on our results and the discrepancies observed in the literature, it seems, as it has been demonstrated for invertebrates and vegetation [43], [81]–[83], that the relevance of chemical extractants for assessing the bioavailability of metals in field situations is also a great matter of concern for vertebrate wildlife. The availability of TMs in soils may not be the main factor governing TM transfer in food webs because bioavailability to herbivorous, omnivorous and carnivorous organisms cannot be predicted by a single chemical extract from a polluted soil. This phenomenon is probably due to the variety of exposure routes [6], [84], [85] and to the fact that factors affecting transfer to secondary consumers are not only related to the partitioning of metals in soils, but also to the parameters that modulate bioavailability along food webs, notably the levels and the chemical storage forms of TM in dietary items as well as the availability of these items [86], [87].

3.4.2 INFLUENCE OF LANDSCAPE ON ANIMAL TM CONCENTRATIONS

Our results show that landscape composition represents a significant variable that, together with the soil TM concentration and the age of the animals, contribute to explain internal TM concentrations in snails and small mammals. The density, availability and diversity of dietary items are likely to vary among landscape types and it has been proven, for both snails and small mammals, that diet varies according to the availability of dietary items [48], [51], [57], [88]. The feeding and foraging behaviour of animals can therefore change according to the landscape type. Resulting variations in diet composition could affect the amount and the bioavailability of metals transferred along food webs, leading to variable levels of bioaccumulation in the organism considered. Indeed, it has been shown that TM bioavailability for an organism depends not only on its own digestive characteristics but also on the amount and the sequestration of metals within the food [87], [89]–[92]. It has been emphasised that metal bioavailability can be affected by foraging and feeding behaviour of the organism [6], [8], [93]. The different patterns of Cd and Pb accumulation between landscapes found in the present work could thus be partially related to such food chain effects. This was also suggested by Hendrickx et al. [86], who observed site-specific TM accumulation in invertebrates (spiders and amphipods) and hypothesized that biological characteristics of sites, via alterations of trophic webs, can modify TM transfer.

Using individual-based models, Schipper et al. [10] modelled the influence of environmental heterogeneity in Dutch floodplains (soil contamination, habitat availability and suitability) on metal exposure using four species of small mammals, including the bank vole. These authors concluded that environmental heterogeneity governed only a minor part of the variation in metal exposure and that intra-species differences in exposure should mainly be due to inter-individual variations in species traits. In our study, soil contamination and landscape explained a non-negligible amount (around 30%) of the variation in internal Cd concentration. This could be due to the fact that we examined a larger pollution range and we considered not only the land use of the sampling point, but the influence of the landscape (a complex mosaic of land uses) surrounding the sampling point. The examination of exposure heterogeneity by combining these two scales of perception (i.e., heterogeneity within and between landscapes) could improve the understanding of ecological factors affecting transfer of TMs in ecosystems. Multi-scale approaches have been shown to be relevant within the context of biological contaminant (e.g. parasite) transfer in eco-epidemiology [94]–[96] and also appear promising for studies on chemical transfer within a landscape ecotoxicology framework.

Our data revealed a high inter-individual variability in TM concentrations, and although we considered age, soil TMs and landscape, more than half of the variance remained unexplained. Therefore, our results are also partly in accordance with those of Schipper et al. [10] concerning the strong influence of intra-species variability on exposure and TM accumulation. This phenomenon suggests that individual characteristics and behaviour govern a large part of the variations in internal concentrations.

3.4.3 SPATIAL SCALE OF CORRELATION BETWEEN ANIMAL AND SOIL TM CONCENTRATIONS

Because internal and soil TM concentrations at the sampling point scale were correlated, we argue that the present study provides evidence for a spatially explicit relationship between TM concentrations in animals and in soils. Our results stress that this spatial correlation is due to both the levels of metals in soils and the landscape composition around the habitats where the animals were sampled. The improvement of the correlation between internal and external concentrations using an increasing buffer size was not straightforward, in contrast to what was expected. Marinussen and van der Zee (1996) modelled exposure to and accumulation of Cd in fictitious organisms with various sizes of home-range (10 to 400 m^2) and showed that home-range size greatly affected exposure [26]. We assume that the range of the strongest correlation between internal and soil metal concentrations depends on numerous biological and ecological parameters. In fact, accumulation abilities, inter-individual variability, spatio-temporal variations in exposure, and heterogeneity in levels of soil contamination may affect the strength of the relationship between accumulated and soil TM concentrations.

3.4.4 OVERALL INTERPRETATION OF THE RESULTS WITH THE EXAMPLE OF PB

The movements of small mammals are known to be variable among landscapes because of the spatial heterogeneity of suitable habitats, their connectivity and the characteristics of the ecological barriers [49], [97]. Presence and survival of animals in heterogeneous landscapes are related to both site-specific characteristics, which define habitat quality on a local

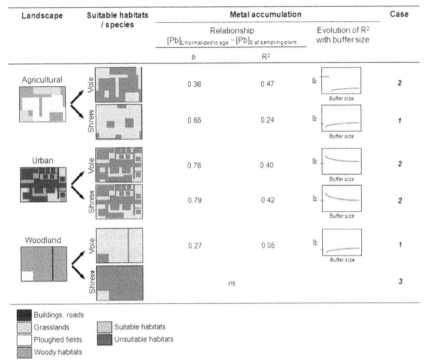

Landscape	Suitable habitats / species	Metal accumulation		Case
		Relationship $[Pb]_{L\ normalized\ to\ age} \sim [Pb]_{S\ at\ sampling\ point}$	Evolution of R^2 with buffer size	
		b	R^2		
Agricultural	Vole	0.38	0.47		2
	Shrew	0.65	0.24		1
Urban	Vole	0.76	0.40		2
	Shrew	0.79	0.42		2
Woodland	Vole	0.27	0.05		1
	Shrew	ns			3

Buildings, roads
Grasslands   Suitable habitats
Ploughed fields   Unsuitable habitats
Woody habitats

FIGURE 6: Summary and overall interpretation of the data with the example of Pb in small mammals. The data concerning Pb in the bank vole and in the greater white-toothed shrew is summarised, thus illustrating our overall interpretation of the data. Data concerning metal accumulation are R^2 values and slopes of the regressions between Pb concentrations in the liver (normalized to age) and total Pb concentrations in soil measured at the sampling point in addition to patterns of evolution of the R^2 values with buffer size. Maps of landscapes are based upon observed situations within the study area. Maps of suitable habitats for each species are hypothesized on the basis of habitat preferences described in the literature. Three patterns linking internal TM concentrations to soil TM concentrations have been uncovered, these patterns are named "Case 1, 2 and 3".

scale, and to factors acting at the landscape and/or meta-population scale [98]. Landscape features, animal habitat preferences and resource requirements govern the spatial repartition of suitable/unsuitable patches over an area [22]. This can result in different exposure patterns among landscapes for a given species, and can lead to different inter-species exposure within the same landscape type [4], [25]. Within this context, we propose a synthetic interpretation of the results we obtained from small mammals, using

Pb as an example (Figure 6). We hypothesize that relationships between internal and soil TMs are modulated by both landscape and the ecological characteristics of the species. In other words, we propose that the exposure of each species and the resulting bioaccumulation are dependent upon its ecological characteristics (such as spatial behaviour and diet) which are likely to be landscape-specific.

We have built hypothetical maps of suitable and unsuitable habitats on the basis of habitat preference data found in the literature. The bank vole is a forest specialist species. It can live in other woody habitats (for instance woodlots and hedgerows) in heterogeneous landscapes (agricultural, urban or others) where it is confined to woody patches and rarely moves to surrounding habitats unless it needs to cross them to reach other suitable woody habitats [49], [99]–[101]. The greater white-toothed shrew is more eurytopic. This species is mostly present in inhabited areas and open landscapes such as agricultural fields [48], [97], [101], [102]. Therefore, within each landscape, habitat suitability differs between voles and shrews.

We propose the existence of three different cases (Figure 6). When almost the entire area is suitable (case 1), home-ranges can overlap several habitats that differ in metal contamination and individuals can easily move between more or less contaminated patches. Therefore, internal concentrations show high inter-individual variability and are poorly related to soil contamination, as evidenced by the relatively low R^2 values and slopes of the relationships). Moreover, the strength of the relationship between internal and soil TMs does not substantially vary regardless the considered scale for soil TM concentration (the R^2 value was roughly similar between the sampling point and buffers as well as among buffers). This is the case for shrews in agricultural lands and voles in woodlands. Conversely, when animals are confined to suitable isolated patches within a hostile matrix (case 2), the absence or the presence of high local contamination in occupied patches would greatly condition animal exposure, since individuals cannot easily move or forage in surrounding habitats. Internal concentrations therefore would more strongly depend on soil TM concentration in occupied patches, resulting in better correlations between internal and soil contamination compared to case 1. In this case, represented by voles in agricultural lands and voles and shrews in urban areas, we found globally strong relationships between TMs in animals and in soils (high R^2 values and/or slopes). Moreover, we observed better correlations on local scales (at the sampling point or at a buffer size lower than 100 m). Finally, in landscapes characterised in a large part by unsuitable habitats (case 3), animals can be found during migration and dispersal or in margins, and it may be assumed that the main duration of exposure has occurred in another place. In such cases (represented by shrews in woodlands and shrublands), we captured few individuals, all of which were non-reproductive adults, and we failed to detect relationships between TMs in animals and in soils.

We observed that landscape composition influences bioaccumulation with species-specific patterns. Thus we conclude that landscape features modulate animal exposure to pollutants, and this effect might differ according to species ecological characteristics. The spatial heterogeneity of soil contamination could more or less affect exposure, depending on the species and the landscape concerned.

3.5 CONCLUSIONS

Apart from animal age and metal concentration in soil at the sampling point, the landscape around the habitat of capture influences the internal TM concentrations. Certain landscapes are therefore more at risk than others depending on the considered organisms and metals. In some cases, transfer is high even at low levels of soil contamination and result in elevated accumulated TM levels in animals (i.e., an elevated biota to soil accumulation factor). In other cases, low internal TM levels were found at low soil TM concentrations, but internal concentrations sharply increased along the pollution gradient. Both phenomena were also found to co-occur. We propose that a landscape gathering both phenomena would the most at risk for wildlife.

The lack of relationship between internal levels in animals and $CaCl_2$-extractable TM concentrations in soil, as well as the differences in relationships between animal and soil TM concentrations between studies, provide two major insights within a risk assessment framework. First, the use of chemical extracts may not be relevant when assessing TM bioavailability to herbivorous/granivorous and carnivorous species. Second, predictions of internal TM using general accumulation models, such as the use of global regression regardless of the environmental characteristics of a site (e.g. landscape) and the intra-species variability should be used cautiously.

Exposure of wildlife to contaminants and subsequent TM accumulation is determined by several parameters acting at different biological organisation levels that are integrative variables of several processes. Based on our results, we suggest that TM accumulation in snails and small mammals is governed by ecological (diet, habitat preferences, mobility, etc) and physiological (assimilation and excretion of TM) characteristics of animals. Our results strongly suggest that availability in soil does not fully determine transfer in food webs. Species ecology and landscape are other key factors that determine organism exposure. Our findings lead us to hypothesize that ecological characteristics, such as food web structure and the way that organisms exploit their environment (home-range size, migration, feeding behaviour, habitat preferences, etc), both dependent on landscape features, mainly explain TM transfer in food webs. The present study points out the need for further investigation to develop a field of "landscape ecotoxicology" and elucidate the underlying mechanisms behind landscape effects, and additionally highlights the interest for a multiscale approach in ecotoxicology.

REFERENCES

1. Smith R, Pollard SJT, Weeks JM, Nathanail CP (2006) Assessing harm to terrestrial ecosystems from contaminated land. Soil Use and Management 21: 527–540. doi: 10.1079/sum2005345.
2. Fairbrother A, Wenstel R, Sappington K, Wood W (2007) Framework for metals risk assessment. Ecotoxicology and Environmental Safety 68: 145–227. doi: 10.1016/j.ecoenv.2007.03.015.
3. Carlsen TM, Coty JD, Kercher JR (2004) The spatial extent of contaminants and the landscape scale: An analysis of the wildlife, conservation biology, and population modeling literature. Environmental Toxicology and Chemistry 23: 798–811. doi: 10.1897/02-202.
4. Ares J (2003) Time and space issues in ecotoxicology: Population models, landscape pattern analysis, and long-range environmental chemistry. Environmental Toxicology and Chemistry 22: 945–957. doi: 10.1002/etc.5620220501.

5. Woodbury PB (2003) Dos and don'ts of spatially explicit ecological risk assessments. Environmental Toxicology and Chemistry 22: 977–982. doi: 10.1002/etc.5620220504.

6. Smith PN, Cobb GP, Godard-Codding C, Hoff D, McMurry ST, et al. (2007) Contaminant exposure in terrestrial vertebrates. Environmental Pollution 150: 41–64. doi: 10.1016/j.envpol.2007.06.009.

7. Fritsch C, Cosson RP, Cœurdassier M, Raoul F, Giraudoux P, et al. (2010) Responses of wild small mammals to a pollution gradient: host factors influence metal and metallothionein levels. Environmental Pollution 158: 827–840. doi: 10.1016/j.envpol.2009.09.027.

8. Peakall D, Burger J (2003) Methodologies for assessing exposure to metals: speciation, bioavailability of metals, and ecological host factors. Ecotoxicology and Environmental Safety 56: 110–121. doi: 10.1016/s0147-6513(03)00055-1.

9. Clifford PA, Barchers DE, Ludwig DF, Sielken RL, Klingensmith JS, et al. (1995) An approach to quantifying spatial components of exposure for ecological risk assessment. Environmental Toxicology and Chemistry 14: 895–906. doi: 10.1002/etc.5620140523.

10. Schipper AM, Loos M, Ragas AMJ, Lopes JPC, Nolte BT, et al. (2008) Modeling the influence of environmental heterogeneity on heavy metal exposure concentrations for terrestrial vertebrates in river floodplains. Environmental Toxicology and Chemistry 27: 919–932. doi: 10.1897/07-252.1.

11. Vermeulen F (2009) Spatially explicit exposure assessment of persistent pollutants through the food chain of the European hedgehog (Erinaceus europaeus). Antwerp: University of Antwerp, Belgium. 211 p.

12. Fowler D, Smith RI, Leith ID, Crossley A, Mourne RW, et al. (1998) Quantifying fine-scale variability in pollutant deposition in complex terrain using 210Pb inventories in soil. Water Air and Soil Pollution 105: 459–470. doi: 10.1007/978-94-017-0906-4_42.

13. Hope BK (2006) An examination of ecological risk assessment and management practices. Environment International 32: 983–995. doi: 10.1016/j.envint.2006.06.005.

14. Fowler D, Skiba U, Nemitz E, Choubedar F, Branford D, et al. (2004) Measuring aerosol and heavy metal deposition on urban woodland and grass using inventories of 210Pb and metal concentrations in soil. Water Air and Soil Pollution 4: 483–499. doi: 10.1023/b:wafo.0000028373.02470.ba.

15. Rieuwerts J, Farago M (1996) Heavy metal pollution in the vicinity of a secondary lead smelter in the Czech Republic. Applied Geochemistry 11: 17–23. doi: 10.1016/0883-2927(95)00050-x.

16. Magiera T, Zawadzki J (2007) Using of high-resolution topsoil magnetic screening for assessment of dust deposition: Comparison of forest and arable soil datasets. Environmental Monitoring and Assessment 125: 19–28. doi: 10.1007/s10661-006-9235-4.

17. Ettler V, Vanek A, Mihaljevic M, Bezdicka P (2005) Contrasting lead speciation in forest and tilled soils heavily polluted by lead metallurgy. Chemosphere 58: 1449–1459. doi: 10.1016/j.chemosphere.2004.09.084.

18. Kapusta P, Sobczyk L, Rozen A, Weiner J (2003) Species diversity and spatial distribution of enchytraeid communities in forest soils: effects of habitat characteristics

and heavy metal contamination. Applied Soil Ecology 23: 187–198. doi: 10.1016/s0929-1393(03)00064-7.

19. Vermeulen F, Van den Brink NW, D'Havé H, Mubiana VK, Blust R, et al. (2009) Habitat type-based bioaccumulation and risk assessment of metal and As contamination in earthworms, beetles and woodlice. Environmental Pollution 157: 3098–3105. doi: 10.1016/j.envpol.2009.05.017.

20. Fritsch C, Giraudoux P, Cœurdassier M, Douay F, Raoul F, et al. (2010) Spatial distribution of metals in smelter-impacted soils of woody habitats: Influence of landscape and soil properties, and risk for wildlife. Chemosphere 81: 141–155. doi: 10.1016/j.chemosphere.2010.06.075.

21. Lidicker WZJ (2008) Levels of organization in biology: on the nature and nomenclature of ecology's fourth level. Biological Reviews 83: 71–78. doi: 10.1111/j.1469-185x.2007.00032.x.

22. Burel F, Baudry J (2003) Landscape ecology - Concepts, methods and applications. Enfield, New Hampshire, USA: Sciences Publishers, Inc. 362 p.

23. Gaines K, Boring C, Porter D (2005) The development of a spatial explicit model to estimate radiocaesium body burdens in raccoons (Procyon lotor) for ecological risk assessment. Science of the Total Environment 341: 15–31. doi: 10.1016/j.scitotenv.2004.09.017.

24. Kapustka LA (2008) Limitations of the current practices used to perform ecological risk assessment. Integrated Environmental Assessment and Management 4: 290–298. doi: 10.1897/ieam_2007-084.1.

25. Barnthouse LW, Munns WRJ, Sorensen MT, editors. (2008) Population-level ecological risk-assessment. Boca Raton, Florida, USA: CRC Press Taylor & Francis Group. 346 p.

26. Marinussen MPJC, van der Zee SEATM (1996) Conceptual approach to estimating the effect of home-range size on the exposure of organisms to spatially variable soil contamination. Ecological Modelling 87: 83–89. doi: 10.1016/0304-3800(94)00207-x.

27. Hope BK (2001) A case study comparing static and spatially explicit ecological exposure analysis methods. Risk Analysis 21: 1001–1010. doi: 10.1111/0272-4332.216169.

28. Hope BK (2000) Generating probabilistic spatially-explicit individual and population exposure estimates for ecological risk assessments. Risk Analysis 20: 573–589. doi: 10.1111/0272-4332.205053.

29. Purucker ST, Welsh CJE, Stewart RN, Starzec P (2007) Use of habitat-contamination spatial correlation to determine when to perform a spatially explicit ecological risk assessment. Ecological Modelling 204: 180–192. doi: 10.1016/j.ecolmodel.2006.12.032.

30. Topping CJ, Sibly RM, Akçakaya HR, Smith GC, Crocker DR (2005) Risk Assessment of UK Skylark Populations Using Life-History and Individual-Based Landscape Models. Ecotoxicology V14: 925–936. doi: 10.1007/s10646-005-0027-3.

31. Kooistra L, Leuven RSEW, Nienhuis PH, Wehrens R, Buydens LMC (2001) A Procedure for Incorporating Spatial Variability in Ecological Risk Assessment of Dutch River Floodplains. Environmental Management 28: 359–373. doi: 10.1007/s0026702433.

32. Kooistra L, Huijbregts MAJ, Ragas AMJ, Wehrens R, Leuven R (2005) Spatial variability and uncertainty in ecological risk assessment: A case study on the potential risk of cadmium for the little owl in a Dutch river flood plain. Environmental Science and Technology 39: 2177–2187. doi: 10.1021/es049814w.

33. Cairns J Jr, Nlederlehner BR (1996) Developing a field of landscape ecotoxicology. Ecological Applications 6: 790–796. doi: 10.2307/2269484.

34. Shore RF, Rattner BA (2001) Ecotoxicology of Wild Mammals. London, UK: John Wiley & Sons. 730 p.

35. Talmage S, Walton B (1991) Small mammals as monitors of environmental contaminants. Review of Environmental Contaminants and Toxicology 119: 47–145. doi: 10.1007/978-1-4612-3078-6_2.

36. Berger B, Dallinger R (1993) Terrestrial snails as quantitative indicators of environmental metal pollution. Environmental Monitoring and Assessment 25: 65–84. doi: 10.1007/bf00549793.

37. de Vaufleury AG, Pihan F (2000) Growing snails used as sentinels to evaluate terrestrial environment contamination by trace elements. Chemosphere 40: 275–284. doi: 10.1016/s0045-6535(99)00246-5.

38. Wijnhoven S, Leuven R, van der Velde G, Jungheim G, Koelemij E, et al. (2007) Heavy-metal concentrations in small mammals from a diffusely polluted floodplain: importance of species- and location-specific characteristics. Archives of Environmental Contamination and Toxicology 52: 603–613. doi: 10.1007/s00244-006-0124-1.

39. van den Brink N, Lammertsma D, Dimmers W, Boerwinkel M-C, van der Hout A (2010) Effects of soil properties on food web accumulation of heavy metals to the wood mouse (Apodemus sylvaticus). Environmental Pollution 158: 245–251. doi: 10.1016/j.envpol.2009.07.013.

40. Baker S, Herrchen M, Hund-Rinke K, Klein W, Kordel W, et al. (2003) Underlying issues including approaches and information needs in risk assessment. Ecotoxicology and Environmental Safety 56: 6–19. doi: 10.1016/s0147-6513(03)00046-0.

41. Van Gestel CAM (2008) Physico-chemical and biological parameters determine metal bioavailability in soils. Science of the Total Environment 406: 385–395. doi: 10.1016/j.scitotenv.2008.05.050.

42. Peijnenburg WJGM, Jager T (2003) Monitoring approaches to assess bioaccessibility and bioavailability of metals: Matrix issues. Ecotoxicology and Environmental Safety 56: 63–77. doi: 10.1016/s0147-6513(03)00051-4.

43. Harmsen J (2007) Measuring bioavailability: from a scientific approach to standard methods. Journal of Environmental Quality 36: 1420–1428. doi: 10.2134/jeq2006.0492.

44. Meers E, Samson R, Tack FMG, Ruttens A, Vandegehuchte M, et al. (2007) Phytoavailability assessment of heavy metals in soils by single extractions and accumulation by Phaseolus vulgaris. Environmental and Experimental Botany 60: 385–396. doi: 10.1016/j.envexpbot.2006.12.010.

45. ISO (2008) Soil quality - Requirements and guidance for the selection and application of methods for the assessment of bioavailability of contaminants in soil and soil materials. ISO 17402:2008. Geneva, Switzerland: International Organization for Standardization.

46. Baur A, Baur B (1993) Daily movement patterns and dispersal in the land snail Arianta arbustorum. Malacologia 35: 89–98.

47. Le Louarn H, Quéré J, Butet A (2003) Les Rongeurs de France - Faunistique et biologie; INRA, editor. Paris, France. 256 p.

48. Lugon-Moulin N (2003) Les Musaraignes - Biologie, Ecologie, Répartition en Suisse; Porte-Plumes, editor. Ayer, Switzerland. 308 p.

49. Kozakiewicz M, Van Apeldoorn R, Bergers P, Gortat T, Kozakiewicz A (2000) Landscape approach to bank vole ecology. Polish Journal of Ecology 48: 149–161.

50. Cameron RAD, Williamson P (1977) Estimating migration and effects of disturbance in mark-recapture studies on snail Cepaea nemoralis L. Journal of Animal Ecology 46: 173–179. doi: 10.2307/3954.

51. Williamson P, Cameron RAD (1976) Natural diet of landsnail Cepaea nemoralis. Oikos 27: 493–500. doi: 10.2307/3543468.

52. Wolda H, Zweep A, Schuitema KA (1971) The role of food in the dynamics of populations of the landsnail Cepaea nemoralis. Oecologia 7: 361–381. doi: 10.1007/bf00345860.

53. Mahtfeld K (2000) Impact of introduced gastropods on molluscan communities, northern North Island. Wellington, New Zealand: Department of Conservation.

54. Kerney MP, Cameron RAD (2006) Guide des escargots et limaces d'Europe. Paris, France: Delachaux et Niestlé. 370 p.

55. Vonproschwitz T (1994) Oxychilus cellarius (Müller) and Oxychilus draparnaudi (Beck) as predators on egg clutches of Arion lusitanicus mabille. Journal of Conchology 35: 183–184.

56. Sadowska ET, Baliga-Klimczyk K, Chrzascik K, Koteja P (2008) Laboratory model of adaptive radiation: a selection experiment in the bank vole. Physiological and Biochemical Zoology 81: 627–640. doi: 10.1086/590164.

57. Abt KF, Bock WF (1998) Seasonal variations of diet composition in farmland field mice Apodemus spp. and bank voles Clethrionomys glareolus. Acta Theriologica 43: 379–389.

58. Frangi J-P, Richard D (1997) Heavy metal soil pollution cartography in northern France. Science of the Total Environment 205: 71–79. doi: 10.1016/s0048-9697(97)00184-8.

59. Sterckeman T, Douay F, Proix N, Fourrier H, Perdrix E (2002) Assessment of the contamination of cultivated soils by eighteen trace elements around smelters in the North of France. Water Air and Soil Pollution 135: 173–194.

60. Douay F, Pruvot C, Roussel H, Ciesielski H, Fourrier H, et al. (2008) Contamination of urban soils in an area of Northern France polluted by dust emissions of two smelters. Water Air and Soil Pollution 188: 247–260. doi: 10.1007/s11270-007-9541-7.

61. Sterckeman T, Douay F, Proix N, Fourrier H (2000) Vertical distribution of Cd, Pb and Zn in soils near smelters in the North of France. Environmental Pollution 107: 377–389. doi: 10.1016/s0269-7491(99)00165-7.

62. Douay F, Pruvot C, Waterlot C, Fritsch C, Fourrier H, et al. (2009) Contamination of woody habitat soils around a former lead smelter in the North of France. Science of the Total Environment 407: 5564–5577. doi: 10.1016/j.scitotenv.2009.06.015.

63. Charissou I (1999) Identification des restes trouvé dans les pelotes de réjection de rapaces. Supplément scientifique de la revue Epops 44: 1–31.

64. Chaline J, Baudvin H, Jammot D, Saint Girons M-C (1974) Les proies des rapaces. Paris: Petits mammifères et leur environnement. 143 p.

65. Kozakiewicz M (1976) The weight of eye lens as the proposed age indicator of the bank vole. Acta Theriologica 21: 314–316.

66. Williamson P (1979) Age determination of juvenile and adult Cepaea. Journal of Molluscan Studies 45: 52–60.

67. AFNOR (1996) Qualité des sols - Méthodes chimiques - sols sédiments, mise en solution totale par attaque acide - NF X31-147. Paris, France: Association Française de Normalisation.

68. Grafen A, Hails R (2002) Modern Statistics for the Life Sciences: Oxford University Press. 351 p.

69. Sánchez-Chardi A, Ribeiro CAO, Nadal J (2009) Metals in liver and kidneys and the effects of chronic exposure to pyrite mine pollution in the shrew Crocidura russula inhabiting the protected wetland of Doñana. Chemosphere 76: 387–394.

70. Williamson P (1980) Variables affecting body burdens of lead, zinc and cadmium in a roadside population of the snail Cepaea hortensis Müller. Oecologia 44: 213–220.

71. R Development Core Team (2006) R: A language and environment for statistical computing. Vienna, Austria: Foundation for Statistical Computing. http://www.R-project.org.

72. Sánchez-Chardi A, Nadal J (2007) Bioaccumulation of metals and effects of landfill pollution in small mammals. Part I. The greater white-toothed shrew, Crocidura russula. Chemosphere 68: 703–711. doi: 10.1016/j.chemosphere.2007.01.042.

73. Sánchez-Chardi A, Lopez-Fuster MJ, Nadal J (2007) Bioaccumulation of lead, mercury, and cadmium in the greater white-toothed shrew, Crocidura russula, from the Ebro Delta (NE Spain): Sex- and age-dependent variation. Environmental Pollution 145: 7–14. doi: 10.1016/j.envpol.2006.02.033.

74. Notten MJM, Oosthoek AJP, Rozema J, Aerts R (2005) Heavy metal concentrations in a soil-plant-snail food chain along a terrestrial soil pollution gradient. Environmental Pollution 138: 178–190. doi: 10.1016/j.envpol.2005.01.011.

75. Damek-Poprawa M, Sawicka-Kapusta K (2004) Histopathological changes in the liver, kidneys, and testes of bank voles environmentally exposed to heavy metal emissions from the steelworks and zinc smelter in Poland. Environmental Research 96: 72–78. doi: 10.1016/j.envres.2004.02.003.

76. Fritsch C, Cœurdassier M, Gimbert F, Crini N, Scheifler R, et al. (2011) Investigations of adaptation to metallic pollution in land snails (Cantareus aspersus and Cepaea nemoralis) from a smelter-impacted area. Ecotoxicology 20: 739–759. doi: 10.1007/s10646-011-0619-z.

77. Dallinger R, Lagg B, Egg M, Schipflinger R, Chabicovsky M (2004) Cd accumulation and Cd-metallothionein as a biomarker in Cepaea hortensis (Helicidae, Pulmonata) from laboratory exposure and metal-polluted habitats. Ecotoxicology 13: 757–772. doi: 10.1007/s10646-003-4474-4.

78. Rogival D, Scheirs J, Blust R (2007) Transfer and accumulation of metals in a soil-diet-wood mouse food chain along a metal pollution gradient. Environmental Pollution 145: 516–528. doi: 10.1016/j.envpol.2006.04.019.

79. Milton A, Cooke JA, Johnson MS (2003) Accumulation of Lead, Zinc, and Cadmium in a Wild Population of (Clethrionomys glareolus) from an Abandoned Lead

Mine. Archives of Environmental Contamination and Toxicology 44: 405–411. doi: 10.1007/s00244-002-2014-5.

80. Loos M, Ragas AMJ, Tramper JJ, Hendriks AJ (2009) Modeling zinc regulation in small mammals. Environmental Toxicology and Chemistry 28: 2378–2385. doi: 10.1897/09-028.1.

81. Migeon A, Richaud P, Guinet F, Chalot M, Blaudez D (2009) Metal accumulation by woody species on contaminated sites in the north of France. Water Air and Soil Pollution 204: 89–101. doi: 10.1007/s11270-009-0029-5.

82. Bleeker EAJ, van Gestel CAM (2007) Effects of spatial and temporal variation in metal availability on earthworms in floodplain soils of the river Dommel, The Netherlands. Environmental Pollution 148: 824–832. doi: 10.1016/j.envpol.2007.01.034.

83. Hobbelen PHF, Koolhaas JE, van Gestel CAM (2006) Bioaccumulation of heavy metals in the earthworms Lumbricus rubellus and Aporrectodea caliginosa in relation to total and available metal concentrations in field soils. Environmental Pollution 144: 639–646. doi: 10.1016/j.envpol.2006.01.019.

84. Cœurdassier M, Gomot-de Vaufleury A, Lovy C, Badot P-M (2002) Is the cadmium uptake from soil important in bioaccumulation and toxic effects for snails? Ecotoxicology and Environmental Safety 53: 425–431. doi: 10.1016/s0147-6513(02)00004-0.

85. Scheifler R, de Vaufleury A, Coeurdassier M, Crini N, Badot PM (2006) Transfer of Cd, Cu, Ni, Pb, and Zn in a soil-plant-invertebrate food chain: A microcosm study. Environmental Toxicology and Chemistry 25: 815–822. doi: 10.1897/04-675r.1.

86. Hendrickx F, Maelfait J-P, Bogaert N, Tojal C, Du Laing G, et al. (2004) The importance of biological factors affecting trace metal concentration as revealed from accumulation patterns in co-occurring terrestrial invertebrates. Environmental Pollution 127: 335–341. doi: 10.1016/j.envpol.2003.09.001.

87. Vijver MG, Van Gestel CAM, Lanno RP, Van Straalen NM, Peijnenburg WJGM (2004) Internal metal sequestration and its ecotoxicological relevance: A review. Environmental Science and Technology 38: 4705–4712. doi: 10.1021/es040354g.

88. Mason CF (1970) Food, feeding rates and assimilation in woodland snails. Oecologia 4: 358–373. doi: 10.1007/bf00393394.

89. Wallace WG, Lopez GR, Levinton JS (1998) Cadmium resistance in an oligochaete and its effect on cadmium trophic transfer to an omnivorous shrimp. Marine Ecology Progress Series 172: 225–237. doi: 10.3354/meps172225.

90. Wallace WG, Luoma SN (2003) Subcellular compartmentalization of Cd and Zn in two bivalves. II. Significance of trophically available metal (TAM). Marine Ecology Progress Series 257: 125–137. doi: 10.3354/meps257125.

91. Hispard F, De Vaufleury A, Cosson RP, Devaux S, Scheifler R, et al. (2007) Comparison of transfer and effects of Cd on rats exposed in a short experimental snail–rat food chain or to CdCl2 dosed food. Environment International 34: 381–389.

92. Monteiro MS, Santos C, Soares A, Mann RM (2008) Does subcellular distribution in plants dictate the trophic bioavailability of cadmium to Porcellio dilatatus (Crustacea, Isopoda)? Environmental Toxicology and Chemistry 27: 2548–2556. doi: 10.1897/08-154.1.

93. Burger J, Diaz-Barriga F, Marafante E, Pounds J, Robson M (2003) Methodologies to examine the importance of host factors in bioavailability of metals. Ecotoxicology and Environmental Safety 56: 20–31. doi: 10.1016/s0147-6513(03)00047-2.

94. Giraudoux P, Delattre P, Takahashi K, Raoul F, Quéré JP, et al. (2002) Transmission ecology of Echinococcus multilocularis in wildlife: what can be learned from comparative studies and multi-scale approaches? In: Craig P, Pawlowski Z, editors. Cestode zoonoses: Echinococcosis and Cystercosis An emergent and global problem. Amsterdam, The Netherlands: NATO Sciences Series, IOS press. pp. 251–262.

95. Kitron U, Clennon JA, Cecere MC, Gurtler RE, King CH, et al. (2006) Upscale or downscale: applications of fine scale remotely sensed data to Chagas disease in Argentina and schistosomiasis in Kenya. Geospatial Health 1: 49–58.

96. Danson FM, Graham AJ, Pleydell DRJ, Campos-Ponce M, Giraudoux P, et al. (2003) Multi-scale spatial analysis of human alveolar echinococcosis risk in China. Parasitology 127: S133–S141. doi: 10.1017/s0031182003003639.

97. Vuilleumier S, Fontanillas P (2007) Landscape structure affects dispersal in the greater white-toothed shrew: Inference between genetic and simulated ecological distances. Ecological Modelling 201: 369–376. doi: 10.1016/j.ecolmodel.2006.10.002.

98. Burel F, Butet A, Delettre YR, de la Pena NM (2004) Differential response of selected taxa to landscape context and agricultural intensification. Landscape and Urban Planning 67: 195–204. doi: 10.1016/s0169-2046(03)00039-2.

99. Pollard E, Relton J, Hedges V (1970) A study of small mammals in hedges and cultivated fields. Journal of Applied Ecology 7: 549–557. doi: 10.2307/2401977.

100. Paillat G, Butet A (1996) Spatial dynamics of the bank vole (Clethrionomys glareolus) in a fragmented landscape. Acta Oecologica – International Journal of Ecology 17: 553–559.

101. Wijnhoven S, van der Velde G, Leuven R, Smits A (2006) Modelling recolonisation of heterogeneous river floodplains by small mammals. Hydrobiologia 565: 135–152. doi: 10.1007/s10750-005-1910-x.

102. Millãn de la Peña N, Butet A, Delettre Y, Paillat G, Morant P, et al. (2003) Response of the small mammal community to changes in western French agricultural landscapes. Landscape Ecology V18: 265–278.

Tables 5 and 6 are missing from this version of the article. To see these tables, as well as other supplemental information, please visit the original version of the article as cited on the first page of this chapter.

CHAPTER 4

HEAVY METAL CONTAMINATION OF SOIL AND SEDIMENT IN ZAMBIA

YOSHINORI IKENAKA, SHOUTA M. M. NAKAYAMA,
KAAMPWE MUZANDU, KENNEDY CHOONGO,
HIROKI TERAOKA, NAOHARU MIZUNO,
AND MAYUMI ISHIZUKA

4.1 INTRODUCTION

Africa is a continent located in the southern hemisphere and known for its rich diversity of wildlife including birds, amphibians, reptiles and large mammals. However, in recent years, there have been concerns about significant environmental problems caused by the mining of rare and major metals and metallurgical activities in African countries by domestic and foreign corporations (Oelofse, 2008). Environmental pollution due to the rapid progress of economic development in Africa can cause various problems and heavy metals are some of the major contaminants in these countries (Akiwumi and Butler, 2008; Norman et al., 2007; Rashad and Barsoum, 2006). Humans and wildlife can be exposed to heavy metals by drinking water and inhaling air or soil contaminated by mining activities and the metal industry (Nakayama et al., 2010).

This chapter was originally published under the Creative Commons Attribution License. Ikenaka Y, Nakayama SMM, Muzandu K, Choongo K, Teraoka H, Mizuno N, and Ishizuka M. Heavy Metal Contamination of Soil and Sediment in Zambia. African Journal of Environmental Science and Technology *4,11 (2010), 729-739.*

Mining activities are considered to have the potential for causing heavy metal pollution and associated diseases (Lacatusu et al., 2009; Kodom et al., 2010). Therefore, many researchers worldwide have focused on and reported assessments of heavy metal concentrations (Zhai et al., 2008; Higueras et al., 2004; Razo et al., 2004; von Braun et al., 2002). However, currently most data on heavy metals in African countries are the result of regional investigations that have been limited to the area around the source of the heavy metals (Aguilar et al., 2002). Surveys of heavy metals across the whole country and comprehensive analyses which include economic activities, are needed to clarify the impact of these chemicals on humans and wildlife and are essential for the protection and management of the environment in African countries.

The Republic of Zambia is an African country that is rich in mineral resources such as copper (Cu), cobalt (Co), zinc (Zn) and lead (Pb) (Stockwell et al., 2001). Mining is the most important industry in Zambia. In 1997, 3% of the world's annual Cu production and 20% of the annual Co production were mined in Zambia and most of this ore was smelted within the country (Stockwell et al., 2001). The core mining areas in Zambia are Kabwe town and the Copperbelt. However, heavy metal pollution is one of the most important environmental issues in Zambia and causes serious effects on humans and animals (Nwanko and Elinder, 1979; Syakalima et al., 2001).

In this study, we suggested that road soils and river and lake sediments are useful to assess widespread environmental pollution. Actually, in polluted aquatic systems, sediments have been increasingly recognized as the most important sink for contaminants and as a reservoir and possible future source of pollutants (Ikenaka et al., 2005a, 2005b). These data can provide basic information on the accumulation and transportation of these pollutants into both human life and ecosystems. Thus, we collected the sediments of the three largest river basins in Zambia, the Zambezi River (including Lake Kariba), the Kafue River (including Lake Itezhi-tezhi) and the Luangwa River. The purpose of the present study was to evaluate the spatial distribution of heavy metals in the main areas of Zambia and to understand the characteristics of pollution in each area, using road soil and sediment.

4.2 MATERIALS AND METHODS

4.2.1 SOIL SAMPLING SITES

We collected 47 soil samples from various cities and towns in Zambia including Lusaka (n = 7), Kabwe (n = 3), and the Eastern (n = 10), western (n = 5), southern (n = 12) and Northern (n = 10) areas (Figure 1 and Table 1).

Lusaka, the capital and largest city of Zambia, is located in the southern part of the central plateau of the country and is the center of economic and industrial activity in the country. We collected two soil samples from the sides of major roads (Lusaka 1, 4) and five samples from industrial areas (Lusaka 2, 3, 5, 6, and 7) (Table 1). Kabwe is located about 130 km North of Lusaka and is one of the main areas of mining activity in Zambia. We collected three samples in Kabwe (Kabwe 1, 2 and 3) (Table 1). In the Eastern area, we collected samples along the T4 road (Chongwe, Chinyunyu, Kachalola, Nyimba, Petauke, Sinda, Chipata, and Mumbwe) and in the South Luangwa National Park (Mfuwe) which is a large wildlife preserve. In the Western area, we collected samples from three towns, Namwala, Mumbwa, and Itezhi-tezhi (ITT). In the Southern area, we collected samples along the T1 road (at Kafue (n=3), Mazabuka, Monze, Muzoka, Choma, Kalomo, Zimba, Livingstone), and at Mambova and Siavonga. The northern area including the Copperbelt is one of the main areas of mining activity in Zambia. In the northern area, we collected samples along the T3 road (Kapiri Mposhi, Luanshya, Ndola, Kitwe, and Chingola (n=3)), and from rural towns along the Kafue River (Mpongwe, Shingwa, and Masaiti).

4.2.2 SEDIMENT SAMPLING SITES

River sediment samples were collected from three rivers that flow through Zambia, the Zambezi (n = 8), Luangwa (n = 5), and Kafue rivers (n = 8)

(Figure 1). Samples were also collected from the tributaries of the Kafue and Luangwa rivers. Of Kafue River tributaries, we collected sediments from the Mushishima (n = 3) and Kakosa streams (n = 1) and one sample was collected from a Luangwa River tributary, the Chongwe Stream.

4.2.3 SAMPLING PROCEDURE

Soil and sediment samples were collected during the dry season between May and September 2008. Approximately 500 g of soil or sediment was collected from each site at a depth of 0 - 5 cm and stored in a plastic bottle. At least three composite soil samples were collected from each sampling point. The soil samples were passed through a 2 mm sieve before extraction. The surface sediment samples were collected using an Ekman grab sampler. Each sediment sample was air-dried in the laboratory at room temperature and was passed through a 2 mm sieve before extraction. The dry weight of each sample was measured after 12 h of drying in an oven at 105°C.

4.2.4 REAGENTS

Sulfuric acid (poisonous metal analysis grade, 96%), nitric acid (atomic absorption spectrometry grade, 60%), perchloric acid (atomic absorption spectrometry grade, 60%), standard solutions of each heavy metal (Cr, Co, Ni, Cu, Zn, As, Cd, and Pb: chemical analysis grade, 100 mg/L in 0.1 M nitric acid; Sr and Hg: chemical analysis grade, 1000 mg/L in 0.1 M nitric acid) were purchased from Kanto Chemical Corp., Tokyo, Japan. Ammonium chloride, hydrochloric acid (special grade: 36%) and lanthanum chloride solution (atomic absorption spectrometry grade, 100 g La/L solution) were purchased from Wako Pure Chemical Industries Ltd., Osaka, Japan.

4.2.5 EXTRACTION AND ANALYSIS OF HEAVY METALS

All laboratory equipment used for the heavy metal analysis was washed in 3% HNO_3 and rinsed at least twice with distilled water. One gram of each

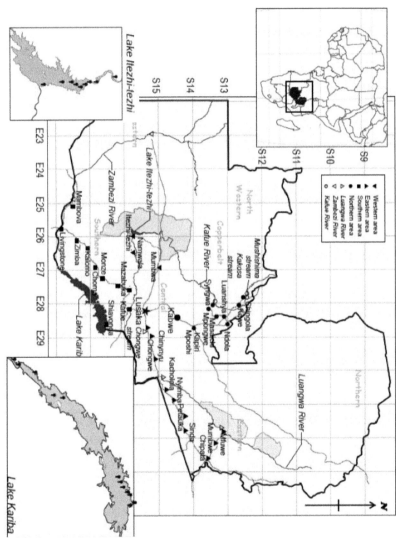

FIGURE 1: Sampling sites for soils and sediments in Zambia.

soil or sediment sample was placed into a 200 mL flask. Then, 0.2 mL of sulfuric acid, 1 mL of nitric acid and 5 mL of perchloric acid were added. The soil and acid mixture was heated to 180°C for 3 h on a hotplate. After cooling, 1 g of ammonium chloride and 20 mL of 0.5 N HCl were added. Samples were reheated to 180°C for one hour and evaporated to approximately 10 mL. After cooling, the extracts were filtered into 100 mL plastic bottles through an ashless 5B filter paper (Advantec, Tokyo, Japan) and 1 mL of lanthanum chloride was added. A reagent blank was also prepared using same the process.

The concentrations of nine of the elements (Cr, Co, Ni, Cu, Zn, As, Sr, Cd, and Pb) were determined using an Analyst™ 800 atomic absorption spectrophotometer (AAS) (Perkin Elmer Instruments, USA) with either an acetylene flame (Cu and Zn) or an argon non-flame (Cr, Co, Ni, As, Sr, Cd and Pb) after preparation of the calibration standards. The concentration of Hg was determined using a mercury analysis system MA-2000 (Nippon Instruments Corp., Tokyo, Japan) after preparation of the calibration standards. The overall recovery rates (mean ± SD) for Cr, Co, Cu, Zn, Cd, Pb and Ni were 91 ± 3.0, 92 ± 3.4, 89 ± 5.6, 91 ± 2.3, 111 ± 8.3, 90 ± 3.5 and $92 \pm 4.2\%$, respectively. The heavy metal concentration in soil or sediment was calculated in mg/kg dry weight (wt). Recommended values for Cr, Ni, Cu, Zn, As, Cd, Hg and Pb in soil Ikenaka et al. were referenced from ICRCL (International Committee on the Redevelopment Contaminated Land, 1987) in the UK.

TABLE 1: Sampling locations.

Sampling points	Latitude	Longitude	Remarks
Lusaka 1	15°26'50.9"S	28°16'13.9"E	Road side
Lusaka 2	15°24'41.1"S	28°16'04.5"E	Industrial area
Lusaka 3	15°23'00.1"S	28°16'08.1"E	Industrial area
Lusaka 4	15°23'52.6"S	28°18'30.4"E	Road side
Lusaka 5	15°22'38.4"S	28°22'15.9"E	Industrial area
Lusaka 6	15°23'23.0"S	28°14'14.9"E	Industrial area
Lusaka 7	15°23'00.1"S	28°16'08.1"E	Industrial area
Kabwe 1	14°28'15.6"S	28°25'23.5"E	Mining
Kabwe 2	14°26'29.0"S	28°26'49.0"E	Mining
Kabwe 3	14°27'25.3"S	28°25'48.8"E	Mining

4.2.6 STATISTICAL ANALYSIS

Each sample was classified using a cluster analysis according to the Euclidean distance based on the composition ratio of each heavy metal. Significant differences ($p < 0.05$) for each sample group were analyzed using either the Mann-Whitney U test or Tukey's test. The significance of correlations was analyzed using the Pearson product-moment correlation coefficient ($p < 0.05$). Statistical analyses were performed using JMP 7.0.1 (SAS Institute, Cary, NC, USA).

4.3 RESULTS AND DISCUSSION

4.3.1 CONCENTRATION AND DISTRIBUTION OF HEAVY METALS IN ZAMBIA

Table 2 shows the heavy metal concentrations in soil samples in each area. A cluster analysis was performed to identify the accumulation pattern in each soil sample using the relative proportions of the ten heavy metals (Figure 2). The results of the cluster analysis divided the 47 individual soil samples into three major groups. Cluster 1 mainly included Kabwe samples. Cluster 2 mainly included the Northern area samples, in particular, the Copperbelt. Cluster 3 mainly included rural towns in the Eastern, Western and Southern areas. Figure 3 shows the concentration ratios of each heavy metal in these three clusters. It shows that each cluster had a characteristic composition. High ratios of Zn (57%) and Pb (32%) were found in Cluster 1, while a high Cu ratio (63%) was observed in Cluster 2. The major components of Cluster 3 were Cr (21%) and Cu (22%), but no individual component exceeded 50%. Interestingly, in Cluster 3, positive correlations were observed between the population of each town and the Zn and Pb concentrations (Figure 4).

These results show that heavy metal pollution in Zambia includes very high Zn, Pb, and Cd concentrations compared with the values recommended by the UK ICRCL (Tables 2, 3 and 4). Furthermore, heavy metal pollution in Zambia has strong regional differences (Table 2, Figures 2 and 3).

4.3.2 KABWE (CLUSTER 1)

In Kabwe, concentrations of Zn, As, Cd, Pb were significantly higher than in other areas (Table 2). Mining and smelting of Pb, Zn and Cd are the major industries around Kabwe. A previous study also showed that mining around Kabwe was responsible for heavy metal pollution, especially by Pb (Tembo et al., 2006). That paper indicated that the heavy metal concentrations decreased with increasing distance from the mine, confirming that mining activities are the main cause of soil contamination. Lead (Pb) toxicity causes many diseases including hematological, gastrointestinal and neurological dysfunctions, and nephropathy (Lockitch, 1993). It is reported that children have a greater susceptibility to Pb toxicity because intestinal absorption of Pb is five times greater in children than in adults. The Blacksmith Institute (2007) reported that blood Pb concentrations of 200 g/dL or more have been recorded in children in some neighborhoods in Kabwe. These records also showed average blood Pb levels for children in Kabwe ranging between 50 and 100 g/dL. On average, children's blood Pb levels in Kabwe were 5 to 10 times the permissible WHO/EPA maximum of 10 g/dL. It was observed that children who played in the soil were most susceptible to Pb pollution caused by mines and smelters.

In the present study, positive correlation coefficients were observed between Pb and Cd, Zn and Cd, As and Pb and As and Zn (Table 5). Because Cd and As are frequently found with Pb (Tembo et al., 2006; Ratnaike, 2003), the Cd and As pollution observed is also related to Pb and Zn mining. The maximum values of Cd and As in Kabwe were 18.7 and 51.5 mg/kg dry-wt, respectively. These concentrations of Cd and As could have the potential for poisoning, as the trigger values for Cd and as are reported to be 3.0 and 10.0 mg/kg (ICRCL, 1987; UK, 1987), respectively. Therefore, in Kabwe, pollution by Cd and As should not be ignored. In Chenzhou City, one of the oldest and largest mining cities in China, high concentrations of Cd were reported in paddy soil (0.35 – 48.3 mg/kg) and rice (0 – 4.4 mg/kg) which exceeded the WHO guideline (Zhai et al., 2008). In that

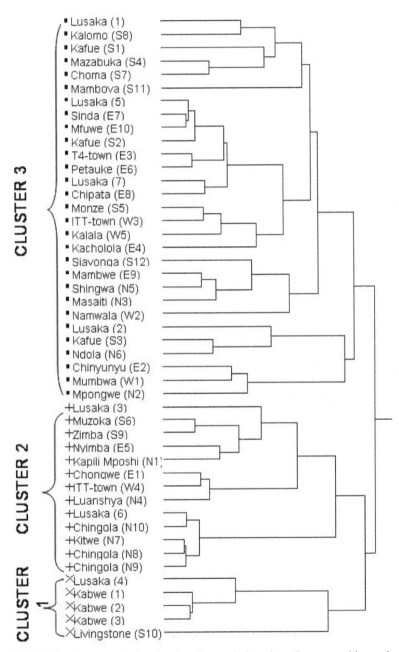

FIGURE 2: Cluster analysis of each soil sample based on the composition ratios of ten heavy metals.

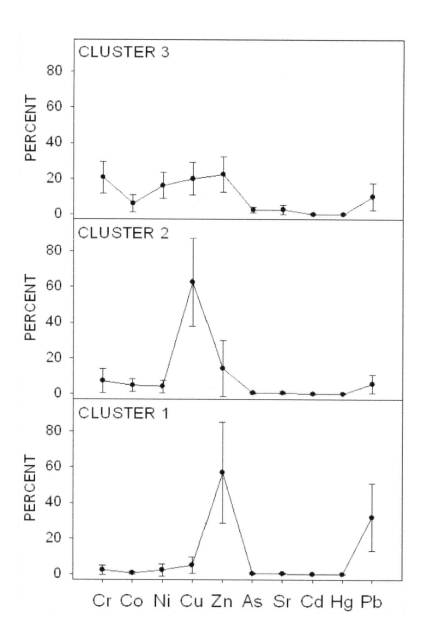

FIGURE 3: Concentration ratios of each heavy metal in the three clusters.

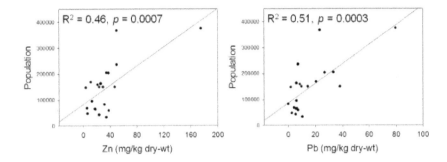

FIGURE 4: Correlation between the populations of each town and concentrations of Zn and Pb.

paper, the calculated total dietary Cd intake in the vicinity of the mine was similar to the intake of patients with Itai-itai disease. Kabwe soil could pose a risk to human health, particularly children playing around the mining area, not only due to the high Pb levels but also high Cd and As levels.

4.3.3 COPPERBELT AREA (CLUSTER 2)

High Co and Cu concentrations were observed in the northern area, particularly in the Copperbelt. Furthermore, the Co and Cu concentrations were strongly positively correlated (R2 = 0.93). It has been reported that the bedrock within the area contains sulfidic mineralization, rich in Co and Cu, embedded in carbonate rich shale and argillite (Mendelsohn, 1961). The high concentrations of Co and Cu observed in the northern area were considered to be the result of mining and smelting activities.

TABLE 2: Mean (± S.D.) and median values and concentration range (mg/kg dry-wt) of heavy metals in soils.

	Cr(mg/kg)	Co(mg/kg)	Ni(mg/kg)	Cu(mg/kg)	Zn(mg/kg)	As(mg/kg)	Sr(mg/kg)	Cd(mg/kg)	Hg(mg/kg)	Pb(mg/kg)
Lusaka (n = 7)										
Average	39±11A	11±9 B	20±7 BC	343±686 AB	147±278 B	4±1 B	9±10 A	0.11±0.10 B	0.02±0.01 B	48±49 B
Median	39	9	21	31	48	4	3	0.08	0.01	26
Range	27–52	3–31	7–31	19–1874	17–777	1–5	2–30	0.03–0.33	0.00–0.04	8–134
Kabwe (n = 3)										
Average	39±18 A	46±43 AB	47±9 A	572±597 AB	16991±20614 A	32±26 A	13±4 A	7.12±9.98 A	0.1±0.1 A	7076±8644 A
Median	33	35	51	330	8259	41	14	1.36	0.06	3398
Range	26–59	8–93	36–53	135–1252	2180–40534	3–52	9–16	1.36–18.65	0.03–0.22	880–16951
Eastern area (n = 10)										
Average	31±16A	11±14B	18±6BC	37±48B	32±27B	2±0B	3±2A	0.05±0.02B	0.00±0.00B	13±12B
Median	29	6	17	16	24	2	2	0.04	0.00	9
Range	14–61A	0–41	6–29	6–164	6–95	1–3	0–8	0.02–0.07	0.00–0.01	2–42
Western area (n = 5)										
Average	32±25 A	13±15 AB	23±7 BC	36±44 AB	22±8 B	3±1 B	3±4 A	0.04±0.03 B	0.01±0.00 B	5±3 B
Median	26	4	24	16	24	2	1	0.04	0.01	5
Range	7–72	0–32	14–33	9–113	9–32	2–4	0–10	0.03–0.09	0.01–0.01	0–9
Southern area (n = 12)										
Average	17±9A	7±9B	15±8C	39±40B	42±41B	3±1B	3±6A	0.06±0.02B	0.01±0.00B	27±35B
Median	15	4	12	27	32	3	2	0.06	0.01	18
Range	9–39	1–33	8–33	10–137	6–161	1–4	0–21	0.03–0.1	0.00–0.01	4–132

TABLE 2: *Cont.*

	Cr(mg/kg)	Co(mg/kg)	Ni(mg/kg)	Cu(mg/kg)	Zn(mg/kg)	As(mg/kg)	Sr(mg/kg)	Cd(mg/kg)	Hg(mg/kg)	Pb(mg/kg)
Northern area (n = 10)										
Average	33±29A	94±102A	29±10B	1646±2415A	99±113B	3±3B	23±41A	0.14±0.1B	0.02±0.02B	52±58B
Median	19	67	27	495	60	2	2	0.10	0.01	36
Range	7–78	1–307	11–45	10–7057	4–372	1–9	0–113	0.03–0.34	0.00–0.06	2–184

Different letter (A,B, C) indicate significant differences (Tukey's tet, $p < 0.05$).

TABLE 3: Mean (± S.D.) and median values and concentration range (mg/kg dry-wt) of heavy metals in river sediments.

	Cr (mg/kg)	Co (mg/kg)	Ni (mg/kg)	Cu (mg/kg)	Zn (mg/kg)	As (mg/kg)	Sr (mg/kg)	Cd (mg/kg)	Hg (mg/kg)	Pb (mg/kg)
Zambezi River (n = 8)										
Average	7±7B	0.9±1.51 B	13±11 B	6±4 B	6±8 B	3±1 B	1±1 A	0.047±0.008 A	0.004±0.002 B	1±1 B
Median	5	0	11	5	4	2	0	0.5	0.00	1
Range	3–24	0–4	3–39	4–15	1–25	1–3	0–3	0.04–0.06	0.00–0.01	0–5
Luangwa River (n = 5)										
Average	21±11 AB	3.5±3.5 B	17±8 B	16±18 B	14±12 AB	1±1 B	2±3 A	0.039±0.011 A	0.003±0.003 B	3±2 AB
Median	15	2	14	7	11	1	1	0.03	0.00	2
Range	14–39	1–7	9–27	6–48	2–33	1–3	0–8	0.03–0.05	0.00–0.01	1–6
Kafue River (n = 8)										
Average	24±8 AB	277±256 A	42±21 A	4745±4464 A	49±43 A	7±4 A	13±15 A	0.122±0.118 A	0.026±0.027 B	22±22 A
Median	23	256	46	5254	35	77	10	0.09	0.02	11
Range	8–38	6.5–748	2–75	23–12906	6–134	1–12	0–42	0.041–0.406	0.003–0.087	5–63
Rivers in Lusaka (n = 3)										
Average	36±39 A	14±20 AB	14±10 B	19±15 AB	65±42 A	3±2 AB	10±13 A	0.11±0.00 A	0.1±0.08 A	19±16 AB
Median	16	3	13	12	45	4	3	0.11	0.07	13
Range	10–81	2–37	4–24	9–36	37–113	0–5	3–25	0.11–0.12	0.04–0.19	6–37

Different letters (A, B,C) indicate significant differences (Tukey's test, p < 0.05).

TABLE 4: Maximum values (mg/kg dry-wt) for each heavy metal and sampling site.

	Maximum value	Maximum place	Recommended value*
Cr	99	Lake Kariba	5-200
Co	748	Mushishima Stream	-
Ni	98	Lake Kariba	40-70
Cu	12906	Mushishima Stream	50
Zn	40534	Kabwe	300
As	52	Kabwe	10
Sr	113	Ndola	-
Cd	19	Kabwe	3
Hg	0.2	Kabwe	0.5-1
Pb	16951	Kabwe	200

Recommended value by the ICRCL in UK (1987).

TABLE 5. Correlation coefficients (R2) for each heavy metal.

	Cr	Co	Ni	Cu	Zn	As	Sr	Cd	Hg	Pb
Cr	1.00									
Co	0.01	1.00								
Ni	0.54	0.20	1.00							
Cu	0.00	0.93	0.13	1.00						
Zn	0.00	0.00	0.02	0.00	1.00					
As	0.05	0.05	0.15	0.04	0.69	1.00				
Sr	0.24	0.16	0.34	0.15	0.00	0.04	1.00			
Cd	0.00	0.00	0.02	0.00	0.98	0.59	0.00	1.00		
Hg	0.02	0.09	0.07	0.08	0.41	0.35	0.10	0.43	1.00	
Pb	0.00	0.00	0.02	0.00	0.99	0.69	0.00	0.98	0.41	1.00

Bold type indicates a significant correlation between each heavy metal determined by the Pearson product-moment correlation coefficient ($p < 0.05$).

The heavy metal concentrations in river sediments are shown in Table 3. Average concentrations of Cu, Co, and Pb along the Kafue River, including Lake ITT, are shown in Figure 5. The concentration of Pb did not change significantly along the Kafue River. However, the Co and Cu concentrations increased significantly after the confluence with the

Mushishima Stream, which drains one of the key mining areas in the Copperbelt. Although, the Co and Cu concentrations decrease down the Kafue River, the Cu concentration observed in Lake ITT was still higher than the UK ICRCL guideline value (50 mg/kg dry-wt). This indicates that copper waste discharged into the upper reaches of the Kafue River was transported to and remained in such areas as Lake ITT, which is more than 450 km downstream from the major mining area. Pettersson and Ingri (2001) also found that dissolved concentrations of Cu (0.88 μmol/L) and Co (0.02 μmol/L) in water of the Kafue River about 100 km downstream from the mining areas in the Copperbelt were elevated compared to average world river concentrations (0.024 μmol/L for Cu and 0.003 μmol/L for Co). Their results support the findings of the present study with high Cu concentrations in the lower reaches of the Kafue River, including Lake ITT sediments (Figure 5).

Copper (Cu) is an essential trace element, but high concentrations of Cu may cause increased oxidative damage to lipids, proteins and DNA (Gaetke and Chow, 2003). Chronic Cu toxicity causes liver cirrhosis and tubular necrosis in the kidney (Gaetke and Chow, 2003; Barceloux, 1999). The high Cu concentrations measured in this study could affect humans, livestock and wildlife living in the area. A study by Syakalima et al. (2001) found that the Kafue Lechwe (Kobus leche kafuensis) which depends on the Kafue River in the Lochinvar and Blue Lagoon national parks had high Cu concentrations in the liver (42.8 ± 22.1 and 75.7 ± 26.4 mg/kg from the two national parks, respectively), which has potential to cause adverse effects. This report and our results suggest that the accumulation of Cu in the lower reaches of the Kafue River might even affect wildlife in National Parks.

4.3.4 LUSAKA AND OTHER AREAS (CLUSTER 3)

Unlike Kabwe and the Copperbelt, no individual heavy metal component was dominant in Lukasa (Table 2 and Figure 3), and no particular point source was identified, but complex sources have been assumed. For example, in Lusaka, higher concentrations of Cu (357 and 1,874 mg/kg dry-wt at Lukasa 3 and 6, respectively) were observed in industrial areas, but high concentrations of Zn (777 mg/kg dry-wt) and Pb (133 mg/kg dry-wt) were found in soil beside major roads (Lusaka 4). High concentrations of Pb (1521.8 ppm) and Cu (1197.6 ppm) were reportedly found in central Transylvania, Romania, which is known as a pollution source due to chemical and metallurgical activities (Suciu et al., 2008). Fatoki (1996) showed that there was an association between the Zn concentration and the distance from road traffic. High concentrations of Pb in topsoil have been reported in the vicinity of a battery factory in Nigeria (Onianwa and Fakayode, 2000). These results indicate that human activities such as the metal industry and combustion of fossil fuels might be major sources of heavy metal contamination in Lusaka. As Lusaka is the capital city and has 10% of the country's population, a number of people are at risk from this pollution.

In other areas, mainly the Eastern, Western and Southern areas, the levels of heavy metals were relatively low. However, we found positive correlations between the population of each town and concentrations of Zn ($R^2 = 0.46$) and Pb ($R^2 = 0.51$) (Figure 4). Previous reports indicate that these pollutants are commonly found in automobile and waste incinerator exhaust emissions (Bradl, 2005). Our findings indicated that the main sources of heavy metals in rural areas of Zambia are human activities such as combustion of fossil fuels.

4.4 CONCLUSION

In Zambia, the major sources of heavy metals are the mining areas, Kabwe and the Copperbelt, and heavy metals are then transported within each area by rivers. Even sediments in national park areas are polluted with high concentrations of Cu and moderate levels of Pb, indicating that mining is the source of the heavy metals. In areas geographically distant from mines, the heavy metal concentrations are moderate or low, but metals were detected in almost areas of Zambia. The findings are summarized as follows:

1. Heavy metal pollution in Zambia has strong regional differences.
2. Kabwe is highly polluted by Pb, Zn, Cu, Cd and As.

3. The Copperbelt area is highly polluted by Cu and Co.
4. Other sampling sites showed relatively low concentrations of heavy metals. However, the heavy metal pollution is currently increasing and caused by human activities. Furthermore, sources in each area were not only mining but other human activities such as metal industries and combustion of fossil fuels.
5. High concentrations of heavy metals, especially Cu, were found in the aquatic environment in the Copperbelt area. This also affected concentrations in Lake ITT, located 450 km downstream on the Kafue River, and in the Kafue National Park.

REFERENCES

1. Aguilar A, Borrel A, Reijnders PJH (2002). Geographical and temporal variation in levels of organochlorine contaminants in marine mammals. Marine Environ. Res., 53: 425-452.
2. Akiwumi FA, Butler DR (2008). Mining and environmental change in Sierra Leone, West Africa: a remote sensing and hydrogeomorphological study. Environ. Monit. Assess., 142: 309-318.
3. Barceloux GD (1999). Copper. Clin. Toxicol, 37:217-230.
4. Blacksmith Institute (2007). The World's Worst Polluted Places, The top ten of the dirty thirty. pp. 1-69.
5. Bradl HB (2005). Heavy metals in the environment: Origin, interaction and remediation. London: Elsevier Academic Press.
6. Von Braun CM, Von Lindern HI, Khristoforova KN, Kachur HA, Yelpatyevsky VP, Elpatyevskaya PV, Spalinger MS (2002). Environmental lead contamination in the Rundnaya Pristan – Dalnegorsk mining and smelter district, Russian Far East. Environ. Res., Section A, 88: 164-173.
7. Fatoki SO (1996). Trace zinc and copper concentration in roadside surface soils and vegetation – measurement of local atmospheric pollution in Alice, South Africa. Environ. Int., 22: 759-762.
8. Gaetke ML, Chow KC (2003). Copper toxicity, oxidative stress, and antioxidant nutrients. Toxicology, 189: 147-163.
9. Higueras P, Oyarzun R, Oyarzun J, Maturana H, Lillo J, Morata D (2004). Environmental assessment of copper-gold-mercury mining in the Andacollo and Punitaqui districts, northern Chile. Appl. Geochem., 19: 1855-1864.
10. Ikenaka Y, Eun H, Watanabe E, Kumon F, Miyabara Y (2005a). Estimation of sources and inflow of dioxins and polycyclic aromatic hydrocarbons from the sediment core of Lake Suwa, Japan. Environ. Pollut., 138: 530-538.
11. Ikenaka Y, Eun H, Watanabe E, Miyabara Y (2005b). Sources, distribution, and inflow pattern of dioxins in the bottom sediment of Lake Suwa, Japan. Bull. Environ. Contaminat. Toxicol., 75: 915-921.

12. Interdepartmental Committee on the Redevelopment of Contaminated Land (ICRCL) (1987). Guidance on Assessment and Redevelopment of Contaminated Land, 2nd Edition. ICRCL Central Directorate on Environmental Protection, Department of the Environment Circular 59/83, London.

13. Kodom K, Wiafe-Akenten J, Boamah D (2010). Soil Heavy Metal Pollution along Subin River in Kumasi, Ghana; Using X-Ray Fluorescence (XRF) Analysis. Conference Information: 20th International Congress on X-Ray Optics and Microanalysis, X-ray Optics and Microanalysis, Proceedings Book Series: AIP Conf. Proc., 1221: 101-108.

14. Lacatusu R, Citu G, Aston J, Lungu M, Lacatusu AR (2009) Heavy metals soil pollution state in relation to potential future mining activities in the Rosia Montana area. Carpathian J. Earth Environ. Sci., 4: 39-50.

15. Lockitch G (1993). Perspectives on lead toxicity. Clin. Biochem., 26: 371-381.

16. Mendelsohn F (1961). The Geology of the Northern Rhodesian Copperbelt. London: Macdonald & Co.

17. Nakayama MMS, Ikenaka Y, Muzandu K, Choongo K, Oroszlany B, Teraoka H, Mizuno N, Ishizuka M. (2010). Heavy Metal Accumulation in Lake Sediments, Fish (Oreochromis niloticus and Serranochromis thumbergi), and Crayfish (Cherax quadricarinatus) in Lake Itezhitezhi and Lake Kariba, Zambia. Arch. Environ. Contaminat. Toxicol., 59(2):291-300.

18. Norman R, Mathee A, Barnes B, van der Merwe L, Bradshaw D, the South African Comparative Risk Assessment Collaborating Group (2007). Estimating the burden of disease attributable to lead exposure in South Africa in 2000. S. Afr. Med. J., 97: 773-780.

19. Nwankwo JN, Elinder CG (1979). Cadmium, lead and zinc concentrations in soils and in food grown near a zinc and lead smelter in Zambia. Bull. Environ. Contam. Toxicol., 22: 625-631.

20. Oelofse S (2008). Mine water pollution–acid mine decant, effluent and treatment: a consideration of key emerging issues that may impact the state of the environment. Emerging Issues Paper: Mine Water Pollut., 2008.

21. Onianwa CP, Fakayode OS (2000). Lead contamination of topsoil and vegetation in the vicinity of a battery factory in Nigeria. Environ. Geochem. Health, 22: 211-218.

22. Pettersson UT, Ingri J (2001). The geochemistry of Co and Cu in Kafue River as it drains the Copperbelt mining area, Zambia. Chem. Geol., 177: 399-414.

23. Rashad S, Barsoum MD (2006). Chronic kidney disease in the developing world. New Engl. J. Med., 354: 997-999.

24. Ratnaike NR (2003). Acute and chronic arsenic toxicity. Postgrad. Med. J., 79: 391-396.

25. Razo I, Carrizales L, Castro J, Barriga DF, Monroy M (2004). Arsenic and heavy metal pollution of soil, water and sediments in a semi-arid climate mining area in Mexico. Water Air Soil Pollut., 152: 129-152.

26. Stockwell LE, Hillier JA, Mills AJ, White R (2001) World mineral statistics 1995-99, Keyworth, Nottingham. British Geological Survey, 2001.

27. Suciu I, Cosma C, Todic M, Bolboac SD, Jäntschi L (2008). Analysis of Soil Heavy Metal Pollution and Pattern in Central Transylvania. Int. J. Mol. Sci., 9: 434-453.

28. Syakalima M, Choongo K, Nakazato Y, Onuma M, Sugimoto C, Tsubota T, Fukushi H, Yoshida M, Itagaki T, Yasuda J (2001). An investigation of heavy metal exposure and risks to wildlife in the Kafue flats of Zambia. J. Vet. Med. Sci., 63: 315-318.

29. Tembo DB, Sichilongo K, Cernak J (2006). Distribution of copper, lead, cadmium and zinc concentrations in soils around Kabwe Town in Zambia. Chemosphere, 63: 497-501.

30. Von Braun CM, Von Lindern HI, Khristoforova KN, Kachur HA, Yelpatyevsky VP, Elpatyevskaya PV, Spalinger MS (2002). Environmental lead contamination in the Rundnaya Pristan – Dalnegorsk mining and smelter district, Russian Far East. Environ. Res., Section A, 88: 164-173.

31. Zhai L, Liao X, Chen T, Yan X, Xie H, Wu B, Wang L (2008). Regional assessment of cadmium pollution in agricultural lands and potential health risk related to intensive mining activities: A case study in Chenzhou City, China. J. Environ. Sci., 20: 696-703.

CHAPTER 5

HUMAN EXPOSURE PATHWAYS OF HEAVY METALS IN A LEAD-ZINC MINING AREA, JIANGSU PROVINCE, CHINA

CHANG-SHENG QU, ZONG-WEI MA, JIN YANG, YANG LIU, JUN BI, AND LEI HUANG

5.1 INTRODUCTION

Heavy metals are known to be persistent in the human body, with excretion half-lives that last for decades. Heavy metals can lead to a wide range of toxic effects, such as carcinogenicity, mutagenicity and teratogenicity [1], [2], [3], [4]. Heavy metal contamination is a major environmental concern on a global scale, particularly in China, with its rapid economic development [5], [6], [7]. Due to large-scale production and consumption and lack of regulations, heavy metals such as lead (Pb), zinc (Zn), cadmium (Cd), mercury (Hg), and chromium (Cr) are emitted into the environment in large quantities through wastewater irrigation, solid waste disposal, sludge application, vehicular exhaust and atmospheric deposition [8]. As a result, heavy metals are present in industrial, municipal and urban runoff, and

This chapter was originally published under the Creative Commons Attribution License. Qu C-S, Ma Z-W, Yang J, Liu Y, Bi J, and Huang L. Human Exposure Pathways of Heavy Metals in a Lead-Zinc Mining Area, Jiangsu Province, China. PLoS ONE 7,11 (2012), doi:10.1371/journal.pone.0046793.

they continuously accumulate in the environment in China [9], [10], [11], [12], [13]. Since 2005, health related incidents caused by heavy metal pollution have risen sharply in China, with major accidents attracting nationwide attention [14].

Heavy metals have been found widely in various environmental media (including soil, water, air and food) around the world [5], [15], [16]. These species may enter the human body through inhalation of dust, direct ingestion of soil and water, dermal contact of contaminated soil and water, and consumption of vegetables grown in contaminated fields. Various studies have been conducted to evaluate population health risks due to heavy metal exposure through various exposure pathways, especially soil and food chain [17], [18], [19], [20], [21], [22]. Since it is difficult to identify the key exposure route because of lack of multipathway risk analysis, media or pathway-specific approach may fail to ensure public safety [23]. Therefore, it is necessary to assess the aggregate exposure to metals concerning about different environmental media and pathways. Previous studies have also demonstrated the importance of conducting multipathway risk assessment to identify the dominant pathway of potential concern [24], [25], [26], [27].

Human health risk assessment, as formalized in 1983 [28], has been recognized as an important tool for estimating the nature and probability of adverse health effects in humans who may be exposed to chemicals and for presenting risk information to the decision maker. The US Environmental Protection Agency (EPA) hazard quotients (HQ) are widely used to characterize non-carcinogenic health effects posed by heavy metals by comparing the exposure level with a reference dose [15], [22]. However, risk assessment is a complex process that is inherently linked with uncertainty [29]. Monte Carlo simulation is commonly used to quantify uncertainty in human health risk estimates [22]. It is a probabilistic approach that works with probability distributions rather than deterministic values of each parameter to estimate a risk.

For metal toxicity monitoring and environmental risk assessment, the identification of heavy metals from biological samples such as blood, urine or hair is useful for identifying exposure. Because the sampling is less invasive, more convenient to store and transport, and less hazardous to handle, hair has been used widely in biomonitoring environmental and occupational exposures of various pollutants [30], [31], [32], [33], [34]. Furthermore, hair sample can be a useful assessment tool in characterizing long-term exposure of the measured contaminant, whereas blood and urine often reflect most recent exposures. Hair has been used in biomonitoring of heavy metals on large cohorts in Brazil [35], determining geological source and exposure through fish consumption in Lake Victoria [36], characterizing human exposure in an abandoned mine in Portugal [32] and examining the residential exposure in an e-waste recycling area in southeastern China [31]. We focus our study on using metal content in hair as a biomonitoring indicator to identify whether differences of long-term exposure existed among residents in different villages.

The current study considered human exposure to heavy metals via drinking water, dietary intake, dermal contact and inhalation in the populations living near a lead-zinc mining area in Jiangsu Province, China. In addition to Pb and Zn, eight other major associated elements were included in this study: Ag (silver), Cd, Cr, Cu (copper), Ni (nickel), Se (selenium), Tl (thallium) and Hg (mercury). These ten metals are all priority pollutant metals (PP metals) as set by the USEPA. The aims of this study are: (1) to evaluate the potential health impacts of these metals on the general population in the mining area, (2) to provide a better understanding of each exposure pathway and (3) to help local governments to prioritize pollution control and health intervention policies to protect local population.

5.2 MATERIALS AND METHODS

5.2.1 ETHICS STATEMENT

This study was approved by the review board of Nanjing University. All participants were informed about the objectives and methods of the study before the investigation. And written consent was obtained from all participants.

5.2.2 STUDY AREAS

The studied Qixia lead-zinc mining area, with reserves of four million tons of lead and zinc, is an important mining region in Jiangsu province, which is one of the fastest-developing provinces in China. The mine has been excavated for 60 years, and the nearby area is suffering serious environmental deterioration from the wastewater and solid waste discharge. Three nearby villages at the leeward side of the mine were sampled in this study to identify the potential health risks and its sources. Village 1 (V1) is 100 m from the mine, village 2 (V2) is 600 m from the mine, and village 3 (V3) is approximately 1 km from the mine. There are 32, 38 and 30 households in V1, V2 and V3 respectively, and they have similar population structures, living conditions and lifestyle. The geographical locations of the three sampled villages are shown in Figure 1.

5.2.3 SAMPLING

The drinking water of the 3 villages is tap water, which is provided by the same water supply company. We collected 10 samples from local families at random to represent the whole area. The water samples were collected in precleaned polyethylene bottles. After acidification with HCl, the samples were placed in an ice bath, then transported to the laboratory and kept at −20°C until they were analyzed.

5 soil samples were collected at the edges of residential streets and crop fields in each village. All soil samples were taken from the upper 5 cm of ground and stored in polyethylene bags. Each sample (1 kg approximately) consisted of 5 subsamples collected in an area of 100 m2, pooled and homogenized to form a representative sample. The samples were air-dried, and stones and coarse plant roots or residues were removed. They were then thoroughly mixed, crushed, passed through a 2-mm mesh sieve and a 0.149-mm mesh sieve, and then stored in polyethylene bottles at ambient temperature prior to chemical analysis.

From January to March 2011, air sampling was conducted simultaneously in the 3 villages. 20 samples in each village and total 60 indoor and

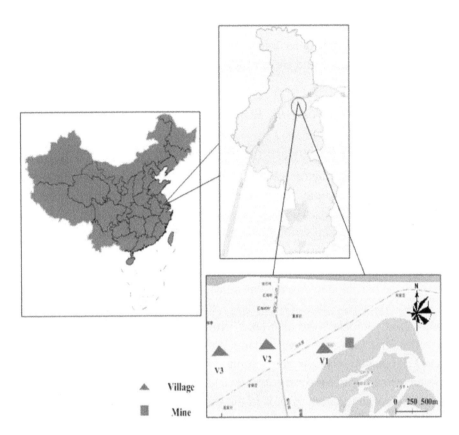

FIGURE 1: Locations of the study areas.

outdoor samples were collected at different sites in and around the study area. The indoor air samples were obtained from the rooms of randomly chosen families. The air samples were collected onto filter membranes using a PM10 air sampler (Tianhong instruments Co. Ltd, Wuhan, China) run for 3 h at a flow rate of 100 L/min. For the detection of trace Hg, the sulfhydryl cotton enrichment method was used.

In each village, we selected 4 fields at different directions, and locally produced vegetable samples were harvested in double in a quadrant of 0.25 m² randomly selected from each site. In total, 24 vegetable samples were collected in the study area. Food produced out of town, including rice, pork and eggs, was collected from the markets. The samples were cut into small pieces after thoroughly washing with tap water and deionized water. After drying to constant weight at 80°C in an oven, the food samples were ground and sieved for acid digestion.

5.2.4 HUMAN EXPOSURE SURVEY

A questionnaire-based survey was conducted in the studied villages to determine key risk factors such as dietary behaviors, daily activities and lifestyle of local people. We invited 50 local residents to participate the survey in each village. All participants were selected randomly. In V1, the research staff obtained written informed consent from 42 participants who are permanent residents of this village, and there were 42 participants in V2 and 36 participants in V3. The environmental samples were collected from these participants' immediate surroundings.

To characterize the exposure level and metal accumulation in human body, hair samples were also taken from the above survey participants. Twenty-nine, 30 and 13 participants provided usable samples in V1, V2 and V3, respectively. The hair samples were analyzed for the 10 metals to characterize the in vivo exposure levels and potential differences among the different indigenous groups in the study area.

5.2.5 ANALYTICAL METHODS

Except for Hg, the USEPA 200.8 method was used to determine the trace metals in the water and soil samples, which were measured with inductively coupled plasma-mass spectrometry (ICP-MS) (Agilent 7500i, Agilent Scientific Technology Ltd., USA). For the airborne particle samples, an atomic absorption spectrometer (AAS, Vario 6, Jena Co., Ltd., Germany) was used to analyze the filters loaded with heavy metals according to the standard examination method for ambient air in China. Using the USEPA 6020A method, the concentrations of metals in the food and hair samples were determined with ICP-MS. The determination of Hg in the water, soil, food and hair samples was conducted using thermal decomposition, amalgamation and atomic absorption spectrophotometry (TDA/AAS) (AMA

254, LECO Co., Ltd., USA) according to the USEPA 7473 method. Trace Hg in the air, enriched in sulfhydryl cotton, was determined using cold vapor atomic absorption spectrophotometry (CVAAS) (JKG-205, Jilin Scientific Technology Co., Ltd., China) according to the NIOSH 6009 method. The detection limit of the measurements was defined as the concentration value, which is numerically equal to three times the standard deviation of 10 replicate blank measurements. Reagent blanks and standard reference materials were used in the analysis for quality assurance and quality control. The recoveries of the elements ranged from 90% to 110%.

5.2.6 RISK CALCULATION

The average daily intake (ADI) of metals by the human subjects was calculated using the following equation, which is recommended by the USEPA [37]. The equation links the time-averaged dose to the exposure medium concentration.

$$ADI = (C \times IR \times EF \times ED) / (BW \times AT) \tag{1}$$

where ADI is the average daily intake or dose through ingestion or inhalation (mg/kg-day); C is the chemical concentration in the exposure medium (mg/L, mg/kg, or mg/m^3); IR is the ingestion rate (L/day, kg/day, or m^3/day); EF is the exposure frequency (day/year); ED is the exposure duration (year); BW is the body weight (kg) and AT is the time period over which the dose is averaged (day).

$$ADI_D = (C \times SA \times AF \times ABS \times EF \times ED) / (BW \times AT) \tag{2}$$

For exposure dose through dermal contact (ADI_D) calculation, SA is the exposed skin surface area (cm^2), AF is the adherence factor (mg/cm^2/day), and ABS is the dermal absorption factor.

To calculate the exposure dose of each pathway, the ingestion rate of drinking water and food were obtained through the above questionnaire-based exposure survey. In addition, the ingestion rate of soil, inhalation rate of air, the adherence factor (AF), and the dermal absorption factor (ABS) were obtained from open database and literature [38], [39].

The human health risks posed by heavy metal exposure are usually characterized by the hazard quotient (HQ) [37], the ratio of average daily intake to the reference dose (RfD) or the reference concentration in air (RfC) for an individual pathway and chemical. A quotient under 1 is assumed to be safe. When the HQ value exceeds 1, there may be concerns for potential health risks associated with overexposure. The RfD and RfC values used in this study were recommended by open chemical databases and shown in Table 1. To assess the overall potential for health effects posed by more than one metal, summing HQs across metals can serve as a conservative assessment tool to estimate high-end risk rather than low-end risk to protect the public. In this study, the total HQ was used as a screening value to identify whether there is significant risk caused by metals, and any difference of total health risk existed among the 3 villages.

TABLE 1: Metal reference doses (RfD).

Metal	RfD (mg/kg-d)	Source	RfC (mg/m³)	Source
Ag	5.0E–3	IRIS[a]	–	
Cd	1.0E–3	IRIS	1.0E–5	ATSDR[b]
Cr	1.5E+0	IRIS	–	
Cu	4.0E–2	HEAST[c]	–	
Ni	2.0E–2	IRIS	9.0E–5	ATSDR
Pb	1.4E–4	Oak Ridge[d]	–	
Se	5.0E–3	IRIS	2.0E–2	Cal EPA[e]
Tl	3.0E–6	IRIS	–	
Zn	3.0E–1	IRIS	–	
Hg	1.6E–4	Cal EPA	3.0E–4	IRIS

[a]*Integrated Risk Information System, U.S. EPA;* [b]*The Agency for Toxic Substances and Disease Registry, U.S.;* [c]*Health Effects Assessment SUmmary Tables, U.S. EPA;* [d]*Oak Ridge National Laboratory, U.S.;* [e]*California Environmental Protection Agency, U.S.*

To accommodate the uncertainties associated within the calculation process, Monte Carlo simulation technique was used based on Crystalball software (Oracle Corporation, Vallejo, US) and considering 10,000 iterations. Before this process, distribution characteristics of each exposure parameter were tested according to the exposure survey results. A probabilistic distribution of the exposure dose and HQ values was then obtained as simulation result.

5.3 RESULTS

5.3.1 HEAVY METAL LEVELS IN DIFFERENT ENVIRONMENTAL MEDIA

The analysis results show that the tap water in the studied area is safe (unpublished data), as all the trace heavy metal concentrations are far less than the national drinking water quality criteria.

Indoor air samples were obtained from 13, 15, and 12 households in V1, V2, and V3, respectively. The chemical analysis results (unpublished data) show that Cu, Zn and Pb are the main pollutants. V1 was the most seriously polluted among the studied villages. The average metal levels there were 6.1 $\mu g/m^3$ for Cu, 2.5 $\mu g/m^3$ for Zn and 2.1 $\mu g/m^3$ for Pb, with a detection rate of nearly 100%. The chemical analysis results for the outdoor air show that Zn and Pb have the highest concentrations. In V1, the median metal levels were 4.2 $\mu g/m3$ for Zn and 2.4 $\mu g/m^3$ for Pb, with a detection rate of 93.8% and 100% respectively.

The chemical analysis results of the soil samples show that 10 PP metals were detected in all samples (unpublished data). In general, V1 is the most seriously polluted, followed by V2, and the soil quality is good in V3. In V1, the soil Pb concentration is as high as 2507 mg/kg, and the average soil zinc concentration as high as 9281 mg/kg, which is 100 times greater than that of V3.

It is well known that vegetables absorb metals from the soil. The self-produced vegetables contained higher metal concentrations in V1, especially of Zn and Pb. The average Zn concentrations in pakchoi in V1, V2 and V3 were 10.48, 5.24 and 2.34 mg/kg, respectively. The average Pb concentrations reached 0.24 and 0.08 mg/kg in V1 and V2, respectively, whereas Pb was not detected in V3. Different from the vegetables, the metals could not be detected in most of the imported egg, pork and rice samples.

TABLE 2: Demographic, lifestyle, and dietary characteristics of the local residents.

Characteristics	Mean	SD	Unit
Weight	62.58	10.23	kg
Activity			
Indoors	8.47	3.99	h
Outdoors	5.84	3.98	h
Sleep	9.24	2.14	h
Dietary			
Rice	360.29	71.92	g/d
Flour	6.29	17.38	g/d
Pakchoi	171.76	82.74	g/d
Cabbage	63.53	38.66	g/d
Spinach	30.44	19.36	g/d
Celery	24.26	11.49	g/d
Pork	56.85	25.01	g/d
Egg	45.44	25.97	g/d
Water ingestion	1.88	0.62	L/d
Exposure duration	35.35	18.68	a

SD: Standard Deviation

5.3.2 EXPOSURE FACTORS

The demographic, activity and lifestyle questionnaire results are listed in Table 2. The periods that local residents have lived in this mining area range from 2 to 35 years. Local residents spend more time indoors than outdoors. The people in this rice-producing region mainly feed on rice, pakchoi, cabbage and pork. The local people almost do not eat flour, whereas the national average level was 140 g/person/day [40]. However, the consumption rate of rice for the local people was 360 g/person/day, much greater than the national average level as 238 g/person/day [40]. The consumption rate of vegetables, pakchoi dominantly, for the local people was 310 g/person/day, higher than the national average level as 276 g/person/day [40]. In this area, the vegetables are mostly self-grown, whereas other foodstuffs are mainly purchased from the market. Additionally, no difference of demographic, activity and lifestyle characteristics among these 3 villages was found at such spatial scale.

TABLE 3: Hazard quotients of different metals in each village.

Metal	V1		V2		V3	
	Mean	SD	Mean	SD	Mean	SD
Ag	0.0009	0.0005	0.0003	0.0001	0.0002	0
Cd	3.32	1.91	1.49	0.843	1.81	0.812
Cr	0.0001	0.0008	0.001	0.0005	0.0009	0.0005
Cu	0.189	0.046	0.164	0.039	0.18	0.046
Ni	0.42	0.15	0.41	0.15	0.35	0.15
Pb	16.2	6.96	3.63	1.33	3.01	0.92
Se	0.018	0.015	0.018	0.016	0.019	0.016
Tl	0.17	0.062	0.058	0.023	0.097	0.13
Zn	0.30	0.076	0.29	0.073	0.28	0.073
Hg	2.84	1.06	2.28	0.65	1.83	0.50

5.3.3 HEALTH RISKS

The health risks for the local population due to exposure to metals were evaluated using Monte Carlo simulation. The cumulative probability distribution of the total HQ values in the three villages is presented in Figure 2. For V1, the HQ ranged from 6.8 to 81.3; for V2, the HQ ranged from 2.9 to 19.3; and for V3, the HQ ranged from 3.3 to 15.8. The mean HQ values of V1, V2 and V3 were 23.4, 8.3 and 7.5, respectively. Obviously, the HQ of V1 was much higher than those of the other two.

The heavy metal HQ values from dietary sources, dust inhalation (both indoor and outdoor air), water ingestion and dermal contact, and soil in-

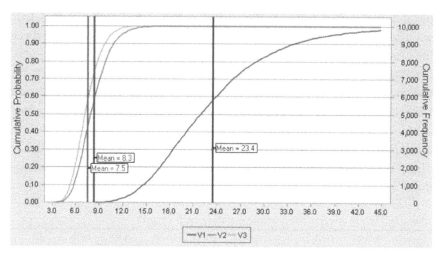

FIGURE 2: Cumulative probability distribution of the total HQs in the three villages. The health risks were evaluated by means of a Monte Carlo simulation based on Crystalball software for 10,000 iterations.

gestion and dermal contact were aggregated. The Monte Carlo simulated health risk assessment results are shown in Table 3. There is a significant difference of Pb, Cd and Hg HQs among the three villages. In V1, the average HQ value of Pb is as high as 16.2, indicating that the chronic daily intake dose exceeds the safe reference dose by 15 fold. The HQ values of Pb in V2 and V3 reach 3.6 and 3.0, respectively. The chronic daily intake dose of Cd and Hg also exceeds the corresponding limits, and the HQ in V1 was greater than those in V2 and V3.

As Figure 3 shows, soil ingestion was the primary pathway to Pb exposure for V1, accounting for 40.3% of the total HQ. Food consumption contributed nearly one-third to the total HQ, and among different food sources, pakchoi were the most significant, contributing 77%. Inhalation contributed nearly one-fifth to the total HQ. Furthermore, indoor air contributed twice as much as did outdoor air. For the other two villages, the proportion of inhalation increased, whereas those of food ingestion and soil ingestion decreased sharply. This result is consistent with trend of Pb levels in the soil with increasing distance from the mine.

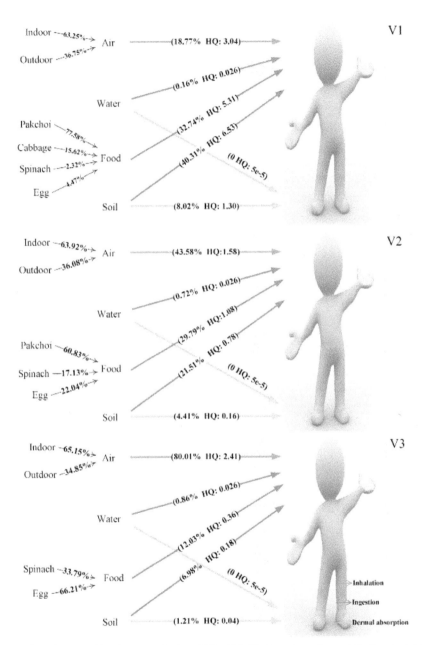

FIGURE 3: Multipathway analysis of HQ (Pb). Each pathway's contribution to total Pb exposure of local residents in the three studied villages was calculated based on average HQ values.

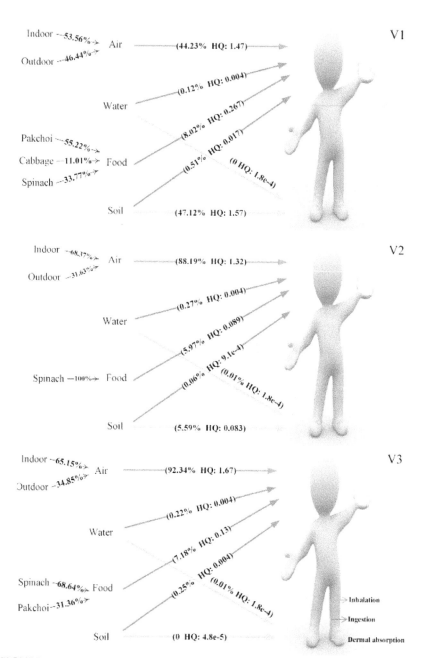

FIGURE 4: Multipathway analysis of HQ (Cd). Each pathway's contribution to total Cd exposure of local residents in the three studied villages was calculated based on average HQ values.

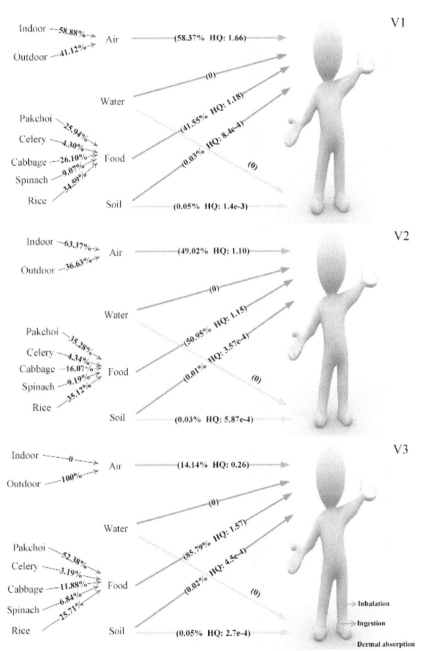

FIGURE 5: Multipathway analysis of HQ (Hg). Each pathway's contribution to total Hg exposure of local residents in the three studied villages was calculated based on average HQ values.

For Cd (Figure 4), the exposure to and uptake of this non-essential element was mainly from soil dermal contact, which contributed 47.1% to the total HQ in V1. Inhalation also contributed 44.3% to the total risk, and indoor air contributed slightly more than outdoor air. However, in V2 and V3, the total HQ decreased with increasing distance from the mine, and the contribution of soil dermal contact decreased sharply with an evident increase of inhalation. The ingestion of water and food is safe for local residents.

The local residents' average daily exposure to Hg also exceeded the safe reference dose, and the HQ value of the residents in V1 was the highest. Taking V1 as an example, the multipathway analysis (Figure 5) showed that the people of that village are exposed to Hg mainly through inhalation, with indoor air contributing the most. Food also contributed 41.6% to the total risk, mainly due to self-produced broad-leaf vegetables, such as pakchoi and cabbage, whereas the risk posed by rice from the market was not trivial. Different from Pb and Cd, soil pollution only contributes a minor fraction. V2 and V3 had similar profiles of metal levels.

5.3.5 HEAVY METAL LEVELS IN HAIR

The metal concentrations in the hair samples are shown in Table 4. Nine metals were detected in the local residents' hair samples. The concentration of Zn was the highest of the detected metals. The geometric mean of Zn reached 240 mg/kg, which is much greater than those of the other metals. The hair samples from the residents in V1 contained higher Pb (8.14 mg/kg) and Hg (0.86 mg/kg) concentrations than those from V2 and V3 (P<0.05). This result is consistent with the variation of the HQ values. For the other metals, no significant differences or strong relationships with the HQ were found among the three villages.

5.4 DISCUSSION

The environmental sample analysis results showed widespread heavy metal contamination in the air, especially of Pb, Zn and Cu. The nearest village (V1) to the mine had the heaviest pollution. Due to its respiratory, neurological, and potential carcinogenic effects, Pb is regulated by the National Ambient Air Quality Standards with seasonal mean levels not to exceed 1.0 μg/m3 and annual mean levels not to exceed 0.5 μg/m³ [41]. The proposed WHO guideline for Pb is 0.5 μg/m³ [42]. The average Pb concentrations in the three villages all exceeded the limit. The highest Pb concentration occurred in the indoor air in V1, which reached 5.9 μg/m³, and this concentration was approximately 6-fold that of the national limit.

TABLE 4: Heavy metal levels in hair (mg/kg).

Metals	LOD	V1		V2		V3	
		Mean (SD)	Detected Proportion	Mean (SD)	Detected Proportion	Mean (SD)	Detected Proportion
Ag	0.05	--	0	--	0	--	0
Cd	0.01	0.15 (0.42)	90.0%	0.08 (0.12)	63.3%	0.14 (0.17)	91.7%
Cr	0.05	1.60 (1.05)	100%	2.68 (2.47)	100%	2.46 (5.20)	100%
Cu	0.05	9.12 (7.88)	100%	11.13 (13.38)	100%	8.86 (3.52)	100%
Ni	0.05	0.44 (0.75)	86.7%	0.73 (0.89)	100%	0.71 (0.90)	91.7%
Pb	0.05	8.14 (12.30)	100%	2.26 (4.19)	100%	3.98 (4.69)	100%
Se	0.25	0.58 (0.37)	46.7%	0.77 (0.94)	76.7%	0.81 (0.86)	75.0%
Tl	0.05	0.07 (0.01)	6.7%	0.09 (0.01)	6.7%	0.10 (0.05)	41.7%
Zn	0.25	270.3 (441.1)	100%	241 (262)	100%	201.4 (126.1)	100%
Hg	0.01	0.86 (5.00)	100%	0.59 (1.55)	100%	0.55 (1.03)	75.0%

LOD: Limit of Detection

Although there was no significant difference between indoor and out-door air in general, the highest metal concentration was detected in in-door air. This might be attributed to various indoor heavy metal pollution emissions. Some indoor particles come from the infiltration of outdoor air or are derived from material inside the house such as household plant material, paint chips, flakes of construction materials, fuel burning and smoking [43], [44], [45]. A previous study found that resuspension has a significant impact on indoor particle concentrations. For example, simply walking into a room can increase the particle concentration by 100% for some supermicron particle sizes [43].

Distance from the mine was a key determinant of soil pollution, as there are significant spatial differences among these three villages. This finding was consistent with the conclusions of previous studies [46]. There was heavy Pb, Zn, Cu and Cd contamination in V1, the nearest village to the mine and the most polluted. The main body of this large mine is underground, whereas the sharp decline of heavy metal concentrations with increasing distance to the mining area suggests that it is the dominant source of metals such as Pb and Zn in the soil. This result was consistent with the previous studies, revealing the fly ash emitted from the ventilation outlet of the mine factory was the pollution source in this area [47]. Additionally, the run-off loss of solid ore deposited on open ground and dust emission from the transportation are likely important pollution sources too.

In addition, the relatively higher Pb and Zn concentrations in the self-produced vegetables were consistent with the concentrations in the top soil. In general, the highest metal concentrations were found in green leafy vegetables [18]. The deposition from dust onto the foliage of these vegetables would also add to the heavy metal content because metals firmly attach to the surfaces [48]. In this study, self-produced pakchoi and Chinese cabbage, especially in V1, had much higher Pb concentrations than did the other vegetables. In contrast, eggs, pork and rice bought from market were safer, as they had much lower metal concentrations than those of the vegetables. It was evident that outside food was safer than the self-produced food.

The calculated HQ values for all three villages exceeded the limits, indicating heavy metal pollution may pose high potential health risk to the people within the vicinity of the lead-zinc mining area. The factory itself declared to comply with strict environmental protection requirements with the goal of zero pollutants emission. The local environmental protection agency had set up hygienic buffer zone with a 200 meters safety protection distance around the mine. However, this regulation was not strictly followed, for the mine factory was only 100 meters from V1. The HQ in V1, which reached 23.4, was significantly higher than those of the other two villages. The average HQs of V2 and V3 also reached 8.3 and 7.5, respectively, indicating that the 200-m safety protection distance failed to protect the public's health. Furthermore, the estimated risks were mainly due to Pb, Cd and Hg. In V1, the average HQ value of Pb was as high as 16.2, accounting for 65% of the total HQ. Pb also contributed nearly half of the average HQ in the other two villages. The chronic daily intake dose of Cd and Hg also exceeded the safe level. Despite Zn having the highest concentration among the metals, its HQ value was below the limit, indicating that Zn does not pose a health risk to the local residents, as Zn is an essential element for the human body, and only superfluous ingestion can cause adverse health effects. It's worth noting that, different from soil exposure, which declined sharply when the distance to the mine increased, as to the pathway of inhalation, V3 had higher Pb and Cd HQ when compared to V2 in average. However, there were no significant differences ($P>0.05$). More environment examinations maybe needed in the future to assess health risk posed by Pb and Cd air pollution.

With regard to each metal, different exposure pathways made different contributions in different locations. Pb posed the highest estimated risk, which was mainly from soil ingestion, indoor air inhalation and pakchoi ingestion. However, in V1, the general population was primarily exposed to Cd via soil dermal contact and inhalation. As to Hg, food ingestion and inhalation contributed most to the HQ. In this study, the average indoor and outdoor air concentrations in V1 reached 734 and 543 ng/m³ respectively, and were much higher than the national background level, 6 ng/ m3 [49]. Accordingly, Hg intake through inhalation is the main and most important Hg exposure source. This finding was in accord with a previous study in Guizhou Province, where atmospheric Hg concentration around power plants ranged from 556.7 to more than 1000 ng/m³, and the local residents have the potential of developing adverse health problems because of over exposure [50]. Additionally, Hg content in rice in the study area was 0.03 mg/kg, which was lower than that of Guizhou in China (0.48 mg/kg), but much higher than that in Taiwan (0.001 mg/kg) and Shanghai (0.0047–0.0056 mg/kg) [51], [52], [53]. The mean concentration of vegetables measured in the study area range from 0.02 to 0.05 mg/kg, which was a bit lower than those measured at Nanning (0.024–0.077 mg/kg), but much higher than those measured at Shanghai (0.0005–0.006 mg/kg) and Zhongshan (0.0023–0.010 mg/kg) in China [54], [55], [56]. In general, the study area had higher Hg concentrations in rice and vegetables among the areas compared, verifying that intake of rice and vegetables is another important Hg exposure source for the residents in the study area.

Generally, the inhalation of indoor air usually contributed more than that of outdoor air because of the relatively heavier indoor air pollution and the amount of time local people spent on indoor activities. The contributions of dermal contact and ingestion of soil declined sharply with increasing distance from the mine, especially for Pb and Cd. This result was in accordance with the spatial trend of metal levels in the soil. Self-grown pakchoi was an important risk source for local residents, specifically in V1, due to its relatively high metal concentrations and large proportion in the dietary structure. These results show that dermal contact, the ingestion of soil, the inhalation of indoor air and vegetable consumption are important in reducing the overall health risk. Human exposure to soil, air and self-produced vegetables indicates the necessity of future policy responses and interventions. Local residents, especially in V1, should reduce their consumption of self-grown vegetables, maintain clean indoor air, and avoid contact with the polluted soil and dust.

Except for Ag, 9 PP metals were detected in the human hair samples. Among these detected metals, the Zn content in the hair of the sampled individuals was found to be the highest and exceeded 200 mg/kg. However, no differences were found among various locations. Consistent with the higher HQ value of the residents in V1, their hair samples exhibited higher, statistically significant Pb (8.14 mg/kg) and Hg (0.86 mg/kg) levels ($P<0.05$) than did the other two villages. The average Pb concentration in V1 was lower than that of an electronic waste recycling area in Taizhou (49.5 mg/kg), Jiangsu Province, China, but higher than that of the control sites in Ningbo (2.53 mg/kg) and Shaoxing (6.61 mg/kg) in a previous study [31]. The highest Pb concentration in hair found in V1, V2 and V3 reached 48.0, 19.7 and 18.0 mg/kg respectively, which exceeded the suggested upper limit of normal value of hair Pb (10.0 mg/kg) [57]. The HQ and corresponding Pb and Hg concentrations in the hair declined with increasing distance from the mine. It can be stated that the heavy metal pollution from the lead-zinc mine increased the exposure level of local residents, especially in V1. Human hair could be a useful biomonitoring tool to assess the extent of Pb and Hg exposure to residents in metal-polluted areas. However, for Zn, Cu, and Cd, higher levels in the hair were not detected in V1, where higher estimated exposure levels were found. This showed that hair was not biomonitoring indicator of exposure for all elements. More studies should be done to better justify use of hair as bio-monitoring indicator for other metals considered in this area.

According to the above analysis, it can be concluded that mining activity may pose high potential risk to the public, and current safety protection distance failed to protect the public's health. Most of the estimated risks came from soil, self-produced vegetables and indoor air inhalation. In summary, our results suggest that the exposure of local inhabitants to heavy metals in the mining area is through multipathways. This study highlights the importance of site-specific multipathway risk assessment. However, it should also be pointed out that our risk model used conservative assumptions and we were not able to conduct a human health impacts analysis. More detailed and in-depth health investigation is necessary in the future to examine whether adverse health outcomes occur, and provide decision-making support for pollution control in this metal-polluted area accordingly.

REFERENCES

1. Tong S, von Schirnding YE, Prapamontol T (2000) Environmental lead exposure: a public health problem of global dimensions. B World Health Organ 78: 1068–1077.
2. Järup L, Hellström L, Alfvén T, Carlsson MD, Grubb A, et al. (2000) Low level exposure to cadmium and early kidney damage: the OSCAR study. Occup Environ Med 57: 668. doi: 10.1136/oem.57.10.668.
3. Thomas LDK, Hodgson S, Nieuwenhuijsen M, Jarup L (2009) Early kidney damage in a population exposed to cadmium and other heavy metals. Environ Health Persp 117: 181. doi: 10.1289/ehp.11641.
4. Putila JJ, Guo NL (2011) Association of arsenic exposure with lung cancer incidence rates in the United States. PLoS ONE 6: e25886 doi:25810.21371/journal.pone.0025886.
5. Wang H, Stuanes AO (2003) Heavy metal pollution in air-water-soil-plant system of Zhuzhou City, Hunan Province, China. Water Air Soil Pollut 147: 79–107. doi: 10.1023/a:1024522111341.

6. Nriagu JO, Pacyna JM (1988) Quantitative assessment of worldwide contamination of air, water and soils by trace metals. Nature 333: 134–139. doi: 10.1038/333134a0.

7. Wang QR, Dong Y, Cui Y, Liu X (2001) Instances of soil and crop heavy metal contamination in China. Soil Sediment Contam 10: 497–510. doi: 10.1080/20015891109392.

8. Wang X, Sato T, Xing B, Tao S (2005) Health risks of heavy metals to the general public in Tianjin, China via consumption of vegetables and fish. Sci Total Environ 350: 28–37. doi: 10.1016/j.scitotenv.2004.09.044.

9. Cheng S (2003) Heavy metal pollution in China: Origin, pattern and control - a state-of-the-art report with special reference to literature published in Chinese journals. Environ Sci Pollut Res 10: 192–198. doi: 10.1065/espr2002.11.141.1.

10. Huang XF, Hu JW, Li CX, Deng JJ, Long J, et al. (2009) Heavy-metal pollution and potential ecological risk assessment of sediments from Baihua Lake, Guizhou, PR China. Int J Environ Health Res 19: 405–419. doi: 10.1080/09603120902795598.

11. Wang XQ, He MC, Xie J, Xi JH, Lu XF (2010) Heavy metal pollution of the world largest antimony mine-affected agricultural soils in Hunan province (China). J Soil Sediment 10: 827–837. doi: 10.1007/s11368-010-0196-4.

12. Zeng HA, Wu JL (2009) Sedimentary Records of Heavy Metal Pollution in Fuxian Lake, Yunnan Province, China: Intensity, History, and Sources. Pedosphere 19: 562–569. doi: 10.1016/S1002-0160(09)60150-8.

13. Zhou JM, Dang Z, Cai MF, Liu CQ (2007) Soil heavy metal pollution around the Dabaoshan Mine, Guangdong Province, China. Pedosphere 17: 588–594. doi: 10.1016/S1002-0160(07)60069-1.

14. Gao Y, Xia J (2011) Chromium contamination accident in China: viewing environment policy of China. Environ Sci Technol 45: 8065–8066. doi: 10.1021/es203101f.

15. Granero S, Domingo J (2002) Human health risks. Environ Int 28: 159–164.

16. Nadal M, Bocio A, Schuhmacher M, Domingo J (2005) Trends in the levels of metals in soils and vegetation samples collected near a hazardous waste incinerator. Arch Environ Contam Tox 49: 290–298. doi: 10.1007/s00244-004-0262-2.

17. MacIntosh DL, Spengler JD, Ozkaynak H, Tsai L, Ryan PB (1996) Dietary exposures to selected metals and pesticides. Environ Health Persp 104: 202. doi: 10.1289/ehp.96104202.

18. Hough RL, Breward N, Young SD, Crout NMJ, Tye AM, et al. (2004) Assessing potential risk of heavy metal exposure from consumption of home-produced vegetables by urban populations. Environ Health Persp 112: 215. doi: 10.1289/ehp.5589.

19. Baastrup R, Sørensen M, Balstrøm T, Frederiksen K, Larsen CL, et al. (2008) Arsenic in drinking-water and risk for cancer in Denmark. Environ Health Persp 116: 231. doi: 10.1289/ehp.10623.

20. Albering HJ, van Leusen SM, Moonen EJC, Hoogewerff JA, Kleinjans JCS (1999) Human health risk assessment: A case study involving heavy metal soil contamination after the flooding of the river Meuse during the winter of 1993–1994. Environ Health Persp 107: 37–43. doi: 10.1289/ehp.9910737.

21. Man YB, Sun XL, Zhao YG, Lopez BN, Chung SS, et al. (2010) Health risk assessment of abandoned agricultural soils based on heavy metal contents in Hong Kong, the world's most populated city. Environ Int 36: 570–576. doi: 10.1016/j.envint.2010.04.014.

22. Mari M, Nadal M, Schuhmacher M, Domingo JL (2009) Exposure to heavy metals and PCDD/Fs by the population living in the vicinity of a hazardous waste landfill in Catalonia, Spain: Health risk assessment. Environ Int 35: 1034–1039. doi: 10.1016/j.envint.2009.05.004.

23. Hu J, Dong Z (2011) Development of Lead Source-specific Exposure Standards Based on Aggregate Exposure Assessment: Bayesian Inversion from Biomonitoring Information to Multipathway Exposure. Environ Sci Technol.

24. Chadha A, McKelvey LD, Mangis JK (1998) Targeting lead in the multimedia environment in the continental United States. J Air Waste Manag Assoc 48: 3–15. doi: 10.1080/10473289.1998.10463666.

25. Vyskocil A, Fiala Z, Chénier V, Krajak L, Ettlerova E, et al. (2000) Assessment of multipathway exposure of small children to PAH. Environ Toxicol Phar 8: 111–118. doi: 10.1016/S1382-6689(00)00032-6.

26. Lu C, Kedan G, Fisker-Andersen J, Kissel JC, Fenske RA (2004) Multipathway organophosphorus pesticide exposures of preschool children living in agricultural and nonagricultural communities. Environ Res 96: 283–289. doi: 10.1016/j.envres.2004.01.009.

27. Marin C, Guvanasen V, Saleem Z (2003) The 3MRA risk assessment framework– A flexible approach for performing multimedia, multipathway, and multireceptor risk assessments under uncertainty. Hum Ecol Risk Assess 9: 1655–1677. doi: 10.1080/714044790.

28. NRC (1983) Risk assessment in the federal government: Managing the process. National Research Council.

29. Li J, Liu L, Huang G, Zeng G (2006) A fuzzy-set approach for addressing uncertainties in risk assessment of hydrocarbon-contaminated site. Water Air Soil Pollut 171: 5–18. doi: 10.1007/s11270-005-9005-x.

30. Goullé JP, Mahieu L, Castermant J, Neveu N, Bonneau L, et al. (2005) Metal and metalloid multi-elementary ICP-MS validation in whole blood, plasma, urine and hair: Reference values. Forensic Sci Int 153: 39–44. doi: 10.1016/j.forsciint.2005.04.020.

31. Wang T, Fu J, Wang Y, Liao C, Tao Y, et al. (2009) Use of scalp hair as indicator of human exposure to heavy metals in an electronic waste recycling area. Environ Pollut 157: 2445–2451. doi: 10.1016/j.envpol.2009.03.010.

32. Pereira R, Ribeiro R, Goncalves F (2004) Scalp hair analysis as a tool in assessing human exposure to heavy metals (S. Domingos mine, Portugal). Sc Total Environ 327: 81–92. doi: 10.1016/j.scitotenv.2004.01.017.

33. Ashe K (2012) Elevated mercury concentrations in humans of Madre de Dios, Peru. PloS ONE 7: e33305 doi:33310.31371/journal.pone.0033305.

34. Sanders AP, Flood K, Chiang S, Herring AH, Wolf L, et al. (2012) Towards prenatal biomonitoring in North Carolina: Assessing arsenic, cadmium, mercury, and lead levels in pregnant women. PloS ONE 7: e31354 doi:31310.31371/journal.pone.0031354.

35. Carneiro MF, Moresco MB, Chagas GR, Souza O, Rhoden CR, et al. (2011) Assessment of trace elements in scalp hair of a young urban population in Brazil. Biol Trace Elem Res 143: 815–824. doi: 10.1007/s12011-010-8947-z.

36. Oyoo-Okoth E, Admiraal W, Osano O, Ngure V, Kraak MH, et al. (2010) Monitoring exposure to heavy metals among children in Lake Victoria, Kenya: Envi-

ronmental and fish matrix. Ecotox Environ Safe 73: 1797–1803. doi: 10.1016/j.ecoenv.2010.07.040.

37. USEPA (1989) Risk assessment guidance for Superfund volume I: human health evaluation manual (Part A). EPA/540/1–89/002. Washington, D. C.: Environmental Protection Agency, United States.

38. USEPA (1997) Exposure Factors Handbook. Washington, D. C.: Environmental Protection Agency, United States.

39. Chen SC, Liao CM (2006) Health risk assessment on human exposed to environmental polycyclic aromatic hydrocarbons pollution sources. Sci Total Environ 366: 112–123. doi: 10.1016/j.scitotenv.2005.08.047.

40. Zhai FY, He YN, Ma GS, Li YP, Wang ZH, et al. (2005) Study on the current status and trend of food consumption among Chinese population. Chin J Epidemiol 26: 485–488 (in Chinese)..

41. MEP (2012) National Ambient Air Quality Standard of China. Ministry of Environmental Protection, China.

42. WHO (2000) Air Quality Guidelines for Europe (2nd Edn.). World Health Organization Regional Publications, European Series No.91.

43. Thatcher TL, Layton DW (1995) Deposition, resuspension, and penetration of particles within a residence. Atmos Environ 29: 1487–1497. doi: 10.1016/1352-2310(95)00016-R.

44. Rashed MN (2008) Total and extractable heavy metals in indoor, outdoor and street dust from Aswan City, Egypt. Clean–Soil Air Water 36: 850–857. doi: 10.1002/clen.200800062.

45. Rasmussen P, Subramanian K, Jessiman B (2001) A multi-element profile of house dust in relation to exterior dust and soils in the city of Ottawa, Canada. Sci Total Environ 267: 125–140. doi: 10.1016/S0048-9697(00)00775-0.

46. Benin AL, Sargent JD, Dalton M, Roda S (1999) High concentrations of heavy metals in neighborhoods near ore smelters in northern Mexico. Environ Health Persp 107: 279. doi: 10.1289/ehp.99107279.

47. Chu BB, Luo LQ (2010) Evaluation of heavy metal pollution in soils from Nanjing Qixiashang lead-zinc mines. Rock Miner Anal 29: 5–8 (in Chinese)..

48. Al Jassir M, Shaker A, Khaliq M (2005) Deposition of Heavy Metals on Green Leafy Vegerables Sold on Roadsides of Riyadh City, Saudi Arabia. B Environ Contam Tox 75: 1020–1027. doi: 10.1007/s00128-005-0851-4.

49. Zhang L, Wong M (2007) Environmental mercury contamination in China: sources and impacts. Environ Int 33: 108–121. doi: 10.1016/j.envint.2006.06.022.

50. Qu L, Liu P (2004) Environmental effect of buring coal for power station. J Guizhou Univ Technol 33: 92–94.

51. Chen H, Zhen C (1999) Heavy metal pollution in soils in China: status and countermeasures. AMBIO 28: 130–134.

52. Lin HS, Wong SS, Li GC (2004) Heavy metal content of rice and shellfish in Taiwan. J Food Drug Anal 12: 167–174.

53. Wang Y (1993) The status of mercury content in water, soil and crop in Shanghai. Shanghai Agric Technol 2: 26–27 (in Chinese)..

54. Li J (1999) Investigation on mercury contamination in suburb soil in Nanning. Guangxi Agricul Ecol Sci 1: 80–83 (in Chinese)..

55. Wang Y (1993) The status of mercury content in water, soil and crop in Shanghai. Shanghai Agric Technol 2: 26–27 (in Chinese)..

56. Zhou RD, Wu HG, Huang C, Deng CG (2002) The investigation on the content of mercury, lead, cadmium, arsenic and chrome in vegetable by bargaining in Zhongshan City. Chin J Health Lab Technol 12: 582–583 (in Chinese)..

57. Qin JF (2004) The upper limit of normal value of hair Pb, Cd, As, Hg in Chinese resident. Guangdong Trace Metal Sci 2004: 29–37 (in Chinese).

To see supplemental information for this article, please visit the original version of the article as cited on the first page of this chapter.

PART III

ANALYSIS AND ASSESSMENT OF HEAVY METAL CONTAMINATION

CHAPTER 6

INTEGRATED ASSESSMENT OF HEAVY METAL CONTAMINATION IN SEDIMENTS FROM A COASTAL INDUSTRIAL BASIN, N. E. CHINA

XIAOYU LI, LIJUAN LIU, YUGANG WANG, GEPING LUO, XI CHEN, XIAOLIANG YANG, BIN GAO, AND XINGYUAN HE

6.1 INTRODUCTION

Coastal and estuarine areas are among the most important places for human inhabitants [1]; however, with rapid urbanization and industrialization, heavy metals are continuously carried to the estuarine and coastal sediments from upstream of tributaries [2]–[5]. Heavy metal contamination in sediment could affect the water quality and bioaccumulation of metals in aquatic organisms, resulting in potential long-term implication on human health and ecosystem [6]–[7]. In most circumstances, the major part of the anthropogenic metal load in the sea and seabed sediments has a terrestrial source, from mining and industrial developments along major rivers and estuaries [8]–[10]. The hot spots of heavy metal concentration are often near industrial plants [11]. Heavy metal emissions have been declining in some industrialized countries over the last few decades [12], [13], however, anthropogenic sources have been increasing with rapid in-

This chapter was originally published under the Creative Commons Attribution License. Li X, Liu L, Wang Y, Luo G, Chen X, Yang X, Gao B, and He X. Integrated Assessment of Heavy Metal Contamination in Sediments from a Coastal Industrial Basin, NE China. PLoS ONE 7,6 (2012), doi:10.1371/journal.pone.0039690.

dustrialization and urbanization in developing countries [14], [15]. Heavy metal contaminations in sediment could affect the water quality, the bioassimilation and bioaccumulation of metals in aquatic organisms, resulting in potential long-term affects on human health and ecosystem [16]–[19]. Quantification of the land-derived metal fluxes to the sea is therefore a key factor to ascertain at which extent those inputs can influence the natural biogeochemical processes of the elements in the marine [20], [21]. The spatial distribution of heavy metals in marine sediments is of major importance in determining the pollution history of aquatic systems [22], [23], and is basic information for identifying the possible sources of contamination and to delineate the areas where its concentration exceeds the threshold values and the strategies of site remediation [24]. Therefore, understanding the mechanisms of accumulation and geochemical distribution of heavy metals in sediments is crucial for the management of coastal environment.

China's rapid growth of the economy since 1979 under the reform policies has been accompanied by considerable environmental side effects [25]. China is one of the largest coastal countries in the world. Booming coastal urban areas are increasingly dumping huge industrial and domestic waste at sea [26]. The elevated metal discharges put strong pressure on China's costal and estuarine area. The average annual input of metals by major rivers was approximately 30,000 t between 2002 and 2008 [27]. Chinese government indicates that 29,720 km² of offshore areas of China are heavily polluted [28]. "Hot spots" of metal contamination can be found along the coast of China [29], from the north to the south, especially in the industry-developed estuaries, such as the Liaodong Bay [30] and Yangtze River catchment [31] and Xiamen Bay [32]. In 2002, China enforced Marine Sediment Quality (GB 18668-2002) to protect marine environment (CSBTS, 2002). Therefore, Marine Sediment Quality (GB 18668-2002) is used as a general measure of marine sediment contamination in China.

Jinzhou Bay, surrounded by highly industrialized regions, is considered as one of the most contaminated coastal areas in China [33]. China produces the largest amount of zinc (Zn) in the world, which was 1.95

million tons in 2000 and will grow to 14.9 million tons in 2010 [26], [34]. And the largest zinc smelting plant in Asia was located at the coast of Jinzhou bay. From 1951 to 1980, the amount of Zn, Cu, Pb and Cd discharged from Huludao Zinc Smelter to Jinzhou bay reached 33745, 3689, 3525 and 1433 t respectively [35]. Although several heavy metal contamination studies have conducted in Jinzhou Bay area recently, these studies were focused on coastal urban soils [36], river sediments [37] and seawater [38] separately. Few researches take the coastal stream, estuary and bay as a whole unit to assess the heavy metal contamination of coastal industrial area spatially and temporally. Thus, it is necessary to understand the process of heavy metal contamination and to evaluate the potential ecological risks of heavy metals in the coastal stream, estuary and bay integratively.

In recent decades different metal assessment indices applied to sediment environments have been developed. Caeiro et al [9] classified them in three types: contamination indices, background enrichment indices and ecological risk indices. The geo-accumulation index (I_{geo}) [39] and the potential ecological risk index (RI) [40] are the most popular methods used to evaluate the ecological risk posed by heavy metals in sediments [41]–[44]. RI method considers the toxic-response of a given substance and the total risk index, and can exhibit the actual pollution condition of seriously polluted sediment [45], [46].

Over the last few decades the study of the sediment cores has shown to be an excellent tool for establishing the effects of anthropogenic and natural processes on depositional environments [44], [47]. Sediment cores can be used to study the pollution history of aquatic ecosystem [48], [49]. Within an individual sediment core, differences in pollutant concentrations at different depths reflect how heavy metal input and accumulation changes over time [50], [51].

The purpose of this study is (1) to quantify and explain the spatial distribution of heavy metal contaminants in modern sediments of Jinzhou bay, NE China; (2) to investigate the natural and anthropogenic processes controlling sediment chemistry; and (3) to identify the potential ecological risks of such heavy metals.

6.2 MATERIALS AND METHODS

6.2.1 STUDY AREA

This study was carried out in Jinzhou Bay and its coastal city, Huludao City, in Liaoning Province, northeast of China (Fig. 1). Jinzhou Bay is one of the important bays in the northwest of Liaodong Bay at the northwestern bank of China's Bohai Sea. It is a semi-closed shallow area with an average depth of 3.5 m and an approximate area of 120 km². Huludao city is located at southwestern coast of Jinzhou bay. The city is an important non-ferrous smelting and chemical industry area in northeast China. More than forty different mineral resources have been discovered in the Huludao region, including gold, zinc, molybdenum, lime and manganese. The economy is dominated by some of China's most important industrial enterprises, such as Asia's biggest zinc manufacturing operation, the Huludao Zinc Smelter (HZS), the Jinxi Oil Refinery and Jinhua Chemical Engineering, and Huludao's Massive Shipyard. The Wuli River, Lianshan River and the Cishan River are three main rivers in the city, flowing into Jinzhou Bay. The water, soil and sediment in the city and Jinzhou bay were heavily polluted by industrial activities. Land reclamation from sea by landfill of soils and solid wastes further increase the level of pollution of the sedimentary environments in this area. These anthropogenic activities have created great threat to the public health and the regional biological and geochemical conditions.

No specific permits were required for the described field studies. The studying area is not privately-owned or protected in any way and the field studies did not involve endangered or protected species.

6.2.2 SAMPLING AND ANALYSIS

Twelve samples of river sediment were collected from the two major rivers (Lianshan river and Wuli river) and and four samples were collected

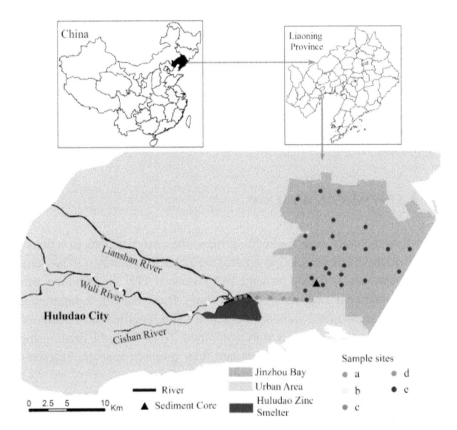

FIGURE 1: The sampling sites in the study area. (a, Lianshan River upstream of Huludao Zinc Smelter (n = 4); b, Wuli River upstream of Huludao Zinc Smelter (n = 5); c, Converged river Downstream of Huludao Zinc Smelter (n = 3); d, Estuary (n = 4); e, Jinzhou Bay (n = 25)).

from their estuary of Huludao City, using a stainless steel shovel. Twenty-five of surface sediments (0–5 cm) and one sediment core were collected in Jinzhou Bay using a stainless gravity corer (40 cm length and 5 cm diameter). The sediment core was sectioned at 2 cm intervals, and each fraction (subsamples) was sliced into 50 ml polyethylene centrifuge tubes with the help of PVC spatula. All the samples were collected in October 2009 in one week.

The samples were oven-dried at 45°C for 3 days, and sieved through a 2-mm plastic sieve to remove large debris, gravel-size materials, plant roots and other waste materials, and stored in closed plastic bags until analysis. Soil was digested with a mixture 5:2:3 of HNO_3–$HClO_4$–HF. The digested solutions were analyzed via an inductively coupled plasma-atomic emission spectroscopy (ICP-AES; Perkin Elmer Optima 3300 DV). All of the soil samples were analyzed for total concentrations of Cu, Zn, Pb, Ni, Mn and Cd.

6.2.3 STATISTICAL ANALYSIS

Statistical methods were applied to process the analytical data in terms of its distribution and correlation among the studied parameters. The commercial statistics software package SPSS version 17.0 for Windows was used for statistical analyses in present study. Basic statistical parameters such as mean, median, standard deviation (SD), coefficient of variation (CV), skewness and kurtosis were computed. To identify the relationship among heavy metals in sediments and their possible sources, Pearson's correlation coefficient analysis were performed.

6.2.4 GEOSTATISTICAL METHODS

Semivariogram is a basic tool of geostatistics and also the mathematical expectation of the square of regional variable $z(x_1)$ and $z(x+h_1)$ increment, namely the variance of regional variable. Its general form is:

$$\gamma(h) = \frac{1}{2N(h)} \sum_{i=1}^{N(h)} [z(x_i) - z(x_i + h)]^2 \tag{1}$$

where r(h) is semivariogram; h is step length, namely the spatial interval of sampling points used for the classification to decrease the individual number of spatial distance of various sampling point assemblages; N(h) is the logarithm of sampling point when the spacing is h; $z(x_i)$ and $z(x_i+h)$ are the values when the variable Z is at the x_i and x_i+h positions respectively. The residual sums of squares (RSS), the determining coefficient (R^2) and F test were used to evaluate the accuracy of the interpolated results.

Kriging, as a geostatistical interpolation method, uses the semivariogram to quantify the spatial variability of regionalized variables, and provides parameters for spatial interpolation. The maps of spatial distribution of heavy metal concentrations were generated by Kriging interpolation with the support of the statistical module of ArcGIS-Geostatistical Analyst.

6.2.4 POTENTIAL ECOLOGICAL RISK

To assess the effect of multiple metal pollutions in the sediments from the river, estuary and Jinzhou bay, potential ecological Risk Index (RI) was used, which was originally developed by Håkanson [40] and is widely used in ecological risk assessments of heavy metals in sediments. According to this methodology, the potential ecological risk index (RI) is defined as

$$RI = \sum Er^i \qquad (2)$$

$$Er^i = Tr^i \cdot C_f^i \qquad (3)$$

$$C_f^i = C_0^i / C_n^i \qquad (4)$$

where RI is calculated as the sum of all risk factors for heavy metals in sediments; Er^i is the monomial potential ecological risk factor; TR^i is the toxic-response factor for a given substance (e.g., $Cu = Pb = Ni = 5$, $Zn = 1$, $Cd = 30$); C_f^i, C_0^i and C_n^i are the contamination factor, the concentration of metals in the sediment and the background reference level, respectively. The background values of Cu, Zn, Pb and Cd are defined as the maximum values of the first category standard of national guideline values of marine sediment quality of China (GB18668-2002) and Ni is defined as the average value of Ni in residual fraction determined, they were 35 mg/kg for Cu, 150 mg/kg for Zn, 60 mg/kg for Pb, 0.5 mg/kg for Cd, and 9 mg/kg for Ni. The concentration of Mn in the sediments showed very weak relationship with the industrial activities, so it was not included in the calculation process of RI.

Still according to Hakanson [40] the following terminology is indicated to be used for the RI value:

- RI <150, low ecological risk for the sediment;
- 150≤ RI <300, moderate ecological risk for the sediment;
- 300≤ Ri <600, considerable ecological risk for sediment;
- RI ≥600, very high ecological risk for the sediment.

6.3 RESULTS AND DISCUSSION

6.3.1 HEAVY METAL IN THE SEDIMENTS

Descriptive statistics of heavy metal concentrations of sediments present in rivers of Huludao city, estuary and Jinzhou bay (Fig. 1) are presented in Table 1, 2, 3. As confirmed by the skewness values (Table 1, 3), the concentrations of elements (except Mn) are characterized by large variability, with positively skewed frequency distributions. This is common for heavy metals, because they usually have low concentrations in the environment, so that the presence of a point source of contamination may cause a sharp increase in local concentration, exceeding the thresholds [24].

TABLE 1: Heavy metal concentrations (mg/kg) of River sediments.

		Minimum	Maximum	Mean	Median	SD	CV%	Skewness	Kurtosis	national guideline values
Cu	a	67.00	186.50	116.50	106.25	50.22	43.11	1.15	2.24	35.00
	b	32.15	85.50	50.61	42.40	24.55	48.51	1.45	1.73	
	c	795.00	2535.00	1533.33	1270.00	899.39	58.66	1.20	--	
Zn	a	471.61	965.31	633.85	549.24	224.08	35.35	1.83	3.51	
	b	153.52	473.93	256.78	199.85	146.71	57.13	1.84	3.50	150.00
	c	1825.96	11010.02	6546.57	6803.73	4597.42	70.23	-0.251	--	
Pb	a	62.64	185.10	112.28	100.69	52.34	46.62	1.18	1.76	
	b	40.15	98.47	57.60	45.89	27.47	47.69	1.90	3.66	60.00
	c	417.58	6090.90	2431.09	784.81	3174.79	130.59	1.70	--	
Ni	a	35.69	109.99	57.41	41.98	35.23	61.36	1.94	3.79	
	b	28.28	35.94	31.49	30.885	3.31	10.51	0.92	0.52	
	c	40.77	87.93	62.58	59.06	23.77	37.98	0.65	--	
Mn	a	520.89	1283.46	812.91	723.65	327.98	40.34	1.47	2.75	
	b	520.21	903.43	710.96	710.11	205.78	28.94	0.004	-5.81	
	c	337.93	1358.55	915.18	1049.07	523.31	57.18	-1.08	--	
Cd	a	25.53	98.78	53.18	44.21	32.61	61.32	1.29	1.30	
	b	8.04	17.75	11.12	9.35	4.49	40.38	1.81	3.29	0.50
	c	136.73	1019.10	503.51	354.71	459.62	91.28	1.30	--	

[a] *Lianshan River upstream of Huludao Zinc Smelter (n = 4);* [b] *Wuli River upstream of Huludao Zinc Smelter (n = 5);* [c] *Converged river downstream of Huludao Zinc SMelter (n = 3). The locations of sampling sites are on Fig. 1.*

TABLE 2: Heavy metal concentrations (mg/kg) of Estuary sediments.

Distance from HZS to the sampling sites	Cu	Zn	Pb	Ni	Mn	Cd
400 m	1510.00	9304.24	1414.14	59.95	744.13	269.34
1200 m	805.00	3898.09	561.21	64.10	677.46	181.90
2000 m	675.00	2750.57	486.43	48.14	514.07	118.08
2800 m	505.00	3221.59	421.28	53.85	682.69	98.29

The locations of sampling sites are Fig. 1.

TABLE 3: Heavy metal concentrations (mg/kg) of Jinzhou Bay sediments (n = 25).

	Min.	Max.	Mean	Median	SD	CV%	Skewness	Kurtosis	national guidelines values
Cu	24.45	327.50	74.11	51.50	68.54	92.48	2.706	7.964	35.00
Zn	168.06	2506.33	689.39	550.58	568.50	82.46	2.183	4.656	150.00
Pb	29.17	523.45	123.98	89.29	114.70	92.51	2.398	5.934	60.00
Ni	26.29	85.99	43.47	41.41	11.92	27.42	1.835	8.867	--
Mn	445.57	1123.03	750.64	774.00	153.82	20.49	0.311	0.137	--
Cd	7.91	105.31	26.81	20.74	22.23	82.92	2.388	6.155	0.50

The locations of sampling sites are in Fig. 1

6.3.2 HEAVY METAL IN RIVER SEDIMENTS.

Among the concentrations of heavy metal of river sediments (Table 1), the low values are from the sediments sampled before the river flowing by the Huludao Zinc Smelter (HZS), and the highest heavy metal concentrations come from the two sediment samples collected after the river flows by the HZS and accepted the wastewater discharged from HZS. Although the Wuli river is the least contaminated river in Huludao City; however the concentration of Cd exceeded the national guideline values of marine sediment quality of China (GB 18668-2002) by 22 times. In the upstream of HZS the sediment contamination of Lianshan river was higher than that of Wuli river. These data showed many other industrial operations at upper reaches of HZS also contributed great heavy metals to the river sediments. According to historical data, wastes from Jinxi Chemical Factory and Jinxi Petroleum Chemical Factory were discharged into Wuli River directly for nearly 40 years till 2000 and caused heavy contamination of river sediments. At the same time, wastewater from other small industrial plants and residents, and nonpoint pollution from soils with runoff or atmospheric deposition contributed additional pollution sources [52], [53]. This is be confirmed by the previous studies about contamination of urban soils [36], [54] and river sediments [37] in Huludao city.

The mean concentrations of Cu, Zn, Pb and Cd in the sediment of converged river downstream of HZS adjacent to the estuary exceeding the national guideline values of marine sediment quality of China (GB 18668-2002) by 42, 42, 40 and 1006 times respectively. These indicated that the Huludao Zinc Smelter (HZS) is the largest source of heavy metals in river sediment adjacent to the estuary. The annual amount of discharged wastewater from HZS is estimated to be more than 8 million tons directly to the Wuli river [35]. This situation lasted for more than 50 years till the water reuse was realized around year 2000 [28].

6.3.3 HEAVY METAL IN THE ESTUARY SEDIMENTS.

Four sediment samples were collected along the estuary, and the distance from HZS to the sampling sites increases from 400 m with an 800 m-interval to the Jinzhou Bay (Fig. 1). All the investigated heavy metals clearly showed the same distribution trend. The maximum concentrations of heavy metals declined as the distance increased from HZS (Table 2). The mean concentration of Cu, Zn, Pb and Cd at estuary were exceeded the national guideline values of marine sediment quality of China (GB 18668-2002) by 24, 31, 11 and 332 times respectively.

Contaminant distributions in the Hudson River estuary were identified two types of trends: increasing trend down-estuary dominated by down-estuary sources such as wastewater effluent, and decreasing trend toward bay dominated by upriver sources, where they are removed and diluted downstream along with the sediment transport [55]. The result of this study showed obvious decreasing trend toward bay. This confirmed that the wastewater discharged from HZS and other industrial plants were the main sources of heavy metals in sediment.

6.3.4 HEAVY METAL IN JINZHOU BAY SEDIMENTS

Twenty-five of surface sediments from Jinzhou Bay were tested and the minimum, maxinum and mean values are all located in Table 3. There were remarkable changes in the concentrations of heavy metals in the sediments of Jinzhou bay (Fig. 1). The heavy metal concentration levels are comparable to those with previous studies [56]–[58], especially for Cu, Pb and Zn, Mean concentrations of Cu, Zn, Pb and Cd were as high as 2.1, 4.6, 2.1 and 53.6 times of the national guideline values of marine sediment quality of China (GB 18668-2002). According to the research results from Institute of Marine Environmental protection, State oceanic Administration of China in 1984, the background value of Cu, Zn, Pb and Cd in Jinzhou bay was 10.0, 48.59, 9.0 and 0.29 mg/kg respectively [35]. This clearly demonstrates an anthropogenic contribution and reveals a serious pollution of sediments in Jinzhou bay. For all metals, total concentrations had a great degree of variability, shown by the large coefficients of variation (CV) from 20.49% of Mn to 92.51% of Pb. The elevated coefficients of variations reflected the inhomogeneous distribution of concentrations of discharged heavy metals. Large standard deviations were found in all heavy metals levels. The results of the K–S test (P<0.05) showed that the concentrations of measured metals were all normally distributed.

The comparison of contaminant concentrations observed in this study with those reported for other regions (Table 4) indicates that levels and ranges of variation of our data are similar to those reported from sites with high anthropogenic impact. It was found that the concentrations of Cd measured in this study were greatly higher than other studies except that of Algeciras Bay was a little close to this study. The levels of Zn and Pd in this study were only lower than that of Izmit Bay and Gulf of Naples respectively. The contents of Cu were relatively higher than other study (except Izmit Bay, Bay of Bengal and Hong Kong). All these show that Jinzhou bay was a highly polluted area in the world.

TABLE 4: Mean concentrations (mg/kg) of heavy metals found in Jinzhou bay compared to the reported average concentrations for other world impacted coastal systems.

Area	Cu	Zn	Pb	Ni	Mn	Cd	Reference
Izmit Bay, Turkey	89.4	754	94.9	52.1	--	6.3	[56]
Ribeira Bay, Brazil	24.6	109	22.9	47	466	0.207	[59]
Sepetiba Bay, Brazil	31.9	567	40	22.3	595	3.22	[59]
Mejillones Bay, Chile	--	29.7	--	20.6	93.8	21.9	[58]
Algeciras Bay, Spain	17	73	24	65	534	0.3	[60]
Taranto Gulf, Italy	47.4	102.3	57.8	53.3	893	--	[11]
Tivoli South Bay, USA	17.6	92.8	26.3	--	--	--	[61]
Bay of Bengal, India	677.7	60.39	25.66	34.03	366.66	5.24	[57]
Gulf of Mannar, India	57	73	16	24	305	0.16	[62]
Gulf of Naples, Italy	27.2	602	221	6.93	1550	0.57	[63]
Hong Kong, China	118.68	147.73	53.56	24.72	523.99	0.33	[64]
This study	74.11	689.39	123.98	43.47	750.64	26.81	

Note: "--" = no data.

6.3.5 CORRELATION BETWEEN HEAVY METALS

Correlation analyses have been widely applied in environmental studies. They provided an effective way to reveal the relationships between multiple variables in order to understand the factors as well as sources of chemical components [50], [65]. Heavy metals in environment usually have complicated relationships among them. The high correlations between heavy metals may reflect that the accumulation concentrations of these heavy metals came from similar pollution sources [66], [67]. Results of Pearson's correlation coefficients and their significance levels ($P<0.01$) of correlation analysis were shown in Table 5. The concentrations of Cu, Zn, Pb, Ni and Cd showed strong positive relationship ($P<0.01$) with each other. This shows that Cu, Zn, Pb, Ni and Cd come from the same source. However, the concentration of Mn showed very weak correlations with the concentrations of the other metals, except Ni. This indicates that Mn have different sources than Cu, Zn, Pb and Cd. Han et al [68] also found the similar result about Mn by multivariate analysis.

TABLE 5 Correlations between heavy metal concentrations.

	Cu	Zn	Pb	Ni	Mn	Cd
Cu	1.000					
Zn	0.919**	1.000				
Pb	0.870**	0.824**	1.000			
Ni	0.499**	0.524**	0.472**	1.00		
Mn	0.115	0.270	0.238	0.515**	1.000	
Cd	0.906**	0.885**	0.972**	0.488**	0.285	1.000

*Levels of significance: **P<0.01*

6.3.6 SPATIAL DISTRIBUTION OF HEAVY METALS IN THE SEDIMENTS OF JINZHOU BAY

Geostatistics is increasingly used to model the spatial variability of con-taminant concentrations and map them using generalized least-squares regression, known as kriging [69]–[71]. The probability map produced based on kriging interpolation and kriging standard deviation integrates information about the location of the pollutant source and transport pro-cess into the spatial mapping of contaminants [71], [72]. There are a lot of studies of the performance of the spatial interpolation methods, but the results are not clear-cut [73]. Some of them found that the kriging method performed better than inverse distance weighting (IDW) [74]; while oth-ers showed that kriging was no better than alternative methods [75]. For example, Kazemi and Hosseini [76] compared the ordinary kriging (OK) and other three spatial interpolation methods for estimating heavy metals in sediments of Caspian Sea, they found that the OK realization smoothed out spatial variability and extreme measured values between the range of observed minimum and maximum values for all of the contaminants.

The spatial distribution of metal concentrations is a useful tool to as-sess the possible sources of enrichment and to identify hot-spot area with high metal concentrations [48], [49]. Semivariogram calculation was conducted and experimental semivariogram of sediment heavy metal concentrations could be fitted with the Gaussian model for Cu, Zn, Pb, Cd, Ni and Mn. The theoretical variation function and experimental variation function exhibits a better fitting (Table 6). The values of R were significant at the 0.01 level by F test, which shows that the semivariogram models well reflect the spatial structural characteristics of sediemnt heavy metals.

TABLE 6: Parameters and F-test of fitted semivariogram models (Gaussian model) for heavy metals in sediment.

	Nugget (C_0)	Sill ($C_0 + C$)	$C/(C_0 + C)$	Range	R^2	RSS	F test
Cu	5200	111500	0.953	0.0744	0.833	1.41E+09	24.94**
Zn	170000	4450000	0..962	0.0831	0.868	1.38E+12	32.88**
Pb	7900	106900	0.926	0.0883	0.837	7.23E+08	25.67**
Ni	10	7460	0.999	0.0277	0.513	3.81E+07	5.27**
Mn	7370	38310	0.808	0.0710	0.946	2.71E+07	87.59**
Cd	220	5550	0.960	0.0935	0.921	8.49E+05	58.29**
RI	890000	22880000	0.961	0.1004	0.927	1.12E+13	63.49**

**Significance at α =0.01 level of F test.

The estimated maps of Cu, Zn, Pb, Cd, Ni and Mn clearly identified that the river, where the Huludao Zinc Smelter (HZS) is located at, is the most important source of heavy metals except Mn (Fig. 2). Among these metals, Cu, Zn, Pb and Cd showed a very similar spatial pattern, with contamination hotspots located at the estuary area, and their concentrations decreased sharply with the distance farther away from the estuary, indicating that they were from the same sources. The concentration of Ni also showed a similar pattern with the concentrations of Cu, Zn, Pb and Cd, but it changed not as sharply as the latter one. However, the concentration of Mn showed a completely different pattern with the others, indicating the industrial activities are not the source of Mn in the Jinzhou bay, and it may be related to geological factor. The mineral constituents of Jinzhou bay consist mainly of hornblende, epidote and magnetite. The percentage of hornblende, which is rich in Mn element, varied from 24.78% up to 64.70% [77]. This confirms that the concentration of Mn comes from the

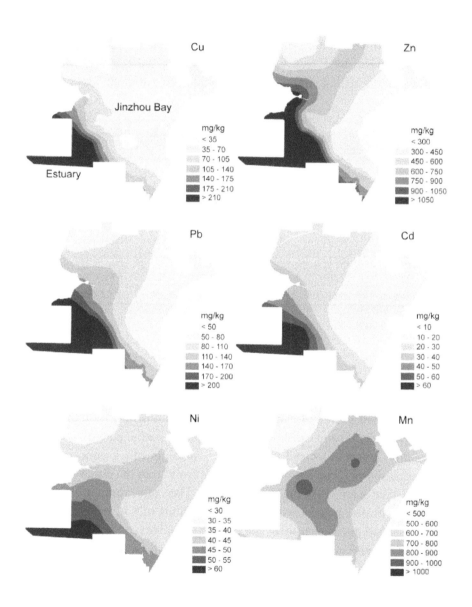

FIGURE 2: Estimated concentration maps for Cu, Zn, Pb, Cd, Ni and Mn (mg/kg).

geological sources. The feature of point sources lies in that the inputs of heavy metals occur over a finite period of time and may have been effectively retained in the sediments near the sources, rather than re-suspended and distributed uniformly throughout the region [66]. Distinct from point sources, metals from non-point sources are more uniformly distributed throughout the area [50]. The results of this study closely correspond to this association.

The Spatial pattern of heavy metal in Jinzhou bay also provided a refinement and reconfirmation of the results in the statistical analysis, in which strong associations were found among Cu, Zn, Pb, Cd and Ni and very weak relations were found between Mn and the other heavy metals except Ni.

TABLE 7: The heavy metal potential ecological risk indexes in sediments.

Sediments	Cu	Zn	Pb	Ni	Cd	RI	Pollution degree
Lianshan River	16.64	4.23	9.36	31.89	3190.804	3252.96	very high
Wuli River	7.23	1.71	4.80	17.50	667.43	698.69	very high
Estuary	162.50	41.10	135.54	34.54	20414.20	20787.87	very high
Jonzhou Bay	10.59	4.60	10.33	24.15	1608.68	1658.34	very high

6.3.7 ASSESSMENT OF POTENTIAL ECOLOGICAL RISK, RI

Almost all the RI values of sampling sites were higher than 600 except the two samples from Wuli river, indicating that the sediments in the rivers of Huludao city and their estuary and Jinzhou bay exhibited very high ecological risk of heavy metals (Fig. 3, Table 7). The RI of sediments in the estuary was as high as 34.6 times of the line value for very high ecological risk level, suggesting that the sediments in the estuary were extremely polluted by heavy metals because of industrial discharge. Cd showed the highest potential ecological risk in the heavy metals, which contributed more than 95% of RI in the sampled sediments.

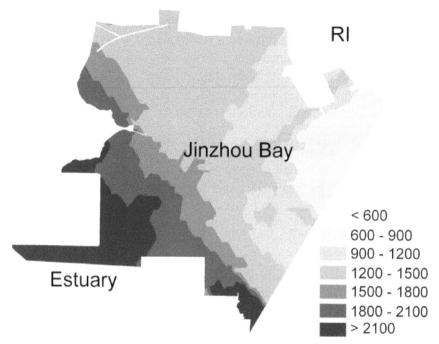

FIGURE 3: The spatial distribution pattern of RI of sediments in Jinzhou Bay.

6.3.8 TEMPORAL DISTRIBUTION OF HEAVY METALS IN THE SEDIMENTS OF JINZHOU BAY

Sediment cores can be used to study the pollution history of aquatic ecosystem [44], [48]. Vertical distribution (0–36 cm) of heavy metals in Jinzhou Bay indicate that the concentration of Cu, Zn, Pb and Cd show similar vertical patterns (Fig. 4). The values of Cu, Zn, Pb and Cd increased sharply from the surface to its highest concentration at the depth of 8 cm and then decreased rapidly at a depth of 20 cm. The values of Cu, Zn, Pb and Cd varied slightly from a depth of 21–36 cm. The concentration of Ni in the sediment core decreased gradually from the surface with small fluctuations while the Mn remained relatively consent throughout the core.

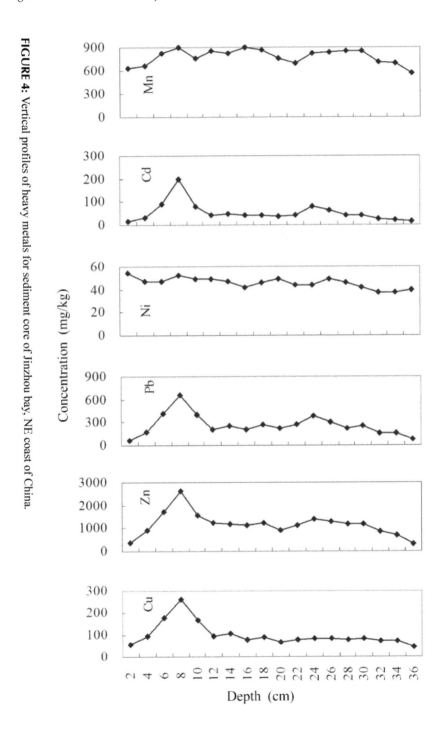

FIGURE 4: Vertical profiles of heavy metals for sediment core of Jinzhou bay, NE coast of China.

According to the sedimentation rate of about 1.0 cm/yr [78], the bottom of the sediment core (at the depth of 36 cm) was polluted about in 1973, 36 years after the set up of Huludao Zinc Smelter (HZS) in 1937. The heavy metals concentrations were already greatly higher than the national guideline values of marine sediment quality of China (GB 18668-2002), especially the concentration of Cd as high as 31 times of the latter. Related to annual Zn yields of HZS between 1973 and 2010 with the concentrations of Cu, Zn, Pb and Cd in sediment core, they showed similar temporal patterns, especially around the year 2000, the peak of Zn yield was followed the highest heavy metal concentrations at the depth of 8 cm of sediment core. This is conclusive evidence that Zn smelting operation was the dominant pollution source of aquatic environment in Jinzhou Bay. With the disposal and reuse of heavy metal wastewater from HZS around 2000, although the Zn yields increased year by year, the sediment pollution was alleviated gradually since 2000. The concentration of Ni in the sediment core decreased gradually from the surface with small fluctuations, the value varied from 54.36 mg/kg at the surface to 37.09 mg/kg at the depth of 32 cm. This also clearly indicated that the content of Ni in the sediment was come from the industrial discharges. Vertical profile of Mn shows some fluctuations in its concentration, and no obvious correlation with the depth of sediment core was observed. This indicates that the concentration of Mn is not under the control of human factors, as the result showed by the spatial pattern of Mn.

6.4 CONCLUSION

This study investigated the concentrations of heavy metals in the sediments from urban-stream, estuary and Jinzhou bay of the coastal industrial city, NE China. The results showed the impact of anthropogenic agents on abundances of heavy metals in sediments. The sediments are found to be extremely contaminated due to many years of random dumping of hazardous waste and free discharge of effluents by industries like Huludao Zinc Smelter (the largest zinc smelting plant in Asia), the Jinxi oil refinery and Jinhua chemical engineering, Huludao's massive shipyard and several arms factories. The potential ecological risk of sediments in lower river reaches, estuary and Jinzhou bay is at very high level, and Cd contributed more than 95% of RI in the sampled sediments. The estuary is the most polluted area, and its RI value was as high as 34.6 times of the line value for very high ecological risk level. The closer the distance to the estuary is, the higher RI values of sediments in Jinzhou bay are. The results of this research updated the information for effective environmental management in the industrial region. This study clearly highlights the urgent need to make great efforts to control the industrial discharges in the coastal area, and the immediate measures should be carried out to minimize the contaminations, and to prevent future pollution problems.

REFERENCES

1. McKinley AC, Dafforn KA, Taylor MD, Johnston EL (2011) High Levels of Sediment Contamination Have Little Influence on Estuarine Beach Fish Communities. PLoS ONE 6(10): e26353. doi:10.1371/journal.pone.0026353.
2. Morton B, Blackmore G (2001) South China Sea. Marine Pollution Bulletin 42: 1236–1263. doi: 10.1016/S0025-326X(01)00240-5.
3. Jha SK, Chavan SB, Pandit GG, Sadasivan S (2003) Geochronology of Pb and Hg pollution in a coastal marine environment using global fallout 137Cs. Journal of Environmental Radioactivity 69: 145–157. doi: 10.1016/S0265-931X(03)00092-4.
4. Muniz P, Danula E, Yannicelli B, Garcia-Alonso J, Medina G, et al. (2004) Assessment of contamination by heavy metals and petroleum hydrocarbons in sediments of Montevideo Harbour (Uruguay). Environment International 29: 1019–1028. doi: 10.1016/S0160-4120(03)00096-5.
5. Xia P, Meng X, Yin P, Cao Z, Wang X (2011) Eighty-year sedimentary record of heavy metal inputs in the intertidal sediments from the Nanliu River estuary, Beibu Gulf of South China Sea. Environmental pollution 159: 92–99. doi: 10.1016/j.envpol.2010.09.014.
6. Fernandes C, Fontainhas-Fernandes A, Peixoto F, Salgado MA (2007) Bioaccumulation of heavy metals in Liza saliens from the Esomriz-Paramos coastal lagoon, Portugal. Ecotoxicology and Environmental Safety 66: 426–431. doi: 10.1016/j.ecoenv.2006.02.007.
7. Abdel-Baki AS, Dkhil MA, Al-Quraishy S (2011) Bioaccumulation of some heavy metals in tilapia fish relevant to their concentration in water and sediment of Wadi Hanifah, Saudi Arabia. African Journal of Biotechnology 10: 2541–2547.
8. Ridgway J, Breward N, Langston WJ, Lister R, Rees JG, et al. (2003) Distinguishing between natural and anthropogenic sources of metals entering the Irish Sea. Applied Geochemistry 18: 283–309. doi: 10.1016/S0883-2927(02)00126-9.
9. Caeiro S, Costa MH, Fernandes F, Silveira N, Coimbra A, et al. (2005) Assessing heavy metal contamination in Sado Estuary sediment: An index analysis approach. Ecological Indicators 5: 151–169. doi: 10.1016/j.ecolind.2005.02.001.

10. Sundaray SK, Nayak BB, Lin S, Bhatta D (2011) Geochemical speciation and risk assessment of heavy metals in the river estuarine sediments – a case study: Mahanadi basin, India. Journal of Hazardous Materials 186: 1837–1846. doi: 10.1016/j.jhazmat.2010.12.081.

11. Buccolieri A, Buccolieri G, Cardellicchio N, Dell'Atti A, Leo AD, et al. (2006) Heavy metals in marine sediments of Taranto Gulf (Ionian Sea, Southern Italy). Marine Chemistry 99: 227–235. doi: 10.1016/j.marchem.2005.09.009.

12. Voet E, Guinée JB, Udo de Haes H (2000) Heavy Metals: A Problem Solved? Methods and Models to Evaluate Policy Strategies for Heavy Metals. Kluwer, Dordrecht, the Netherlands.

13. Hjortenkrans D, Bergback B, Haggerud A (2006) New metal emission patterns in road traffic environments. Environmental Monitoring and Assessment 117: 85–98. doi: 10.1007/s10661-006-7706-2.

14. Govil PK, Sorlie JE, Murthy NN, Sujatha D, Reddy GLN, et al. (2008) Soil contamination of heavy metals in the Katedan Industrial Development Area, Hyderabad, India. Environmental Monitoring and Assessment 140: 313–323. doi: 10.1007/s10661-007-9869-x.

15. Wu SH, Zhou SL, Li XG (2011) Determining the anthropogenic contribution of heavy metal accumulations around a typical industrial town: Xushe, China. Journal of Geochemical Exploration 110: 92–97. doi: 10.1016/j.gexplo.2011.04.002.

16. Snodgrass JW, Casey RE, Joseph D, Simon JA (2008) Microcosm investigations of stormwater pond sediment toxicity to embryonic and larval amphibians: variation in sensitivity among species. Environmental Pollution 154: 291–297. doi: 10.1016/j.envpol.2007.10.003.

17. Besser J, Brumbaugh W, Allert A, Poulton B, Schmitt C, et al. (2009) Ecological impacts of lead mining on Ozark streams: toxicity of sediment and pore water. Ecotoxicology and Environmental Safety 72: 516–526. doi: 10.1016/j.ecoenv.2008.05.013.

18. Ip CCM, Li XD, Zhang G, Wai OWH, Li YS (2007) Trace metal distribution in sediments of the Pearl River. Estuary and the surrounding coastal area, South China. Environmental Pollution 147: 311–323. doi: 10.1016/j.envpol.2006.06.028.

19. Suthar S, Arvind KN, Chabukdhara M, Gupta SK (2009) Assessment of metals in water and sediments of Hindon River, India: Impact of industrial and urban discharges. Journal of Hazardous Materials 178: 1088–1095. doi: 10.1016/j.jhazmat.2009.06.109.

20. IGBP (1995) Global Change. In: Pernetta JC, Milliman JD, editors. Stockholm: ICSU.

21. Cobelo-Garcia A, Prego R, Labandeira A (2004) Land inputs of trace metals, major elements, particulate organic carbon and suspended solids to an industrial coastal bay of the NE Atlantic. Water Research 38: 1753–1764. doi: 10.1016/j.watres.2003.12.038.

22. Birch GF, Taylor SE, Matthai C (2001) Small-scale spatial and temporal variance in the concentration of heavy metals in aquatic sediments: a review and some new concepts. Environmental Pollution 113: 357–372. doi: 10.1016/S0269-7491(00)00182-2.

23. Rubio B, Pye K, Rae JE, Rey D (2001) Sedimentological characteristics, heavy metal distribution and magnetic properties in subtidal sediments, Ria de Pontevedra, NW Spain. Sedimentology 48: 1277–1296. doi: 10.1046/j.1365-3091.2001.00422.x.

24. Sollitto D, Romic M, Castrignanò A, Romic D, Bakic H (2010) Assessing heavy metal contamination in soils of the Zagreb region (Northwest Croatia) using multivariate geostatistics. CATENA 80: 182–194. doi: 10.1016/j.catena.2009.11.005.

25. Liu JG (2010) China's Road to Sustainability. Scinece 328: 50. doi: 10.1126/science.1186234.

26. Pan K, Wang WX (2012) Trace metal contamination in estuarine and coastal environments in China. Science of the Total Environment 421–422: 3–16. doi: 10.1016/j.scitotenv.2011.03.013.

27. NBSC (2010) National Bureau of Statistics of China. China Statistical Yearbook (2001–2009). Beijing.

28. NBO National Bureau of Oceanography of China (2010) Bulletin of Marine Environmental Quality, 2008–2009.

29. Yang ZF, Wang Y, Shen ZY, Niu JF, Tang ZW (2009) Distribution and speciation of heavy metals in sediments from the mainstream, tributaries, and lakes of the Yangtze River catchment of Wuhan, China. Journal of Hazardous Materials 166: 1186–1194. doi: 10.1016/j.jhazmat.2008.12.034.

30. Fang TH, Li JY, Feng HM, Chen HY (2009) Distribution and contamination of trace metals in surface sediments of the East China Sea. Marine Environmental Research 68: 178–187. doi: 10.1016/j.marenvres.2009.06.005.

31. Müller B, Berg M, Yao ZP, Zhang XF, Wang D, et al. (2008) How polluted is the Yangtze river? Water quality downstream from the Three Gorges Dam. Science of the Total Environment 402: 232–247. doi: 10.1016/j.scitotenv.2008.04.049.

32. Chen C, Lu Y, Hong J, Ye M, Wang Y, Lu H (2010) Metal and metalloid contaminant availability in Yundang Lagoon sediments, Xiamen Bay, China, after 20 years continuous rehabilitation. Journal of Hazardous Materials 175: 1048–1055. doi: 10.1016/j.jhazmat.2009.10.117.

33. Zhang YF, Wang LJ, Huo CL, Guan DM (2008) Assessment on heavy metals pollution in surface sediments in Jinzhou Bay. Marine Environmental Science 2: 178–181.

34. Research and Markets website. Business monitor international. China Metals report. Accessed 2012 May 30.

35. Institute of Marine Environmental protection, State oceanic Administration (1984) Studies on the contamination and protection of Jinzhou Bay.

36. 36. Lu CA, Zhang JF, Jiang HM, Yang JC, Zhang JT, et al. (2010) Assessment of soil contamination with Cd, Pb and Zn and source identification in the area around the Huludao Zinc Plan. Journal of Hazardous Materials 182: 743–748. doi: 10.1016/j.jhazmat.2010.06.097.

37. Zheng N, Wang QC, Liang ZZ, Zheng DM (2008) Characterization of heavy metal concentrations in the sediments of three freshwater rivers in Huludao City, Northeast China. Environmental Pollution 154: 135–142. doi: 10.1016/j.envpol.2008.01.001.

38. Wang J, Liu RH, Yu P, Tang AK, Xu LQ, et al. (2012) Study on the Pollution Characteristics of Heavy Metals in Seawater of Jinzhou Bay. Procedia Environmental Sciences 13: 507–1516. doi: 10.1016/j.proenv.2012.01.143.

39. Porstner U (1989) Lecture Notes in Earth Sciences (Contaminated Sediments). Springer Verlag, Berlin, 107–109.

40. Håkanson L (1980) An ecological risk index for aquatic pollution control: A sedimentological approach. Water Research 14: 975–1001. doi: 10.1016/0043-1354(80)90143-8.

41. Selvaraj K, Mohan Ram V, Piotr S (2004) Evaluation of Metal Contamination in Coastal Sediments of the Bay of Bengal, India: Geochemical and Statistical Approaches. Marine Pollution Bulletin 49: 174–185. doi: 10.1016/j.marpolbul.2004.02.006.

42. Verca P, Dolence T (2005) Geochemical Estimation of Copper Contamination in the Healing Mud from Makirina Bay, Central Adriatic. Environment International 31: 53–61. doi: 10.1016/j.envint.2004.06.009.

43. Yi YJ, Yang ZF, Zhang SH (2011) Ecological risk assessment of heavy metals in sediment and human health risk assessment of heavy metals in fishes in the middle and lower reaches of the Yangtze River basin. Environmental Pollution 159: 2575–2585. doi: 10.1016/j.envpol.2011.06.011.

44. Harikumar PS, Nasir UP (2010) Ecotoxicological impact assessment of heavy metals in core sediments of a tropical estuary. Ecotoxicology and Environmental Safety 73: 1742–1747. doi: 10.1016/j.ecoenv.2010.08.022.

45. Huang YL, Zhu WB, Le MH, Lu XX (2011) Temporal and spatial variations of heavy metals in urban riverine sediment: An example of Shenzhen River, Pearl River Delta, China. Quaternary International. doi:10.1016/j.quaint.2011.05.026.

46. Uluturhan E, Kontas A, Can E (2011) Sediment concentrations of heavy metals in the Homa Lagoon (Eastern Aegean Sea): Assessment of contamination and ecological risks. Marine Pollution Bulletin 62: 1989–1997. doi: 10.1016/j.marpolbul.2011.06.019.

47. Rosales-Hoz L, Cundy AB, Bahena-Manjarrez JL (2003) Heavy metals in sediment cores from a tropical estuary affected by anthropogenic discharges: Coatzacoalcos estuary. Estuarine, Coastal and Shelf Science 58: 117–126. doi: 10.1016/S0272-7714(03)00066-0.

48. Karbassi AR, Nabi-Bidhendi GHR, Bayati I (2005) Environmental geochemistry of heavy metals in a sediment core off Bushehr, Persian Gulf. Iranian Journal of Environmental Health Science & Engineering 2: 255–260.

49. Viguri JR, Irabien MJ, Yusta I, Soto J, Gomez J, et al. (2007) Physico-chemical and toxicological characterization of the historic estuarine sediments. A multidisciplinary approach. Environment International 33: 436–444. doi: 10.1016/j.envint.2006.10.005.

50. Shine JP, Ika RV, Ford TE (1995) Multivariate statistical examination of spatial and temporal patterns of heavy-metal contamination in New-Bedford Harbor marine-sediments. Environmental Science and Technology 29: 1781–1788. doi: 10.1021/es00007a014.

51. White HK, Xu L, Lima ANL, Egliton TI, Reddy CM (2005) Abundance, composition and vertical transport of PAHs in marsh sediments. Environmental Science and Technology 39: 8273–8280. doi: 10.1021/es050475w.

52. Berthelsen BO, Steinnes E, Solberg W (1995) Heavy metal concentrations in plants in relation to atmospheric heavy metal deposition. Journal of Environmental Quality 24: 1018–1026. doi: 10.2134/jeq1995.00472425002400050034x.

53. Gray CW, McLaren RG, Roberts AHC (2003) Atmospheric accessions of heavy metals to some New Zealand pastoral soils. Science of the Total Environment 305: 105–115. doi: 10.1016/S0048-9697(02)00404-7.

54. Li LL, Yi YL, Wang YS, Zhang DG (2006) Spatial distribution of soil heavy metals and pollution evaluation in Huludao City. Chinese Journal of Soil Science 37: 495–499.

55. Feng H, Cochran JK, Lwiza H, Brownawell BJ, Hirschberg DJ (1998) Distribution of heavy metal and PCB contaminants in the sediments of an urban estuary: The Hudson River. Marine Environmental Research. 45: 69–88. doi: 10.1016/S0141-1136(97)00025-1.

56. Pekey H (2006) Heavy metal pollution assessment in sediments of the Izmit Bay, Turkey. Environmental Monitoring and Assessment 123: 219–231. doi: 10.1007/s10661-006-9192-y.

57. Raju K, Vijayaraghavan K, Seshachalam S, Muthumanickam J (2011) Impact of anthropogenic input on physicochemical parameters and trace metals in marine surface sediments of Bay of Bengal off Chennai, India. Environmental Monitoring and Assessment 177: 95–114. doi: 10.1007/s10661-010-1621-2.

58. Valdés J, Vargas G, Sifeddine A, Ortlieb L, Guiñez M (2005) Distribution and enrichment evaluation of heavy metals in Mejillones Bay Northern Chile: Geochemical and statistical approach. Marine Pollution Bulletin 50: 1558–1568. doi: 10.1016/j.marpolbul.2005.06.024.

59. Gomes F, Godoy J, Godoy M, Carvalho Z, Lopes R, et al. (2009) Metal concentrations, fluxes, inventories and chronologies in sediments from Sepetiba and Ribeira Bays: A comparative study. Marine Pollution Bulletin 59: 123–133. doi: 10.1016/j.marpolbul.2009.03.015.

60. Alba MD, Galindo-Riaño MD, Casanueva-Marenco MJ, García-Vargas M, Kosore CM (2011) Assessment of the metal pollution, potential toxicity and speciation of sediment from Algeciras Bay (South of Spain) using chemometric tools. Journal of Hazardous Materials 190: 177–187. doi: 10.1016/j.jhazmat.2011.03.020.

61. Benoit G, Wang EX, Nieder WC, Levandowsky M, Breslin VT (1999) Sources and history of heavy metal contamination and sediment deposition in Tivoli South Bay, Hudson River, New York. Estuaries 22: 167–178. doi: 10.2307/1352974.

62. Jonathan M P, Stephen-Pichaimani V, Srinivasalu S, RajeshwaraRao N, Mohan SP (2007) Enrichment of trace metals in surface sediments from the northern part of Point Calimere, SE coast of India. Environmental Geology 55: 1811–1819. doi: 10.1007/s00254-007-1132-9.

63. Romano E, Ausili A, Zharova N, Magno MC, Pavoni B, et al. (2004) Marine sediment contamination of an industrial site at Port of Bagnoli, Gulf of Naples, Southern Italy. Marine Pollution Bulletin 49: 487–495. doi: 10.1016/j.marpolbul.2004.03.014.

64. Zhou F, Guo HC, Hao ZJ (2007) Spatial distribution of heavy metals in Hong Kong's marine sediments and their human impacts: a GIS-based chemometric approach. Marine Pollution Bulletin 54: 1372–84. doi: 10.1016/j.marpolbul.2007.05.017.

65. Al-Khashman OA, Shawabkeh RA (2006) Metals distribution in soils around the cement factory in southern Jordan. Environmental Pollution 140: 387–394. doi: 10.1016/j.envpol.2005.08.023.

66. Facchinelli A, Sacchi E, Mallen L (2001) Multivariate statistical and GIS-based approach to identify heavy metal sources in soils. Environmental Pollution 114: 313–324. doi: 10.1016/S0269-7491(00)00243-8.

67. Manta DS, Angelone M, Bellanca A, Neri R, Sprovieri M (2002) Heavy metals in urban soils: a case study from the city of Palermo (Sicily), Italy. The Science of the Total Environment 300: 229–243. doi: 10.1016/S0048-9697(02)00273-5.

68. Han Y, Du P, Cao J, Posmentier ES (2006) Multivariate analysis of heavy metal contamination in urban dusts of Xi'an, Central China. Science of the Total Environment 355: 176–186. doi: 10.1016/j.scitotenv.2005.02.026.

69. Carlon C, Critto A, Marcomini A, Nathanail P (2001) Risk based characterisation of contaminated industrial site using multivariate and geostatistical tools. Environmental Pollution 111: 417–427. doi: 10.1016/S0269-7491(00)00089-0.

70. Romic M, Romic D (2003) Heavy metals distribution in agricultural topsoils in urban area. Environmental Pollution 43: 795–805.

71. McGrath D, Zhang CS, Carton OT (2004) Geostatistical analyses and hazard assessment on soil lead in Silvermines area, Ireland. Environmental Pollution 127: 239–248. doi: 10.1016/j.envpol.2003.07.002.

72. Saito H, Goovaerts P (2001) Accounting for source location and transport direction into geostatistical prediction of contaminants. Environmental Science & Technology 35: 4823–4829. doi: 10.1021/es010580f.

73. Xie YF, Chen TB, Lei M, Yang J, Guo QJ, et al. (2011) Spatial distribution of soil heavy metal pollution estimated by different interpolation methods: Accuracy and uncertainty analysis. Chemosphere 82: 468–476. doi: 10.1016/j.chemosphere.2010.09.053.

74. Yasrebi J, Saffari M, Fathi H, Karimian N, Moazallahi M, et al. (2009) Evaluation and comparison of ordinary kriging and inverse distance weighting methods for prediction of spatial variability of some soil chemical parameters. Research Journal of Biological Sciences 4: 93–102.

75. Gotway CA, Ferguson RB, Hergert GW, Peterson TA (1996) Comparison of kriging and inverse-distance methods for mapping soil parameters. Soil Science Society of America Journal 60: 1237–1247. doi: 10.2136/sssaj1996.03615995006000040040x.

76. Kazemi SM, Hosseini SM (2011) Comparison of spatial interpolation methods for estimating heavy metals in sediments of Caspian Sea. Expert Systems with Applications 38: 1632–1649. doi: 10.1016/j.eswa.2010.07.085.

77. Compiling Council of Chinese Embayment (1997) Chinese Embayment (Part 2). Beijing: China Ocean Press.

78. Ma JR, Shao MH (1994) Variation in heavy metal pollution of offshore sedimentary cores in Jinzhou Bay. China Environmental Science 14: 22–29.

CHAPTER 7

A DETERMINATION OF METALLOTHIONEIN IN LARVAE OF FRESHWATER MIDGES (*Chironomus riparius*) USING BRDICKA REACTION

IVO FABRIK, ZUZANA RUFEROVA, KLARA HILSCHEROVA, VOJTECH ADAM, LIBUSE TRNKOVA, AND RENE KIZEK

7.1 INTRODUCTION

The on-line monitoring of a specific pollutant can be performed for only a short time interval, thus the alternative methods for long-term monitoring of pollution are developing. Various species of plants and animals, which are sensitive to higher levels of pollutants or which can synthesize easily detectable biomolecules as a response to environmental pollution, have been used to assess the effects of longterm environmental stress [1-8]. Among wide spectrum of biomolecules induced by various stress factors low molecular mass thiols are suitable for assessment of the environmental pollution because of their main physiological functions in scavenging of reactive oxygen species and detoxification of toxic organic and inorganic molecules via binding with free –SH groups [9]. Metallothioneins (MT) as a group of intracellular, low molecular mass, free of aromatic amino acids and rich in cysteine proteins with molecular weight from 6 to 10 kDa can be considered members of forementioned thiols biomark-

*This chapter was originally published under the Creative Commons Attribution License. Fabrik I, Rugerova Z, Hilscherova K, Adam V, Trnkova L, and Kizek R. A Determination of Metallothionein in Larvae of Freshwater Midges (*Chironomus riparius*) Using Brdicka Reaction.* Sensor *8 (2008), 4081-4094. doi:10.3390/s8074081.*

ers [10-14]. These proteins are abundant through whole animal kingdom, and they were also found in higher plants, eukaryotic microorganisms and some prokaryotes. MT can be found mostly in liver, kidney, pancreas and intestines at animal species. Moreover, MT is accumulated in lysosomes and was found also in nuclei [15].

It has been shown that the MT level increases, when an organism is affected by heavy metals ions. This event can be used for monitoring of environmental contamination by heavy metals [16-18]. Besides stress factors MT level strongly depends on animal specie, analysed tissue, age of an animal, eating habits and likely on others, not yet fully understood and identified factors.

Various analytical techniques and methods can be used to determine MT [11,16,18-41]. One of the most sensitive techniques called Brdicka reaction belongs to wide group of electrochemical methods and measures the hydrogen evolution catalyzed by a protein containing free –SH groups in the presence of Co(III) complex. This method was discovered by Brdicka in 1933 [42-44]. Since the discovery Brdicka reaction has been utilized for determination of MT levels in various animal species (fish, mussels, gastropods) [45-50]. A modification of this method to improve its sensitivity and selectivity has been recently proposed [10,51].

The aim of this work was to determine the metallothionein and total thiols content in larvae of freshwater midges (*Chironomus riparius*) using Brdicka reaction.

7.2 MATERIAL AND METHODS

7.2.1 CHEMICALS AND INSTRUMENTS

Rabbit liver MT (MW 7143), containing 5.9 % Cd and 0.5 % Zn, was purchased from Sigma Aldrich (St. Louis, USA). Tris(2-carboxyethyl)phosphine (TCEP) is produced by Molecular Probes (Evgen, Oregon, USA). $Co(NH_3)_6Cl_3$ and other used chemicals were purchased from Sigma Aldrich in ACS purity unless noted otherwise. The stock standard solutions of MT at 10 µg/ml was prepared with ACS water (Sigma-Aldrich, USA), reduced by adding of 1 mM TCEP [52] and stored in the dark at –20 °C. Working standard solutions were prepared daily by dilution of the stock solutions. Deionised water underwent demineralization by reverse osmosis using the instruments Aqua Osmotic 02 (Aqua Osmotic, Tisnov, Czech Republic) and then it was subsequently purified using Millipore RG (Millipore Corp., USA, 18 MΩ) – MiliQ water. The pH value was measured using WTW inoLab Level 3 with terminal Level 3 (Weilheim, Germany), controlled by personal computer program (MultiLab Pilot; Weilheim). The pH-electrode (SenTix-H) was calibrated by set of WTW buffers (Weilheim).

7.2.2 LARVAE OF FRESHWATER MIDGES

The larvae of freshwater midges *Chironomus riparius* were used in this study. The laboratory populations have been cultured according to standard procedures in siliceous sand overlaid by defined media at 20+-2°C, with constant humidity and 16 h light:8 h dark photoperiod (US EPA, 2000). The samples for analysis of thiols were collected after fifteen days of exposure of chironomid larvae to sediment artificially contaminated by cadmium (Cd^{2+}) at concentrations of 50 ng/g or 50 µg/g. The control samples were reared under same conditions without exposure to heavy metal. In the second part of the study, environmentally exposed chironomid larvae (3rd instar) have been collected from eight field sites with various level of pollution by heavy metals. All the sampling sites were located at smaller streams with abundant chironomid populations. Several sites were in a close proximity of large industrial works, which are the potential sources of pollution including heavy metals (Fig. 1). The control samples were the same as in the first part of the study.

7.2.3 HEAVY METALS ANALYSIS

Heavy metals content in sediment samples was evaluated based on Aqua Regia leaching process and after total decomposition of silicate

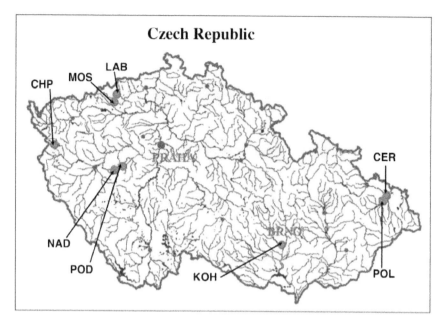

FIGURE 1: Map of sampling sites: CHP – Planenský potok (creek), MOS – Bílina: U Mostu, LAB – Bílina: U Labutí, NAD – Červený potok (creek), POD – Červený potok (creek), KOH – Kohoutovický potok (creek), POL – Polančice, CER – Černý potok (creek).

matrix. Inductively coupled plasma – mass spectrometry (ICP-MS) (Agilent 7500ce, Agilent Technologies, Japan) was used for determination of heavy metals in Aqua Regia leachate and total decomposed sediment samples. Elements (isotopes) suffering from polyatomic interferences were measured in He collision mode using Octopole Reaction System.

7.2.4 PREPARATION OF BIOLOGICAL SAMPLES FOR ELECTROCHEMICAL ANALYSIS

Larvae of freshwater midges (app. 0.2 g) were homogenized using liquid nitrogen. The homogenate was quantitatively transferred to test tube and vortexed (Vortex Genie, USA) for 15 min at room temperature. The vortexed sample was prepared by heat treatment. Briefly, the sample was kept at 99 °C in a thermomixer (Eppendorf 5430, USA) for 15 min with occasional stirring, and then cooled to 4°C. The denatured homogenates were centrifuged at 4°C, 15 000 g for 30 min (Eppendorf 5402, USA). Heat treatment effectively denature and remove high molecular weight proteins from samples [10,12,53].

7.2.5 STATIONARY ELECTROCHEMICAL ANALYSER – ADSORPTIVE TRANSFER STRIPPING DIFFERENTIAL PULSE VOLTAMMETRY BRDICKA REACTION – MT CONTENT

Electrochemical measurements were performed using an AUTOLAB analyser (EcoChemie, The Netherlands) connected to VA-Stand 663 (Metrohm, Switzerland), using a standard cell with three electrodes. The three-electrode system consisted of hanging mercury drop electrode as working electrode, an Ag/AgCl/3 M KCl reference electrode and a glassy carbon auxiliary electrode. For smoothing and baseline correction the software GPES 4.4 supplied by EcoChemie was employed. The Brdicka supporting electrolyte containing 1 mM $Co(NH_3)_6Cl_3$ and 1 M ammonia buffer (NH_3(aq) + NH_4Cl, pH = 9.6) was used and changed after five measurements, surface-active agent was not added. AdTS DPV Brdicka reaction parameters were as follows: initial potential of –0.6 V, end potential –1.6 V, modulation time 0.057 s, time interval 0.2 s, step potential of 1.05 mV, modulation amplitude of 250 mV, Eads = 0 V. Temperature of the supporting electrolyte was 4 °C. For other experimental conditions see in Ref. No. [10].

7.2.6 STATIONARY ELECTROCHEMICAL ANALYSER COUPLED WITH AUTOSAMPLER – DIFFERENTIAL PULSE VOLTAMMETRY BRDICKA REACTION – TOTAL CONTENT OF THIOLS

Electrochemical measurements were performed with 747 VA Stand instrument connected to 746 VA Trace Analyzer and 695 Autosampler (Metrohm, Switzerland), using a standard cell with three electrodes and cooled sample holder (4 °C). A hanging mercury drop electrode (HMDE) with a drop area of 0.4 mm2 was the working electrode. An Ag/AgCl/3M KCl electrode was the reference and glassy carbon electrode was auxiliary electrode. The Brdicka supporting electrolyte mentioned in Section 2.5

was used and changed per one analysis. The DPV parameters were as follows: initial potential of –0.7 V, end potential of –1.75 V, modulation time 0.057 s, time interval 0.2 s, step potential 2 mV, modulation amplitude -250 mV, Eads = 0 V. All experiments were carried out at temperature 4 °C (Julabo F12, Germany). For smoothing and baseline correction the software GPES 4.9 supplied by EcoChemie was employed.

A measurement proceeds as follows: A sample is positioned on the thermostatic sample holder (4 °C). The electrochemical cell is rinsed with distilled water (3 × 25 ml MiliQ water) using three computer controlled pumps. After draining of the water the supporting electrolyte (temperature 4 °C) is pipetted into the washed cell. Further a sample is introduced using the autosampler. The syringe from the autosampler is rinsed and the sample is injected to the cell. The measurement itself consists of a few following processes: At first, the injected sample is accumulated on the surface of hanging mercury drop electrode at open circuit for two minutes. At second, the current responses as function of various potentials are measured. At third, the measured values are processed by 746 VA Trace Analyser and transferred to a personal computer. At fourth, the transferred data is then processed using GPES 4.9 software (Fig. 2).

7.2.7 STATISTICAL ANALYSES

Data were processed using MICROSOFT EXCEL® (USA). Results are expressed as mean ± S.D. unless noted otherwise. Differences with p < 0.05 were considered significant (t-test was applied for means comparison).

7.3 RESULTS AND DISCUSSION

Larvae of freshwater midges (*Chironomus riparius*) were used for monitoring of pollution of environment by heavy metals already more than 25 years ago [54]. Since then midges and mosquito larvae have been employed for assessment of environmental contamination [55-59]. As mentioned in "Introduction" section, level of MT determined in animal blood and tissues is thought to be a marker of heavy metal stress. Brdicka reaction

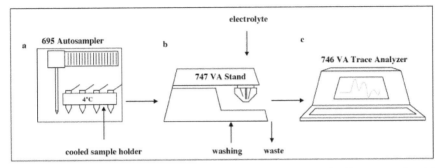

FIGURE 2: Scheme of stationary electrochemical analyser coupled with autosampler: (a) 695 Autosampler with cooled sample holder, (b) 747 VA Stand instrument with potentiostat/ galvanostat using a standard cell with three electrodes and (c) 746 VA Trace Analyzer for data processing. Vessels with washing water and the supporting electrolyte are other parts of instrument.

is very promising analytical method to determine MT due to its very low detection limit [10], however, automated analyser is needed to analyse tens of real samples.

7.3.1 STATIONARY ELECTROCHEMICAL ANALYSER COUPLED WITH AUTOSAMPLER

Automatic autosampler injecting low sample volumes (units of μl) coupled with stationary electrochemical analyser was used to overcome a lack of using of Brdicka reaction to analyse tens even hundreds of real samples (Fig. 2). A measurement is carried out automatically under the control of microprocessor within five minutes. The typical voltammograms of the Brdicka supporting electrolyte and MT (100 μM) are shown in Fig. 3A. The calibration curve (dependence of Cat2 peak height on MT concentration, $y = 1.4961 + 0.0799$, $R^2 = 0.9928$) obtained within the range from 0.25 to 5 μM is shown in Fig. 3B. Relative standard deviation of measurements was 3.6 %. The detection limit for MT estimated as 3 S/N was 5 nM. The detection limits (3 S/N) were calculated according to Long [60], whereas N was expressed as standard deviation of noise determined in the signal domain. The proposed methodology was utilized for determination of MT levels in larvae of freshwater midges exposed to cadmium(II) ions.

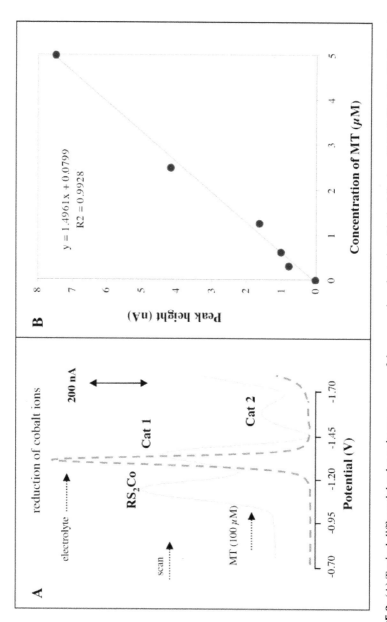

FIGURE 3: (A) Typical differential pulse voltammograms of the supporting electrolyte (dashed line) and MT (100 μM). (B) Dependence of Cat2 peak height on MT concentration.

7.3.2 DETERMINATION OF MT CONTENT IN LARVAE OF FRESHWATER MIDGES EXPOSED TO CADMIUM(II) IONS

Martinez et al. published a paper on study of morphological deformities in larvae of insect from chironomidae family collected in heavy metal contaminated areas [61]. The authors also determined content of various heavy metals as As, Cd, Ni, Pb, Zn and Ni. They found a significant correlation between metal concentrations and deformity rates for all metals except Ni. To our knowledge a correlation between MT in Chironomids and heavy metals exposure has not been studied so far. In our first laboratory experiments, chironomid larvae have been exposed to sediment contaminated by cadmium(II) ions at concentrations of 50 ng/g or 50 µg/g for fifteen days. One may expected that correlation between MT content and concentration of the heavy metal could be assessed. The larvae exposed to cadmium doses showed no visible marks of heavy metal intoxication because they developed and behaved same as control group. At the very end of the exposure the larvae were killed, washed in distilled water and frozen (-20°C) prior to analysis. The samples of larvae were prepared according to procedure mentioned in "Material and Methods" section. The typical DP voltammograms of real samples are shown in Fig. 4A. The characteristic peaks obtained Cat1 and Cat2 were very well separated and developed. To quantify MT in the samples of interest the Cat2 signal was used. However, we have found previously that if we analysed real samples by the automatic electrochemical analyser (Fig. 2), we quantified not only MT content but also content of all heat stable low molecular mass thiols such as glutathione and others [62,63]. The content of the heat stable thiols in the treated larvae is shown in Fig. 4B. The total content of thiols was enhanced with cadmium(II) dose compared to control samples. Moreover, we attempted to determine MT itself by adsorptive transfer stripping differential pulse voltammetry Brdicka reaction (AdTS DPV Brdicka reaction) [10]. We found that the content of MT also increased with increasing dose of the heavy metal ions but more substantially. MT content in control samples was 1.2 µM, in larvae exposed to 50 ng Cd/g it was 2.0 µM and in larvae exposed to 50 µg Cd/g 2.9 µM. Based on the results obtained it follows that the automated analyser is suitable for routine determination of

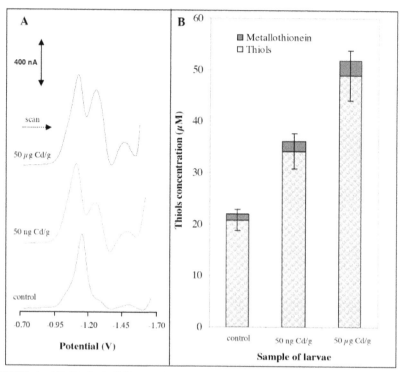

FIGURE 4: (A) Typical differential pulse voltammograms of samples obtained from larvae of freshwater midges exposed to 0 ng Cd/g, 50 ng Cd/g and 50 μg Cd/g. (B) The content of total thiols (stationary electrochemical analyser coupled with autosampler) and metallothionein (stationary electrochemical analyser) measured in the larvae exposed to cadmium(II) ions under controlled experimental conditions.

thiols content in larvae exposed to heavy metals and, thus, to assess heavy metal pollution of environment.

TABLE 1: Content of heavy metals in the sediments from the studied field sites.

Locality	Heavy Metals										
	Cr	Co	Ni	Cu	Zn	As	Mo	Cd	Sn	Pb	Sum
MOS	23.7	28.9	44.0	32.0	153	33.8	1.5	0.7	0.8	23.9	342.3
LAB	74.4	170*	249*	346*	715*	81.7*	17.9	2.8*	5.8*	74.0	1736.6
CER	243*	21.0	38.7	114	417	9.6	8.1	0.9	1.5	152	1005.8

TABLE 1: *Cont.*

Locality	Heavy Metals										
	Cr	Co	Ni	Cu	Zn	As	Mo	Cd	Sn	Pb	Sum
POL	17.3	4.9	14.8	23.9	132	3.9	0.6	0.3	0.5	15.4	213.6
NAD	57.0	24.3	33.6	38.3	336	10.2	118	1.3	3.8	39.2	661.7
POD	51.1	20.3	41.6	46.9	264	19.2	133*	2.5	3.4	41.6	623.6
CHP	66.6	10.0	39.9	43.1	567	7.6	3.3	0.6	1.0	22.7	761.8
KOH	121	23.3	41.7	93.1	667	5.9	4.0	0.6	2.2	86.9*	1045.7

* ... *The highest concentration of the metal.*

7.3.3 DETERMINATION OF THE CONTENT OF HEAVY METALS AND THIOLS IN CHIRONOMID LARVAE FROM THE FIELD STUDY

The field study with sampling of chironomids populations from the environment has been conducted to examine the levels of studied thiols and MT in the wild populations. Chironomid larvae as well as sediment samples have been collected from eight field sites with different levels of pollution by heavy metals called LAB, KOH, CER, CHP, NAD, POD, MOS and POL. Most of these sites have been impacted by undergoing or former industrial activities, thus increased concentration of pollutants including heavy metals can be expected. The metals content in sediment extracts were determined by ICP-MS. We determined content of chromium, nickel, copper, zinc, arsenic, molybdenum, cadmium, tin and lead (Tab. 1). Using the simplified criterion for assessment of environmental contamination (sum of concentration of all heavy metals determined in the locality) we evaluated the localities with highest and lowest contamination. The most polluted locality was LAB followed by KOH, CER, CHP, NAD, POD, MOS and POL. Besides the sum of the heavy metal concentrations sediments samples from the LAB locality contained the highest concentrations of seven out of the ten analyzed metals, particularly cobalt, nickel, copper, zinc, arsenic, cadmium and tin.

Moreover total content of thiols and MT in the chironomid larvae (3rd instar) sampled at the same sites as sediments was determined

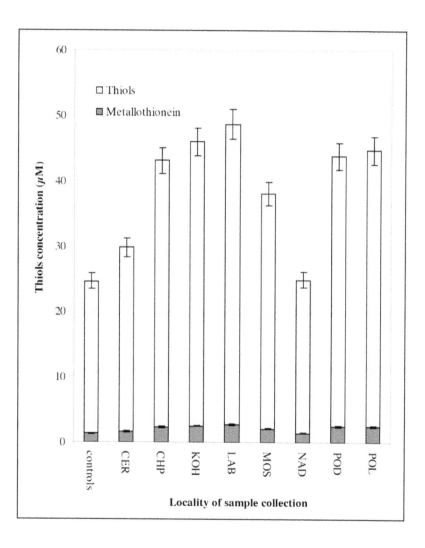

FIGURE 5: The content of total thiols and metallothionein measured in the larvae sampled from eight field sites with various level of pollution by heavy metals.

electrochemically (Fig. 5). The thiols and MT content determined in the larvae sampled was higher in comparison to control ones (from 102 to 198 % of content in control group). However, the content of the target molecules in the larvae samples at CER and NAD localities was lower about 30-40 % compared to other field localities. This difference could be associated with many factors hardly identified due to natural origin of the samples. After exclusion of results obtained from these two localities (CER and NAD) due to the lowest levels of MT, the good correlation (y(sum of content of heavy metals as mg/kg) = 0.0004x(content of MT, μM) + 2.0917; R2 = 0.8294) between metallothionein content and observed contamination expressed as total heavy metal concentration was obtained.

7.4 CONCLUSION

We have shown that the level of thiols in larvae of freshwater midges can be determined by using stationary electrochemical analyser coupled with autosampler. Moreover, we have compared the total thiols content and MT level determined by AdTS DPV Brdicka reaction and found that these values well correlated.

REFERENCES

1. Houtman, C.J.; Booij, P.; van der Valk, K.M.; van Bodegom, P.M.; van den Ende, F.; Gerritsen, A.A.M.; Lamoree, M.H.; Legler, J.; Brouwer, A. Biomonitoring of estrogenic exposure and identification of responsible compounds in bream from Dutch surface waters. Environ. Toxicol. Chem. 2007, 26, 898-907.
2. Dardenne, F.; Smolders, R.; De Coen, W.; Blust, R. Prokaryotic gene profiling assays to detect sediment toxicity: Evaluating the ecotoxicological relevance of a cell-based assay. Environ. Sci. Technol. 2007, 41, 1790-1796.
3. Tripathi, A.K.; Gautam, M. Biochemical parameters of plants as indicators of air pollution. J.Environ.Biol. 2007, 28, 127-132.
4. Ervin, G.N.; Herman, B.D.; Bried, J.T.; Holly, D.C. Evaluating non-native species and wetland indicator status as components of wetlands floristic assessment. Wetlands 2006, 26, 1114-1129.
5. Misik, M.; Micieta, K.; Solenska, M.; Misikova, K.; Pisarcikova, H.; Knasmuller, S. In situ biomonitoring of the genotoxic effects of mixed industrial emissions using the Tradescantia micronucleus and pollen abortion tests with wild life plants: Demonstration of the efficacy of emission controls in an eastern European city. Environ. Pollut. 2007, 145, 459-466.

6. Tomasevic, M.; Rajsic, S.; Dordevic, D.; Tasic, M.; Krstic, J.; Novakovic, V. Heavy metals accumulation in tree leaves from urban areas. Environ. Chem. Lett. 2004, 2, 151-154.

7. Sasmaz, A.; Yaman, M. Distribution of chromium, nickel, and cobalt in different parts of plant species and soil in mining area of Keban, Turkey. Commun. Soil Sci. Plant Anal. 2006, 37, 1845- 1857.

8. Rusu, A.M.; Jones, G.C.; Chimonides, P.D.J.; Purvis, O.W. Biomonitoring using the lichen Hypogymnia physodes and bark samples near Zlatna, Romania immediately following closure of a copper ore-processing plant. Environ. Pollut. 2006, 143, 81-88.

9. Huska, D.; Zitka, O.; Adam, V.; Beklova, M.; Krizkova, S.; Zeman, L.; Horna, A.; Havel, L.; Zehnalek, J.; Kizek, R. A sensor for investigating the interaction between biologically important heavy metals and glutathione. Czech J. Anim. Sci. 2007, 52, 37-43.

10. Petrlova, J.; Potesil, D.; Mikelova, R.; Blastik, O.; Adam, V.; Trnkova, L.; Jelen, F.; Prusa, R.; Kukacka, J.; Kizek, R. Attomole voltammetric determination of metallothionein. Electrochim. Acta 2006, 51, 5112-5119.

11. Kukacka, J.; Vajtr, D.; Huska, D.; Prusa, R.; Houstava, L.; Samal, F.; Diopan, V.; Kotaska, K.; Kizek, R. Blood metallothionein, neuron specific enolase, and protein S100B in patients with traumatic brain injury. Neuroendocrinol. Lett. 2006, 27, 116-120.

12. Kizek, R.; Trnkova, L.; Palecek, E. Determination of metallothionein at the femtomole level by constant current stripping chronopotentiometry. Anal. Chem. 2001, 73, 4801-4807.

13. Adam, V.; Krizkova, S.; Zitka, O.; Trnkova, L.; Petrlova, J.; Beklova, M.; Kizek, R. Determination of apo-metallothionein using adsorptive transfer stripping technique in connection with differential pulse voltammetry. Electroanalysis 2007, 19, 339-347.

14. Studnickova, M.; Turanek, J.; Zabrsova, H.; Krejci, M.; Kysel, M. Rat liver metallothioneins are metal dithiolene clusters. J. Electroanal. Chem. 1997, 421, 25-32.

15. Tsujikawa, K.; Imai, T.; Kakutani, M.; Kayamori, Y.; Mimura, T.; Otaki, N.; Kimura, M.; Fukuyama, R.; Shimizu, N. Localization of Metallothionein in Nuclei of Growing Primary Cultured Adult-Rat Hepatocytes. FEBS Lett. 1991, 283, 239-242.

16. Prusa, R.; Svoboda, M.; Blastik, O.; Adam, V.; Zitka, O.; Beklova, M.; Eckschlager, T.; Kizek, R. Increase in content of metallothionein as marker of resistence to cisplatin treatment. Clin. Chem. 2006, 52, A174-A175.

17. Petrlova, J.; Potesil, D.; Zehnalek, J.; Sures, B.; Adam, V.; Trnkova, L.; Kizek, R. Cisplatin electrochemical biosensor. Electrochim. Acta 2006, 51, 5169-5173.

18. Strouhal, M.; Kizek, R.; Vecek, J.; Trnkova, L.; Nemec, M. Electrochemical study of heavy metals and metallothionein in yeast Yarrowia lipolytica. Bioelectrochemistry 2003, 60, 29-36.

19. Adam, V.; Baloun, J.; Fabrik, I.; Trnkova, L.; Kizek, R. An electrochemical detection of metallothioneins in nanolitres at zeptomole level. Sensors 2008, 8, 2293-2305.

20. Adam, V.; Beklova, M.; Pikula, J.; Hubalek, J.; Trnkova, L.; Kizek, R. Shapes of differential pulse voltammograms and level of metallothionein at different animal species. Sensors 2007, 7, 2419-2429.

21. Adam, V.; Blastik, O.; Krizkova, S.; Lubal, P.; Kukacka, J.; Prusa, R.; Kizek, R. Application of the Brdicka reaction in determination of metallothionein in patients with tumours. Chem. Listy 2008, 102, 51-58.

22. Adam, V.; Fabrik, I.; Nakielna, J.; Hrdinova, V.; Blahova, P.; Krizkova, S.; Kukacka, J.; Prusa, R.; Kizek, R. New perspectives in electrochemical determination of metallothioneins. Tumor Biol. 2007, 28, 79-79.

23. Adam, V.; Krizkova, S.; Fabrik, I.; Zitka, O.; Horakova, Z.; Binkova, H.; Hrabeta, J.; Eckschlager, T.; Kukacka, J.; Prusa, R.; Sykorova, E.; Kizek, R. Metallothioneins as a new potential tumour marker. Tumor Biol. 2007, 28, 43-43.

24. Adam, V.; Petrlova, J.; Potesil, D.; Zehnalek, J.; Sures, B.; Trnkova, L.; Jelen, F.; Kizek, R. Study of metallothionein modified electrode surface behavior in the presence of heavy metal ionsbiosensor. Electroanalysis 2005, 17, 1649-1657.

25. Diopan, V.; Baloun, J.; Adam, V.; Macek, T.; Havel, L.; Kizek, R. Determination of expression of metallothionein at transgenic tobacco plants. Listy Cukrov. Reparske 2007, 122, 325-327.

26. Huska, D.; Krizkova, S.; Beklova, M.; Havel, L.; Zehnalek, J.; Diopan, V.; Adam, V.; Zeman, L.; Babula, P.; Kizek, R. Influence of cadmium(II) ions and brewery sludge on metallothionein level in earthworms (Eisenia fetida) – Bio-transforming of toxic wastes. Sensors 2008, 8, 1039-1047.

27. Kizek, R.; Vacek, J.; Trnkova, L.; Klejdus, B.; Havel, L. Application of catalytic reactions on a mercury electrode for electrochemical detection of metallothioneins. Chem. Listy 2004, 98, 166-173.

28. Krizkova, S.; Zitka, O.; Adam, V.; Beklova, M.; Horna, A.; Svobodova, Z.; Sures, B.; Trnkova, L.; Zeman, L.; Kizek, R. Possibilities of electrochemical techniques in metallothionein and lead detection in fish tissues. Czech J. Anim. Sci. 2007, 52, 143-148.

29. Kukacka, J.; Petrlova, J.; Prusa, R.; Adam, V.; Sures, B.; Beklova, M.; Havel, L.; Kizek, R. Changes of content of glutathione and metallothionein at plant cells and invertebrate treated by platinum group metals. Faseb J. 2006, 20, A75-A75.

30. Petrlova, J.; Krizkova, S.; Zitka, O.; Hubalek, J.; Prusa, R.; Adam, V.; Wang, J.; Beklova, M.; Sures, B.; Kizek, R. Utilizing a chronopotentiometric sensor technique for metallothionein determination in fish tissues and their host parasites. Sens. Actuator B-Chem. 2007, 127, 112-119.

31. Prusa, R.; Blastik, O.; Potesil, D.; Trnkova, L.; Zehnalek, J.; Adam, V.; Petrlova, J.; Jelen, F.; Kizek, R. Analytic method for determination of metallothioneins as tumor markers. Clin. Chem. 2005, 51, A56-A56.

32. Prusa, R.; Kizek, R.; Vacek, J.; Trnkova, L.; Zehnalek, J. Study of relationship between metallothionein and heavy metals by CPSA method. Clin. Chem. 2004, 50, A28-A29.

33. Trnkova, L.; Kizek, R.; Vacek, J. Catalytic signal of rabbit liver metallothionein on a mercury electrode: a combination of derivative chronopotentiometry with adsorptive transfer stripping. Bioelectrochemistry 2002, 56, 57-61.

34. Vajtr, D.; Kukacka, J.; Adam, V.; Kotaska, K.; Houstava, L.; Toupalik, P.; Kizek, R.; Prusa, R. Serum levels of metallothionein in expansive contusions, interleukine-6 and TNF-α during reparative phase of the blood brain barrier damage. Tumor Biol. 2007, 28, 44-44.

35. Szpunar, J. Bio-inorganic speciation analysis by hyphenated techniques. Analyst 2000, 125, 963-988.

36. Lobinski, R.; Chassaigne, H.; Szpunar, J. Analysis for metallothioneins using coupled techniques. Talanta 1998, 46, 271-289.

37. Dabrio, M.; Rodriguez, A.R.; Bordin, G.; Bebianno, M.J.; De Ley, M.; Sestakova, I.; Vasak, M.; Nordberg, M. Recent developments in quantification methods for metallothionein. J. Inorg. Biochem. 2002, 88, 123-134.

38. Vodickova, H.; Pacakova, V.; Sestakova, I.; Mader, P. Analytical methods for determination of metallothioneins. Chem. Listy 2001, 95, 477-483.

39. Sestakova, I.; Vodickova, H.; Mader, P. Voltammetric methods for speciation of plant metallothioneins. Electroanalysis 1998, 10, 764-770.

40. Werner, J.; Palace, V.; Baron, C.; Shiu, R.; Yarmill, A. A real-time PCR method for the quantification of the two isoforms of metallothionein in lake trout (Salvelinus namaycush). Arch. Environ. Contam. Toxicol. 2008, 54, 84-91.

41. Ndayibagira, A.; Sunahara, G.I.; Robidoux, P.Y. Rapid isocratic HPLC quantification of metallothionein-like proteins as biomarkers for cadmium exposure in the earthworm Eisenia andrei. Soil Biol. Biochem. 2007, 39, 194-201.

42. Brdicka, R. Polarographic Studies with the Dropping Mercury Kathode. -Part XXXII. - Activation of Hydrogen in Sulphydryl Group of Some Thio-Acids in Cobalt Salts Solutions. Coll. Czech. Chem. Commun. 1933, 5, 148-164.

43. Brdicka, R. Polarographic Studies with the Dropping Mercury Kathode. -Part XXXI. - A New Test for Proteins in The presence of Cobalt Salts in Ammoniacal Solutions of Ammonium Chloride. Coll. Czech. Chem. Commun. 1933, 5, 112-128.

44. Brdicka, R. Polarographic studies with the dropping mercury kathode.- Part XXXIII.- The microdetermination of cysteine and cystine in the hydrolysates of proteins, and the course of the protein decomposition. Coll. Czech. Chem. Commun. 1933, 5, 238-252.

45. Dragun, Z.; Erk, M.; Raspor, B.; Ivankovic, D.; Pavicic, J. Metal and metallothionein level in the heat-treated cytosol of gills of transplanted mussels Mytilus galloprovincialis Lmk. Environ. Int. 2004, 30, 1019-1025.

46. Raspor, B.; Dragun, Z.; Erk, M.; Ivankovic, D.; Pavicic, J. Is the digestive gland of Mytilus galloprovincialis a tissue of choice for estimating cadmium exposure by means of metallothioneins? Sci. Total Environ. 2004, 333, 99-108.

47. Marijic, V.F.; Raspor, B. Metal exposure assessment in native fish, Mullus barbatus L., from the Eastern Adriatic sea. Toxicol. Lett. 2007, 168, 292-301.

48. Marijic, V.F.; Raspor, B. Metal exposure assessment in native fish, Mullus barbatus, from the Eastern Adriatic Sea. Toxicol. Lett. 2006, 164, S156-S156.

49. Marijic, V.F.; Raspor, B. Age- and tissue-dependent metallothionein and cytosolic metal distribution in a native Mediterranean fish, Mullus barbatus, from the Eastern Adriatic Sea. Comp. Biochem. Physiol. C-Toxicol. Pharmacol. 2006, 143, 382-387.

50. Dragun, Z.; Raspor, B.; Erk, M.; Ivankovic, D.; Pavicic, J. The influence of the biometric parameters on metallothionein and metal level in the heat-treated cytosol of the whole soft tissue of transplanted mussels. Environ. Monit. Assess. 2006, 114, 49-64.

51. Raspor, B.; Paic, M.; Erk, M. Analysis of metallothioneins by the modified Brdicka procedure. Talanta 2001, 55, 109-115.

52. Kizek, R.; Vacek, J.; Trnkova, L.; Jelen, F. Cyclic voltammetric study of the redox system of glutathione using the disulfide bond reductant tris(2-carboxyethyl)phosphine. Bioelectrochemistry 2004, 63, 19-24.

53. Erk, M.; Ivanković, D.; Raspor, B.; Pavičić, J. Evaluation of different purification procedures for the electrochemical quantification of mussel metallothioneins. Talanta 2002, 57, 1211-1218.

54. Lang, C.; Langdobler, B. Melimex, an Experimental Heavy-Metal Pollution Study - Oligochaetes and Chironomid Larvae in Heavy-Metal Loaded and Control Limno-Corrals. Schweizerische Zeitschrift Fur Hydrologie-Swiss Journal of Hydrology 1979, 41, 271-276.

55. Bervoets, L.; Solis, D.; Romero, A.M.; Van Damme, P.A.; Ollevier, F. Trace metal levels in chironomid larvae and sediments from a Bolivian river: Impact of mining activities. Ecotox. Environ. Safe. 1998, 41, 275-283.

56. Pourang, N. Heavy metal concentrations in surficial sediments and benthic macroinvertebrates from Anzali wetland, Iran. Hydrobiologia 1996, 331, 53-61.

57. Gillis, P.L.; Reynoldson, T.B.; Dixon, D.G. Metallothionein-like protein and tissue metal concentrations in invertebrates (Oligochaetes and Chironomids) collected from reference and metal contaminated field sediments. J. Gt. Lakes Res. 2006, 32, 565-577.

58. Bhattacharyay, G.; Sadhu, A.K.; Mazumdar, A.; Chaudhuri, P.K. Antennal deformities of chironomid larvae and their use in biomonitoring of heavy metal pollutants in the River Damodar of West Bengal, India. Environ. Monit. Assess. 2005, 108, 67-84.

59. Mousavi, S.K.; Primicerio, R.; Amundsen, P.A. Diversity and structure of Chironomidae (Diptera) communities along a gradient of heavy metal contamination in a subarctic watercourse. Sci. Total Environ. 2003, 307, 93-110.

60. Long, G.L.; Winefordner, J.D. Limit of Detection. Anal. Chem. 1983, 55, A712-A724.

61. Martinez, E.A.; Moore, B.C.; Schaumloffel, J.; Dasgupta, N. The potential association between mental deformities and trace elements in Chironomidae (diptera) taken from a heavy metal contaminated river. Arch. Environ. Contam. Toxicol. 2002, 42, 286-291.

62. Krizkova, S.; Fabrik, I.; Adam, V.; Kukacka, J.; Prusa, R.; Trnkova, L.; Strnadel, J.; Horak, V.; Kizek, R. Effects of reduced glutathione, surface active agents and ionic strength on detection of metallothioneinS by using of brdicka reaction. Electroanalysis 2008, submitted.

63. Krizkova, S.; Fabrik, I.; Adam, V.; Kukacka, J.; Prusa, R.; Chavis, G.J.; Trnkova, L.; Strnadel, J.; Horak, V.; Kizek, R. Utilizing of adsorptive transfer stripping technique Brdicka reaction for determination of metallothioneins level in melanoma cells, blood serum and tissues. Sensors 2008, 8, 3106-3122.

CHAPTER 8

MULTIVARIATE STATISTICAL ASSESSMENT OF HEAVY METAL POLLUTION SOURCES OF GROUNDWATER AROUND A LEAD AND ZINC PLANT

ABBAS ALI ZAMANI, MOHAMMAD REZA YAFTIAN, AND ABDOLHOSSEIN PARIZANGANEH

8.1 INTRODUCTION

Water is one of essential compounds for all forms of plants and animals [1], thus its pollution is generally considered more important than soil and air. Due to its specific centeracteristics, this liquid bears unique properties. It is the most effective dissolving agent, and adsorbs or suspends many different compounds [2].

More than one billion people in the world do not have suitable drinking water, and two to three billions lack access to basic sanitation services. About three to five millions die annually from water related diseases [3].

This chapter was originally published under the Creative Commons Attribution License. Zamani AA, Yaftian MR, and Parizanganeh A. Multivariate Statistical Assessment of Heavy Metal Pollution Sources of Groundwater Around a Lead and Zinc Plant. Iranian Journal of Environmental Health Science & Engineering 9,29 (2012), doi:10.1186/1735-2746-9-29.

Surface water (fresh water lakes, rivers, streams) and groundwater (borehole water and well water) are the principal natural water resources. Nowadays one of the most important environmental issues is water contamination [4,5]. Heavy metals are among the major pollutants of water sources [6]. Despite this, heavy metals are sensitive indicators for monitoring changes in the marine environment. Due to human industrial activities, the levels of heavy metals in the aquatic environment are seriously increasing and have created a major global concern [7,8]. Some of these metals are essential for the growth, development and health of living organisms, whereas others are non-essential as they are indestructible and most of them are categorized as toxic species on organisms [9]. Nonetheless the toxicity of metals depends on their concentration levels in the environment. With increasing concentrations in environment and decreasing the capacity of soils towards retaining heavy metals, they leach into groundwater and soil solution. Thus, these toxic metals can be accumulated in living tissues and concentrate through the food chain.

Cadmium is regarded as the most serious contaminant of the modern age [10]. Copper is classified as a priority pollutant because of its adverse health effects [11]. Zinc and iron are essential elements and are generally considered to be non-toxic below certain levels [12]. Lead is not an essential trace element in any organism and has no known biological function. It can cause a variety of harmful health effects [13] and is known as a fatal neurotoxicant [14]. Excessive concentrations of cobalt can cause death and various compounds of nickel are carcinogenic [15]. These menaces provoke the studies on the monitoring of these heavy metals in this chain being important for protection of public health.

A variety of techniques including x-ray fluorescence (XRF), neutron activation analysis (NAA), inductively coupled plasma-atomic emission spectrometry (ICP-AES), atomic absorption spectrometry (AAS) and graphite furnace atomic absorption spectrometry (GFAAS) have been used for evaluating the heavy metal concentration in environmental samples [16-20]. Beside their valuable centeracteristics, these techniques suffer from some disadvantages such as heavy capital cost, expensive maintenance, and insufficient sensitivity for very low concentrations of metals. Voltammetric methods are known as sensitive techniques for determination of a variety of chemical species [21]; among these techniques, differential pulse polarography (DPP) bears some advantages for accurate and precise detection and determination of trace amounts of heavy metal ions in environmental samples [22,23].

Evaluation of the contaminants resulted from excavation of zinc and lead mines and development of related industries in Zanjan province-Iran and their negative environmental impacts is critical and important. Lack of a systematic investigation of the probable heavy metals contamination around National Iranian Lead and Zinc Company (NILZ) in Bonab Industrial Estate (BIE), in Zanjan province, promotes to assess the quality of groundwater sources in this industrial zone. These are the main sources of drinking water and irrigation for a part of people who live around NILZ Company. In this research, DPP technique was used to determine the concentrations of seven heavy metals (iron, cobalt, nickel, copper, zinc, cadmium and lead) in water samples and the results were compared with the maximum contaminant levels (MCLs) specified by WHO as well as Institute of Standards and Industrial Research of Iran (ISIRI). The multivariate statistical analysis was conducted to categorize the metals and to distinguish the source of the contaminants.

8.2 MATERIALS AND METHODS

8.2.1 STUDY AREA

Zanjan province (located in north west Iran), has a large metalliferrous site and has been considered as a traditional mining region since antiquity [24]. There are still large reserves of lead and zinc in the area. Both mines and smelting units within the province present a risk of contamination of soils, plants, and surface/groundwater resources through dissemination of particles carrying metals by wind action and/or by runoff from the tailings [25]. Transportation of concentrated ore by trucks for about 110 kilometers from mines in Angouran to NILZ is another anthropogenic source of metal contamination, especially along the roads.

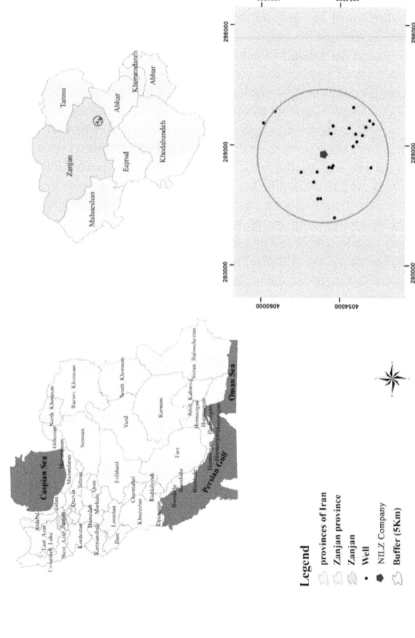

FIGURE 1: Location map of the studied area indicating sampling points.

In this study, Bonab Industrial Estate (BIE) and its neighborhood was selected for detailed study. The research was focused on the environmental impacts of NILZ Company (36° 66′ N, 48° 48′ E) located within BIE, about 12 km east of Zanjan city. The NILZ Company was established in 1992, with a current consumption of about 300,000 tons of raw ore and an annual production of 55000 tons of Pb and Zn [26,27]. The plant is situated over an aquifer, which is the only source of fresh water available in the area, supplying a part of drinking water to Zanjan citizens and its neighboring areas as well as water used for agricultural and industrial consumptions. The tailings from BIE, estimated to be about 2.5 million tons, contain a variety of toxic elements, notably Pb, Zn, and Cd [26]. They are damped in the vicinity of the Estate and are exposed to wind and rain, contributing to soil, surface and groundwater contamination.

8.2.2 SAMPLE COLLECTION AND STORAGE

To examine the extent of the contamination by toxic metals leached from tailings, 23 spring/groundwater samples were collected and analyzed from the studied area. Nineteen groundwater and four spring water stations were selected from the site within a radius of five km from NILZ Company (Figure 1).

Sampling stations were selected, taking into account the direction of groundwater flow (west), direction of prevailing winds (west and south west) and also the density of the population within the studied area. However, limitations on number and distribution of sampling stations are set due to the spatial distribution of available bore wells within the studied area. Table 1 shows the location of sampling stations for this study.

From each station three replicate samples were selected for analysis. Glassware and vessels were treated in 10% (v/v) nitric acid solution for 24 h and were washed with distilled and deionized water. The samples were collected in polypropylene containers, labeled and immediately few drops of HNO_3 (ultra pure grade) to pH < 2 were added to prevent loss of metals, bacterial and fungal growth and then stored in a refrigerator.

TABLE 1: GPS location and some physical properties of sampling wells

Site/ location	GPS location (UTM)		pH	EC (µS /cm)	DO (mg /L)	t (°C)	Depth (m)	Distance[1] (m)
	X	Y						
W1	290458	4054526	7.81	576	6.91	16	60	2418
W2	289862	4052820	7.93	465	7.46	16	75	2958
W3	290843	4051745	7.58	729	8.36	15	50	4349
W4	289758	4052297	7.70	758	7.73	15	35	3308
W5	287326	4051641	8.15	900	7.10	28	spring	3617
W6	290619	4051470	7.41	1607	7.51	14	13	4512
W7	284992	4055674	7.86	500	6.40	17	45	3234
W8	284999	4055445	8.23	1857	6.95	18	32	3247
W9	283572	4054398	8.27	826	7.22	27	spring	4736
W10	286973	4056943	7.51	361	8.39	17	150	2203
W11	289883	4054679	7.45	369	8.07	19	150	1777
W12	291848	4052965	7.53	480	8.87	15	150	4352
W13	290659	4059821	7.87	334	9.03	16	spring	5320
W14	291539	4058941	8.29	384	7.10	26	spring	5073
W15	290371	4052030	7.52	990	9.36	13	25	3971
W16	289285	4052723	7.45	1157	8.22	14	10	2746
W17	288931	4053010	7.43	1073	7.78	14	150	2344
W18	287475	4054510	7.85	815	9.26	18	20	995
W19	287004	4055727	8.08	564	7.59	14	50	1300
W20	287319	4054582	7.20	326	7.20	15	13	1054
W21	286247	4055985	7.35	850	7.70	15	70	2183
W22	287340	4054826	7.50	671	6.97	15	42	927
W23	290294	4053296	7.80	415	8.90	16	70	2837

[1] *Distance from NILZ Company.*

8.2.3 REAGENTS AND STANDARDS

All the chemicals used in this study were mostly reagents of highest grades (Merck) and used without further treatment. The chemicals used were: dimethylglyoxime (>99%), ammonia solution (25%), ammonium chloride

(>99.8), acetic acid (>99.8), hydrochloric acid (37%), nitric acid (65%), pyrocatechol (>99%), and sodium hydroxide (>97%). The heavy metal standards were prepared from stock solutions of 1000 ± 5 mg/L (Merck) by successive dilution with ultra-pure water. Polargraphic mercury was used as electrode in heavy metal determination (Merck).

8.2.4 SAMPLE DIGESTION

Groundwater samples were filtered through 0.45 μm filters. To ensure the removal of organic impurities from the samples and thus preventing interference in analysis, the samples were preserved and digested with concentrated nitric acid. To this end 1 mL of nitric acid was added to water sample in 50 mL volumetric flask.

8.2.5 SAMPLE ANALYSIS IN THE FIELD

The pH, electrical conductivity (EC) and dissolved oxygen (DO) of the samples were immediately measured at sampling stations by using a portable digital pH meter (Hach HQ 40d). Recorded pH and EC of samples varied in the range of 7.2 -8.3 and 326–1857 (μS /cm) respectively (Table 1). The pH values of the samples were within the WHO range (6.5 -8.5) but those of ECs were below the announced value of MCL by WHO (1500 μS /cm), except for samples number W6 and W8.

8.2.6 SAMPLE ANALYSIS

Water samples were analyzed for the presence of iron, cobalt, nickel, copper, zinc, cadmium and lead using a differential pulse polarography (Metrohm 797 VA). Dissolved air was removed from the solutions by degassing with N2 gas (99.999%) for 5–10 min prior to each run. Standard addition method was used for the analysis. The polarography parameters are given in Table 2. Digested samples were analyzed in triplicate and the average concentrations of metals were reported in μg/L.

TABLE 2. Instrument operating parameters for the analysis of the investigated heavy metals

Parameters	Heavy metals		
	Fe[1]	Co and Ni[2]	Cu, Zn, Cd and Pb[3]
Working electrode	HMDE	HMDE	HMDE
Drop size	7	4	4
Stirrer speed	2000 rpm	2000 rpm	2000 rpm
Mode	DP	DP	DP
Purge time	300 s	300 s	300 s
Deposition potential	−300 mV	−0.7 V	−1.15V
Deposition time	60 s	90 s	90 s
Equilibrium time	5 s	10 s	10 s
Pulse amplitude	50 mV	50 mV	50 mV
Start potential	−200 mV	−0.8 V	−1.15 V
End potential	−550 mV	−1.25 V	0.05 V
Voltage step	4 mV	4 mV	6 mV
Voltage step time	0.4 s	0.3 s	0.1 s
Sweep rate	10 mV/s	13 mV/s	60 mV/s
Peak potential	−380 mV	−1.13, -0.97V	−0.10, -0.98, -0.56, -0.38 V

[1]*10 mL sample solution + 100 μL Catechol solution (1M) + 1 mL phosphate buffer; pH =7.0,* [2]*10 mL. sample solution + 100 μL dimethylglyoxime solution (0.1 M) + 0.5 ml NH$_4$Cl pH = 9.5,* [3]*10 mL. sample solution + 1 mL ammonium acetate buffer; pH =4.6.*

8.2.7 STATISTICAL ANALYSIS

SPSS statistical package (Window version 18) and software Excel 2007 are used for data analysis. The analysis of the experimental data was carried out by using one-way ANOVA, Pearson correlation matrix, Cluster Analysis, Principal Component Analysis (PCA) and Factor Analysis (FA) methods [28,29]. Pearson correlation matrix shows a probable common source of the pollutants. Cluster analysis is used for dividing the studied metal ions into the similar classes with respect to their normalized concentration level. PCA is designed to transform the original variables into new, uncorrelated variables (axes), called the principal components. Factor Analysis is similar to Principal Component Analysis method except for the preparation of the observed correlation matrix for extraction and the underlying theory [30].

The one-way ANOVA method allows testing the significant difference of the means. For this test each sampling location was selected as a group and its heavy metal concentration as the corresponding variable. The ANOVA test requires three assumptions, i.e. the random behavior of the occurrence, the homogeneity of variance and the normal distribution behavior of the metal ions in the sample stations. These were tested by using Runs test, Levene statistic and the K-S (Kolmogorov-Smirnov) methods, respectively. It is noteworthy that instead of the ANOVA test, one can use the Kruskal-Wallis test. The latter is a non-parametric test without requirements announced for the ANOVA test [28,29]. In this work both of the methods were tested for a comparison.

The bivariate correlation procedure computes the pair wise associations for a set of variables and displays the results in a matrix. It is useful for determining the strength and direction of the association between two variables. The correlation coefficients computed by bivariate correlation procedure lay in the range -1 (for the cases in which a perfect negative relationship exists) to $+1$ (for a perfect positive relationship). A value of 0 indicates there is no linear relationship among the variables. For normally distributed variables, the Pearson method can be used to calculate the correlation coefficient. For normally distributed variables, the Pearson correlation was used for bivariate correlation, otherwise non-parametric Spearman method was applied.

Cluster analysis is a method for dividing a group of metals into classes so that similar metals, with respect to variable space, are in the same class. In fact, the groups are not known prior to applying this mathematical analysis and no assumption is made about the distribution of the variables [28,29].

The major objective of FA is to reduce the contribution of less significant variables to simplify even more of the data structure given by PCA. This goal can be achieved by rotating the axis defined by PCA and constructing new variables, also called Varifactors [31]. PCA reduces the dimensionality of data by a linear combination of original data to generate new latent variables which are orthogonal and uncorrelated to each other

[32]. The major objective of FA is to reduce the contribution of less significant variables to simplify even more of the data structure coming from PCA. All significance statements reported in this study are at the $P < 0.05$ level.

8.3 RESULTS

8.3.1 EXTENT OF HEAVY METALS CONTAMINATION

The results of analysis of target metal ions i.e. Fe, Co, Ni, Zn, Cd and Pb in samples from 23 studied wells are given in Table 3. It is noteworthy that the reported values are based on three replicate determinations. Table 4 is prepared in order to give a simple comprehensive interpretation on the obtained data, and to compare the concentration of the studied metals in the samples with the MCL values reported by WHO and ISIRI. The results show that Fe, Co, Ni, Cu, Zn, Cd and Pb are detected in 100%, 47.8%, 78.3%, 100%, 100%, 65.2% and 100% of the samples, respectively. The concentration of metals (in μg /L) in the samples were found in the range of 75.90 -339.75 for Fe; ND (not detected) -99.82 for Co; ND −84.15 for Ni; 6.59-65.31 for Cu; 27.79 -2227.80 for Zn; ND −14.87 for Cd; and 0.74 -12.45 for Pb.

8.3.2 COMPARISON OF THE CONCENTRATION OF HEAVY METALS

In order to deduce the frequencies of the concentration of each metal in the samples, the Chi-Square test was applied [29]. Here, the frequency means the number of times a given range of concentrations occurs, and the Chi-Square test is used to examine whether the observed frequencies differ significantly from those which would be expected on the null hypothesis. This test indicates that there is no significant difference between observed frequencies of the heavy metals.

TABLE 3: Metal contents in water samples (μg/ L) from the wells

Sample	Fe	Co	Ni	Cu	Zn	Cd	Pb
W1	129.30 ± 10.91	9.91 ± 0.91	84.15 ± 10.84	44.39 ± 0.26	196.95 ± 11.40	4.55 ± 0.25	11.78 ±2.76
W2	339.75 ± 39.77	60.96 ± 1.95	52.24 ± 4.46	26.25 ± 2.03	169.51 ± 6.68	14.87 ± 0.86	6.10 ± 0.54
W3	143.99 ± 20.16	11.07 ± 1.15	41.59 ± 1.46	15.27 ± 1.44	140.53 ± 13.55	0.68 ± 0.06	1.26 ± 0.64
W4	228.71 ± 4.50	3.51 ± 0.82	42.98 ± 2.41	18.39 ± 1.46	460.19 ± 26.21	1.29 ± 0.45	1.67 ± 0.78
W5	194.11 ± 18.49	1.25 ± 0.15	21.82 ± 1.61	16.65 ± 1.05	353.92 ± 59.79	ND	2.63 ± 0.28
W6	109.83 ± 16.98	ND[2]	20.09 ± 1.23	8.83 ± 0.27	232.38 ± 41.84	ND	1.76 ± 0.42
W7	146.51 ± 3.63	0.43 ± 0.02	14.12 ± 0.91	15.71 ± 1.93	582.66 ± 61.54	3.55 ± 0.37	5.46 ± 0.48
W8	116.20 ± 9.88	ND	ND	27.71 ± 1.90	96.54 ± 6.23	ND	2.51 ± 0.16
W9	288.97 ± 16.40	1.89 ± 0.31	12.15 ± 0.44	65.31 ± 6.00	541.43 ± 22.53	3.41 ± 0.34	12.45 ± 0.73
W10	302.70 ± 20.74	2.24 ± 0.15	6.13 ± 0.22	35.81 ± 3.41	60.36 ± 6.74	0.36 ± 0.05	6.52 ± 0.31
W11	124.80 ± 7.37	ND	5.20 ± 0.21	12.06 ± 0.67	205.36 ± 12.64	ND	2.04 ± 0.19
W12	144.21 ± 15.68	99.82 ± 9.64	9.55 ± 0.41	21.96 ± 0.92	84.54 ± 6.23	0.62 ± 0.20	5.45 ± 0.38
W13	132.58 ± 6.49	ND	ND	9.84 ± 1.11	49.26 ± 2.11	1.23 ± 0.23	3.25 ± 0.24
W14	308.73 ± 22.94	ND	12.21 ± 1.11	13.79 ± 1.09	43.68 ± 2.15	0.83 ± 0.11	6.71 ± 0.24
W15	120.64 ± 24.25	1.52 ± 0.27	8.04 ± 0.28	60.77 ± 3.10	113.99 ± 7.12	ND	6.56 ± 0.31
W16	75.90 ± 10.80	3.19 ± 0.35	7.68 ± 0.47	31.95 ± 2.47	90.04 ± 7.71	ND	6.60 ± 0.24
W17	245.96 ± 29.36	ND	7.21 ± 0.24	21.52 ± 1.66	115.69 ± 15.27	1.43 ± 0.36	3.31 ± 0.27
W18	179.67 ± 22.70	ND	6.37 ± 0.27	8.22 ± 0.72	31.72 ± 4.87	0.73 ± 0.05	0.91 ± 0.07
W19	101.49 ± 6.39	ND	ND	24.87 ± 0.12	27.79 ± 1.51	0.59 ± 0.06	0.74 ± 0.21
W20	84.60 ± 12.10	ND	ND	28.17 ± 1.02	133.82 ± 15.59	0.15 ± 0.06	4.60 ± 0.22

TABLE 3: *Cont.*

Sample	Fe	Co	Ni	Cu	Zn	Cd	Pb
W21	115.60 ± 32.23	ND	ND	6.59 ± 0.92	65.99 ± 4.21	2.25 ± 0.39	2.25 ± 0.32
W22	137.38 ± 23.88	ND	78.66 ± 6.60	15.02 ± 2.10	2227.80 ± 145.12	ND	5.26 ± 0.15
W23	113.57 ± 6.72	ND	5.57 ± 0.85	34.84 ± 3.75	73.67 ± 10.12	ND	2.56 ± 0.66

[1]Average of three determinations. [2]Not Detected.

The random and normal distribution assumptions were checked by Runs and K–S methods, respectively. Another requirement for applying the ANOVA test is that the variances of the groups are equivalent. Based on the statistically verification done by Levene test, the homogeneity of variance was found to be significant for the samples (Levene statistic = 5.696, $P < 0.001$). Although the Levene statistic parameter rejects the null hypothesis, as the group variances are equal the ANOVA test can be yet used. Alternatively, the homogeneity and normal distribution in the data can be achieved by transforming the obtained data to another mathematically presentation which lowers the difference between the data. This can be achieved for example by using the logarithmic form of data. In addition, one can use a non-parametric test. This type of tests does not require to a homogeneity assumption.

The ANOVA method was used under two conditions. In fact, although the homogeneity of the data was not shown, ANOVA was applied to the data. In addition, by transforming the data as their logarithmic form, homogeneity in the observed data was achieved. The non-parametric method Kruskal-Wallis is based on ranks of the data variances. This method was used for the same scope as ANOVA. Both parametric and non-parametric methods used for comparison of the concentrations of heavy metals among sampling sites show a statistically significant difference depending on sampling locations.

TABLE 4: Summery statistics of heavy metal content in water samples ($\mu g/ L$) analysis

	Fe	Co	Ni	Cu	Zn	Cd	Pb
Detected (%)	100	47	78	100	100	65	100
Min. of the detected concentration	75.90	ND[1]	ND[1]	6.59	27.79	ND1	0.74
Max. of the detected concentration	339.75	99.82	84.15	65.31	2227.80	14.87	12.45
Mean of the detected concentration	168.92	17.80	24.21	24.52	265.12	2.44	4.45
Standard deviation[2]	77.95	32.33	25.21	15.50	456.66	3.68	3.14
MCL (based on WHO)	300.003	-	70.00	1000	3000	3.00	10.00
Percentage of samples containing metals > WHO (%)	13.04	-	8.70	0	0	17.39	8.70
MCL (based on ISIRI)[3]	-	-	70.00	1000	3000	3.00	10.00
Percentage of samples containing metals > ISIRI (%)	-	-	8.70	0	0	17.39	8.70

[1]*Not Detected.* [2]*Standard deviation for heavy metal concentration in all samples.* [3]*Institute of standards and industrial research of Iran, 1997.*

8.3.3 BIVARIATE CORRELATIONS OF INVESTIGATED HEAVY METALS

To deduce the probable common source of metals in water samples, the bivariate correlation procedure was used (Table 5). This procedure computes the pair wise associations for a set of metals and displays the results as a matrix. It is useful for determining the value of association of the investigated metals. Because, obtained data was not normally distributed, Spearman method was applied.

8.3.4 CLASSIFICATION OF THE INVESTIGATED HEAVY METALS BY CLUSTER ANALYSIS

Cluster analysis grouped the studied heavy metals into clusters (called groups in this study) on the basis of similarities within a group and dissimilarities between different groups. CA was performed on the data using Ward method and squared Euclidean distance. A dendrogram was produced by cluster analysis, shown in Figure 2. Seven studied heavy metals were classified into five groups based on spatial similarities and dissimilarities.

TABLE 5 Spearman correlation coefficient (r) of heavy metals in the sampling stations

	Fe	Co	Ni	Cu	Zn	Cd	Pb	DO
Co	0.019							
Ni	0.129	0.342						
Cu	0.007	0.008	−0.025					
Zn	0.121	**−0.433**[a]	**0.593**[b]	0.084				
Cd	0.291	0.027	**0.679**[b]	0.046	**0.551**[b]			
Pb	**0.283**[a]	−0.054	0.135	**0.583**[b]	0.168	**0.374**[a]		
DO	−0.079	0.273	**−0.681**[b]	−0.043	**−0.518**[a]	−0.436	−0.250	
Depth	**0.516**[a]	0.519	−0.258	0.003	−0.133	0.155	0.084	0.136
Dist. Ind.	0.266	0.027	0.234	0.029	0.080	0.261	0.241	0.063

Correlation is significant (a) at the 0.05 level and (b) at the 0.01 level.

8.3.5 PRINCIPAL COMPONENT ANALYSIS AND FACTOR ANALYSIS

PCA reduces the dimensionality of data by a linear combination of original data to generate new latent variables which are orthogonal and uncorrelated to each other [32]. Prior to PCA and FA analysis, the raw data was commonly normalized to avoid misclassifications due to the different order of magnitude and range of variation of the analytical parameters [30]. The rotation of the principal components was executed by the Varimax method with Kaiser normalization.

Four principal components are obtained for heavy metals through FA performed on the PCA. This indicates that four main controlling factors influenced the quality of surface water in the study area. Corresponding components, variable loadings, and the variances are presented in Table 6. Only PCs with eigenvalues greater than 1 were considered. PCA of the whole data set yielded 4 data sets explaining 88.92% of the total variance.

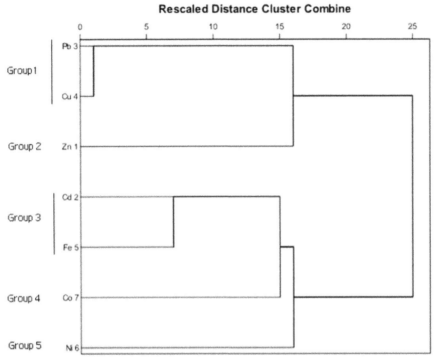

FIGURE 2: Dendrogram of heavy metal concentrations of water samples.

First component which explained 32.02% of the total variance is correlated with Pb and Cu. The second component is due to Zn and Co. The third component is a location for only Ni metal. The latest extracted factor is related to Fe and Cd.

8.4 DISCUSSION

According to results mentioned in Table 4, all of the samples contained Co, Cu and Zn inferior to the values specified by related MCLs. In contrast, in 13.0% and 8.7% of the samples the amounts of Fe and Ni, respectively, were above WHO MCLs. The amount of cadmium found in 34.8% of the samples was lower than the detection limit of the DPP method, but 17.4% contained the metal ion superior than the ISIRI and WHO MCL. This is of concern because cadmium has carcinogenic properties as well as a long biological half life leading to chronic effects as a result of accumulation in liver and renal cortex. It can also cause kidney damage as well as producing acute health effects resulting from over exposure to high concentrations [20].

Due to possible long term effects of chronic exposure, the presence of lead in drinking water is crucially important for public concern. Although all of the samples included this metal, 8.7% of them contained lead ions above the levels proposed by WHO and ISIRI MCL. Overall average concentration of heavy metals in water samples varies as $Zn > Fe > Cu \approx Ni > Co > Pb > Cd$. The results reveal that the amount of heavy metals depends on the sampling locations.

TABLE 6: Rotated component matrix of four-factor model[a]

	Component			
	1	2	3	4
Fe	0.195	0.036	0.267	0.881
Co	−0.129	−0.867	0.047	0.142
Ni	0.044	−0.015	−0.974	−0.027
Cu	0.960	0.151	0.070	0.144
Zn	0.047	0.858	0.088	0.078
Cd	0.042	−0.140	−0.504	0.808
Pb	0.961	0.047	−0.123	0.078
Eigen value	2.241	1.708	1.216	1.059
% of total variance	32.020	24.405	17.366	15.132
% Cumulative of variance	32.020	56.425	73.791	88.924

[a]*Extraction method: Principal component analysis. Rotation method: Varimax with Kaiser Normalization. Rotation converged in 5 iterations.*

As shown in Table 5, a close relationship between the couples Fe/Pb, Cu/Pb, Cd/Pb, Co/Zn, Ni/Zn, Ni/Cd and Zn/Cd states a probable common source of the couples. A further statistical investigation was performed by testing the correlation between the determined concentration of heavy metals and the distance of the sampling site from NILZ Company. The calculated correlation (Cu/Pb and Zn/Cd) can confirm the significant effect of NILZ Company activities as a main source of heavy metal contamination observed in the investigated groundwater samples. In addition, close correlation between Fe and depth of the wells (0.52) suggests that this metal is totally of pedo-geochemical source leached from the upper soil layers.

Cluster analysis allows identification of five clusters or groups of associated metals (Figure 2). On the basis of similarities found for group 1 (Pb, Cu), one can suggest the anthropogenic origin of the contamination sources. The presence of iron in group 3 (Cd, Fe) notifies, probably, mixed anthropogenic and pedo-geochemical source of the metals presented in this group. Therefore Zn, Co and Ni were located in single member groups.

Also according to Table 6, Component 1 is attributed to lead and copper with positive sign. These elements are important byproducts of lead industries indicating its anthropogenic sources. Component 2 reveals 24.4% of the total variances are positively loaded with Zn and negatively loaded with Co. Component 3 shows that 17.4% of the total variance is positively loaded with Ni and it can be represented by oil industries activities near the NILZ Company. Component 4 explains 15.1% of the total variance, is positively loaded with Cd and Fe.

The heavy metal grouping has been explored in the plot of the first three principal components generated from these parameters (Figure 3). The low correlation found for the studied metal ions in the four components defined by FA, suggests both anthropogenic and pedo-geochemical sources for the metal contaminations.

8.5 CONCLUSION

Overall, the present study has shown that the groundwater source within radius of 5 km from National Iranian Lead and Zinc Company (NILZ) in Bonab Industrial Estate (Zanjan province-Iran) is contaminated by iron, cobalt, nickel, copper, zinc, cadmium and lead. This can be considered as a menace for people who daily intake the corresponding waters, planted vegetables and food crops irrigated by the same water source. The higher health risk comes from those elements which are present at higher levels than announced by WHO and ISIRI notably lead, nickel and cadmium. Multivariate statistical techniques have shown correlations and similarities among the investigated heavy metals and classification of these ion

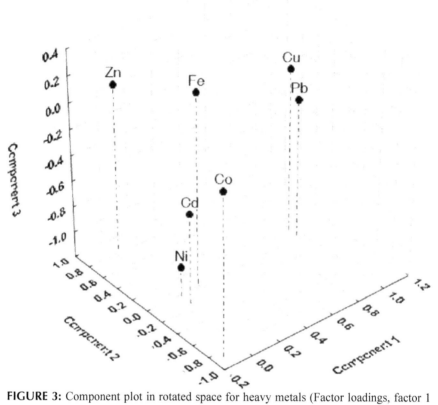

FIGURE 3: Component plot in rotated space for heavy metals (Factor loadings, factor 1 vs. factor 2 vs. factor 3, Rotation: varimax normalized, extraction: principal components).

groups. Cluster analysis has identified five clusters among the heavy metals. The statistical investigations reveal the pollution sources influencing water quality in the study area as anthropogenic (with a very high contribution of NILZ Company) and pedo-geochemical for Fe, Cu. The results suggest a significant risk to the population of Zanjan city and its neighborhoods given the toxicity of the studied metals and the fact that this aquifer by far is the main source of their drinking water and irrigation. This study has also highlighted the need for further research and regular monitoring, in order to determine the permitted levels of metals in the studied aquifer.

REFERENCES

1. Vanloon GW, Duffy SJ: The hydrosphere, in: environmental chemistry: a gold per-spective. 2nd edition. New York: Oxford University Press; 2005.
2. WHO: Water for pharmaceutical use, in quality assurance of pharmaceuticals: a compendium of guidelines and related materials. 2nd edition. Geneva: World Health Organization; 2007.
3. Kumar R, Singh RN: Municipal water and wastewater treatment. New Delhi: Capi-tal Publishing Company; 2006.
4. Vodela JK, Renden JA, Lenz SD, Henney WHM, Kemppainen BW: Drinking water contaminants (Arsenic, cadmium, lead, benzene, and trichloroethylene) 1. Interac-tion of contaminants with nutritional status on general performance and immune function in broiler chickens. Pollution Science 1997, 76:1474-1492.
5. Öztürk M, Özözen G, Minareci O, Minareci E: Determination of heavy metals in fish, water and sediments of Avsar Dam Lake in Turkey. Iran J Environ Health Sci Eng 2009, 6:73-80.
6. Marcovecchio JE, Botte SE, Freije RH: Heavy metals, major metals, trace elements. In handbook of water analysis. 2nd edition. Edited by Nollet LM. London: CRC Press; 2007.
7. Khodabakhshi A, Amin MM, Mozaffari M: Synthesis of magnetic nanoparticles and evaluation efficiency for arsenic removal from simulated industrial wastewater. Iran J Environ Health Sci Eng 2011, 8:189-200.
8. Ghasemi M, Keshtkar AR, Dabbagh R, Jaber Safdari S: Biosorption of uranium in a continuous flow packed bed column using Cystoseira indica biomass. Iran JEnviron Health Sci Eng 2011, 8:65-74.
9. Underwood EJ: Trace elements in humans and animals nutrition. 3rd edition. New York: Academic Press; 1956.
10. Wang LK, Chen JP, Hung YT, Shammas NK: Heavy metals in the environment. London: Taylor and Francis; 2009.
11. Borah KK, Bhuyan B, Sarma HP: Heavy metal contamination of groundwater in the tea garden belt of Darrang district, Assam, India. E-Journal of Chemistry 2009, 6(S1):S501-S507.
12. EPA: Toxicological review of zinc and compounds. Washington D.C: U.S. Environ-mental Protection Agency; 2005.
13. Botkin BD, Keller EA: Environmental science: Earth as a living planet. New York: Wiley; 2005.
14. Thomson RM, Parry GJ: Neuropathies associated with excessive exposure to lead. Muscle Nerve 2006, 33:732-741.
15. Dunnick JK, Elwell MR, Radovsky AE, Benson JM, Hahn FF, Nikula KJ, EBarr B, Hobbs CH: Comparative carcinogenic effects of nickel subsulfide, nickel oxide, or

nickel sulfate hexahydrate chronic exposures in the lung. Cancer Res 1995, 55:5251-5256.

16. Nassef M, Hannigan R, ELsayed KA, Tahawy EMS: Determination of some heavy metals in the environment of Sadat industrial city. Alexandria, Egypt: Proceeding of the 2nd Environmental Physics Conference; 2006.

17. Sanayei S, Norli I, Talebi SM: Determination of heavy metals in Zayandeh Rood river, Isfahan-Iran. World Appl Sci J 2009, 6:1209-1214.

18. Samunding K, Abustan I, Abdulrahman MT, Hasnainisa M: Distribution of heavy metals profile in groundwater system at solid waste disposal site. Eur J Sci Res 2009, 1:58-66.

19. Surindra S, Arvind KN, Maryuri C, Sanjay KG: Assessment of metals in water and sediments of Hindon river, India: impact of industrial and urban discenterges. J Hazard Mater 2009, 171:1088-1095.

20. Momodu MA, Anyakora CA: Heavy metal contamination of groundwater: the Surulere case study. Res J Environ Earth Sci 2010, 2:39-43.

21. Rao DB (Ed): Electrochemistry for environmental protection. New Delhi: Discovery Publishing House; 2001.

22. Ricenterdson DHS: Applications of voltammetry in environmental science. Environ Pollut; (Series B) 1985, 10:261-276.

23. Prasad PR, Rajasekar Reddy S, Chandra Mohan M, Sreedhar NY: Differential pulse polarographic determination of Cr(VI) in various environmental and soil samples using 2,2'- {benzene-1,2-diylbis(nitrilomethylylidene]}diphenol. Int J ChemTech Res 2010, 2:295-302.

24. National Geoscience Database of Iran, http://www.ngdir.ir webcite, (Accessed on 9 August 2012).National Geoscience Database of Iran, http://www.ngdir.ir webcite, (Accessed on 9 August 2012).

25. Chehregani A, Noori M, Lari Yazdi H: Phytoremediation of heavy metal polluted soils: screening for new accumulator plants in angouran mine (Iran) and evaluation of removal ability. Ecotoxicol Environ Saf 2009, 72:1349-1353.

26. Mohammadi H, Eslami A: Quantity and quality of special wastes in Zanjan province. Research report. Zanjan: Department of the Environment; 2007.

27. Parizanganeh AH, Bijnavand V, Zamani AA, Hajabolfath A: Concentration, distribution and comparison of total and bioavailable heavy metals in top soils of Bonab District in Zanjan province. Open J Soil Sci 2012, 2:123-132.

28. Einax GW, Zwanziger HW, Geiss S: Chemometrics in environmental analysis. VCH Weinheim; 1997.

29. Miller JN, Miller JC: Statistics and chemometrics for analytical chemistry. 5th edition. London: Pearson; 2005.

30. Tabachnick BG, Fidell LS: Using multivariate statistics. London: Allyn and Bacon; 2007.

31. Shrestha S, Kazama F: Assessment of surface water quality using multivariate statistical techniques: A case study of the Fuji river basin. Japan Environ Model Software 2007, 22:464-475.

32. Nkansah K, Dawson-Andoh B, Slahor J: Rapid centeracterization of biomass using near infrared spectroscopy coupled with multivariate data analysis: Part 1 Yellow-poplar (Liriodendron tulipifera L.). Bioresour Technol 2010, 101:4570-4576.

CHAPTER 9

ASSESSMENT OF HEAVY METAL CONTAMINATION OF AGRICULTURAL SOIL AROUND DHAKA EXPORT PROCESSING ZONE (DEPZ), BANGLADESH: IMPLICATION OF SEASONAL VARIATION AND INDICES

SYED HAFIZUR RAHMAN, DILARA KHANAM,
TANVEER MEHEDI ADYEL, MOHAMMAD SHAHIDUL ISLAM,
MOHAMMAD AMINUL AHSAN,
AND MOHAMMAD AHEDUL AKBO

9.1 INTRODUCTION

The role of heavy and trace elements in the soil system is increasingly becoming an issue of global concern at private as well as governmental levels, especially as soil constitutes a crucial component of rural and urban environments [1], and can be considered as a very important "ecological crossroad" in the landscape [2]. Agricultural soil contamination with heavy metals through the repeated use of untreated or poorly treated wastewater from industrial establishments and application of chemical fertilizers and pesticides is one of the most severe ecological problems in Bangladesh. Although some trace elements are essential in plant nutrition, plants growing in the close vicinity of industrial areas display increased concentration of heavy metals, serving in many cases as biomonitors of pollution loads [3]. Vegetables cultivated in soils polluted with toxic and heavy metals take up such metals and accumulate them in their edible and non-edible parts in quantities high enough to cause clinical problems both to animals and human beings consuming these metal-rich plants as there is no good mechanism for their elimination from the human body [4–6]. Toxic metals are known to have serious health implications, including carcinogenesis induced tumor promotion, and hence the growing consciousness about the health risks associated with environmental chemicals has brought a major shift in global concern towards prevention of heavy metal accumulation in soil, water and vegetables [7,8]. Heavy metals and trace elements are also a matter of concern due to their non biodegradable nature and long biological half-lives. Wastewater from industries or other sources carries appreciable amounts of toxic heavy metals such as Cd, Cu, Zn, Cr, Ni, Pb, and Mn in surface soil which create a problem for safe rational utilization of agricultural soil [9–13]. Long-term use of industrial or municipal wastewater in irrigation is known to have a significant contribution to the content of trace and heavy elements such as Cd, Cu, Zn, Cr, Ni, Pb, and Mn in surface soil [12]. As a result, excessive accumulation of trace elements in agricultural soils through wastewater irrigation may not only result in soil contamination but also affect food quality and safety [14–16].

The Dhaka Export Processing Zone (DEPZ) being the 2nd EPZ and the largest industrial belt of Bangladesh at present houses 92 industrial units which are categorically the leading pollution creators. These industrial units include cap/accessories/garments; textile/knitting plastic goods; footwear/leather goods; metal products; electronic goods; paper products; chemicals and fertilizers and miscellaneous products [17,18]. Industrial activities discharge untreated or poorly treated industrial wastewater, effluent and even sludge into the surrounding environment that can decrease soil quality by increasing concentrations of pollutants such as heavy metals, resulting in adverse effects on macrophytes, soil fauna and human health [7,8,19,20]. So it is a nerve-racking issue to find out the present status of toxic and heavy metals in this surface soil, redress the affected subsequent environmental problems and adopt a future mitigation strategy. The present work is aimed at investigating seasonal and spatial variation of the contamination levels of different heavy metals (Fe, As, Mn, Cu, Ni, Pb, Zn, Hg, Cr, Cd) in the agricultural soil around DEPZ using different indices such as the index of geoaccumulation, contamination factor, degree of contamination, and the pollution load index. This approach would help adopt an effective effluent management strategy towards control over enhanced metal levels with recycling of effluents for toxic metal separation and soil remediation and reclamation. The data generated in this work may help to work out an effluent management strategy towards control over effective treatment of the DEPZ discharges in terms of toxic and heavy metal contents.

9.2 MATERIALS AND METHODS

9.2.1 GEOLOGICAL AND HYDROLOGICAL SETTING OF THE STUDY AREA

DEPZ, located at about 35 km south-east of Dhaka, the capital city of Bangladesh, was selected as the study area in the present research work. This area belongs to Dhamsona Union under Savar Upazila of Dhaka District. The area is situated at the southwestern fringe of a Pleistocene terrace named the Madhupur Tract, an elevated landscape distinct from the surrounding Fluvio-deltaic plains by the Ganges, Brahmaputra and Meghna River. The major geomorphic units of the area are: the high land, the low lands or floodplains, depressions, swamps, marshes and abandoned channels. Soil of the Modhupur Formation, in general reddish brown in color, contains pre-existing paleosol materials. Hydrology of the study area is governed by rainfall intensity and distribution, permanent or ephemeral water bodies and rivers or canals. The average rainfall distribution pattern in various months shows a distinct conformation with the climatic pattern prevailing, with strong Monsoon influence. The average annual rainfall in dry and wet season in the area is about 25 and 380 mm, respectively. The Bansi-Daleshwari and Turag River comprise the drainage network of the area-Bansi on the west and Turag away on the east.

9.2.2 SAMPLING AND ANALYTICAL PROCEDURE

Agricultural soil samples were collected during February 2010 to April 2011 from the surface layer (15–20 cm) of soil with a stainless steel Ekman Grab Sampler from twenty (20) different locations adjacent to DEPZ (Figure 1) in two different seasons—dry and wet season. The sampler was inspected for possible cross-contamination and cleaned with ambient water for individual sample collection. The difference from one sampling point to another was approximately 100 m. About 250–300 g of the soil was sampled from the central part of the grab sampler by avoiding contact with the inside materials of the sampler and transferred to a pre-cleaned plastic container. Collected soil samples were air dried for several days over pre-cleaned Pyrex petry dishes. Then 2–3 g dry soil samples were digested in about 15 mL of aqua-regia (HCL:HNO_3 = 3:1) for approximately 4–5 hours using a hotplate maintaining a heating temperature of approximately 110 °C. The samples were next placed in a 100 mL Pyrex glass beaker and diluted with distilled water up to 50 mL. The solution was filtered and the filtrates were analyzed in the Analytical Research Division, BCSIR Laboratories, Dhaka by Atomic Absorption Spectrophotometer (AA-6401 F, Shimadzu, Japan). The working standard solutions for each metal were prepared before every analysis. The analytical procedures were verified with National Institute of Standards and Technology (NIST) traceable certified reference standards. Concentrations of Fe, Mn, Cu, Ni, Zn, Cr and Cd were measured by an air acetylene flame AAS, with As determined by hydride vapor generation AAS, Hg by hydride generation with cold vapor AAS and Pb by graphite furnace AAS.

9.2.3 INDEX OF GEOACCUMULATION

The index of geoaccumulation (I_{geo}) actually enables the assessment of contamination by comparing the current and pre-industrial concentrations originally used with bottom sediments [21]; it can also be applied to the assessment of soil contamination. The method assesses the degree of metal pollution in terms of seven enrichment classes (Table 1) based on the

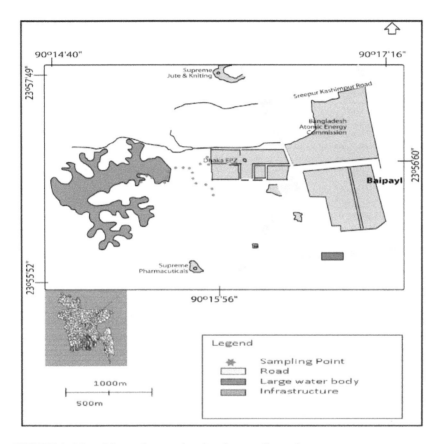

FIGURE 1: Map of the study area showing the sampling points.

increasing numerical values of the index. It is computed using the Equation (1) as:

$$I_{geo} = \log_2 \frac{C_n}{1.5B_n} \tag{1}$$

where C_n is the measured concentration of the element in the pelitic sediment fraction (<2 μm) and B_n is the geochemical background value in fossil argillaceous sediment (average shale). The constant 1.5 allows us to analyze natural fluctuations in the content of a given substance in the environment as well as very small anthropogenic influences.

TABLE 1: Index of geoaccumulation (Igeo) for contamination levels in soil [23].

I_{geo} Class	I_{geo} Value	Contamination Level
0	$I_{geo} \leq 0$	Uncontaminated
1	$0 < I_{geo} < 1$	Uncontaminated/moderately contaminated
2	$1 < I_{geo} < 2$	Moderately contaminated
3	$2 < I_{geo} < 3$	Moderately/strongly contaminated
4	$3 < I_{geo} < 4$	Strongly contaminated
5	$4 < I_{geo} < 5$	Strongly/extremely contaminated
6	$5 < I_{geo}$	Extremely contaminated

In the present paper we applied the modified calculation based on the equation given in [22], where Cn denoted the concentration of a given element in the soil tested, while Bn denoted the concentration of elements in the earth's crust [22]. For some elements like As, Hg and Sb the average concentration in the Earth's crust is much higher than the average concentration in the shale accepted by Muller [21] as a reference value. Here the focus is between the concentration obtained and the concentration of elements in the Earth's crust, because soil is a part of the layer of the Earth's crust and its chemical composition is related to that of the crust.

9.2.4 CONTAMINATION FACTOR AND DEGREE OF CONTAMINATION

The assessment of soil contamination was also carried out using the contamination factor (C_f^i) and degree of contamination (C_d). The C_f^i is the single element index, the sum of contamination factors for all elements examined represents the C_d of the environment and all four classes are recognized [24].

Table 2 shows the different contamination factor class and level. Equation (2) was used as follows:

$$C_f^i = \frac{C_0^i}{C_n^i}$$

$$(1)$$

where C_0^i is the mean content of metals from at least five sampling sites and C_n^i is the pre-industrial concentration of the individual metal.

TABLE 2: Different contamination factor (C_f^i) for soil [24].

C_i^f Value	Contamination Factor level
$C_i^f < 1$	Low contamination factor indicating low contamination
$1 \leq C_i^f < 3$	Moderate contamination factor
$3 \leq C_i^f < 6$	Considerable contamination factor
$6 \leq C_i^f$	Very high contamination factor

The calculated C_d is therefore defined as the sum of the C_i^f for the pollutant species specified by Hakanson [24]. C_d was assessed using Equation (3).

$$C_d = \sum_{i=1}^{n} c_f^i$$

$$(2)$$

The C_d is aimed at providing a measure of the degree of overall contamination in surface layers in a particular sampling site. In the present study we applied a modification of the factor as applied by Krzysztof et al. [25] that used the concentration of elements in the earth's crust as a reference value, similar to the other factors. The C_d was divided into four groups as given in Table 3.

TABLE 3: Different degree of contamination (C_d) for soil [24].

C_d Class	Degree of Contamination Level
$C_d < 8$	Low degree of contamination
$8 \leq C_d < 16$	Moderate degree of contamination
$16 \leq C_d < 32$	Considerable degree of contamination
$32 \geq C_d < 8$	Very high degree of contamination

9.2.5 MODIFIED DEGREE OF CONTAMINATION (MC_D)

Abrahim [26] presented a modified and generalized form of the Hakanson [24] equation for the calculation of the overall degree of contamination at a given sampling or coring site as follows: (a) The modified formula is generalized by defining the degree of contamination (mC_d) as the sum of all the contamination factors (C_f^i) for a given set of estuarine pollutants divided by the number of analyzed pollutants; (b) The mean concentration of a pollutant element is based on the analysis of at least three samples; and (c) The baseline concentrations are determined from standard earth materials The modified equation for a generalized approach to calculating the degree of contamination is given in Equation 4.

$$mC_d = \frac{\sum_{i=1}^{i=n} C_f^i}{n} \qquad (4)$$

where n is the number of analyzed elements and and C_f^i is the contamination factor.

Using this generalized formula to calculate the mC_d allows the incorporation of as many metals as the study may analyse with no upper limit. For the classification and description of the mC_d seven gradations are proposed as shown in Table 4.

An intrinsic feature of the mC_d calculation is that it produces an overall average value for a range of pollutants. As with any averaging procedure, care must however be used in evaluating the final results since the effect of significant metal enrichment spikes for individual samples may be hidden within the overall average result.

TABLE 4: Different modified degree of contamination (mC_d) for soil [26].

mC_d Class	Modified Degree of Contamination Level
$mC_d < 1.5$	Nil to very low degree of contamination
$1.5 \leq mC_d < 2$	Low degree of contamination
$2 \leq mC_d < 4$	Moderate degree of contamination
$4 \leq mC_d < 8$	High degree of contamination
$8 \leq mC_d < 16$	Very high degree of contamination
$16 \leq mC_d < 32$	Extremely high degree of contamination
$mC_d \geq 32$	Ultra high degree of contamination

9.2.6 POLLUTION LOAD INDEX (PLI)

The pollution load index (PLI) was proposed by Tomlinson et al. [27] for detecting pollution which permits a comparison of pollution levels between sites and at different times. The PLI was obtained as a concentration factor of each heavy metal with respect to the background value in the soil. In this study, the world average concentrations of the metals studied reported for shale [28] were used as the background for those heavy metals. According to Angula [29], the PLI is able to give an estimate of the metal contamination status and the necessary action that should be taken. A PLI value of ≥ 100 indicates an immediate intervention to ameliorate pollution; a PLI value of ≥ 50 indicates a more detailed study is needed to monitor the site, whilst a value of <50 indicates that drastic rectification measures are not needed. The formulas applied are as Equation (5).

$$PLI = n\sqrt{cf_1 \times cf_2 \times \cdots\cdots\cdots \times cf_n} \qquad (5)$$

9.3 RESULTS AND DISCUSSION

9.3.1 SEASONAL AND SPATIAL VARIATION OF HEAVY METAL CONTENT

The average concentration of different metals in the agricultural soil of the study area in two seasons is given in Table 5. Average concentration of Fe, As, Mn, Cu, Zn, Cr, Pb, Hg, Ni and Cd in the study area during the dry season was 30,404, 4,073.1, 339, 60, 209, 49.66, 27.6, 486.6, 48.1 and 0.0072 mg/kg, respectively. While average concentration of Fe, As, Mn, Cu, Zn, Cr, Pb, Hg, Ni and Cd in the wet season was 17,103, 2,326.2, 305, 90, 194, 34.2, 23.83, 133.2, 5.5 and 1.04 mg/kg, respectively. So the trend of metals according to mean concentration in the dry season was: As > Fe > Hg > Mn > Zn > Cu > Cr > Ni > Pb > Cd, while in the wet season the trend was: As > Fe > Mn > Zn > Hg > Cu > Ni > Cr > Pb > Cd. The variation of heavy metal concentration in the study area was due to irrigation of land by industrial wastewater and other agronomic practices. The higher standard deviation reveals higher variations in heavy metal distributions from the point source of discharge to the adjacent areas. The low concentration of heavy metals in the soil may be ascribed to its continuous removal by vegetables grown in the designated areas. Among the different metals examined in soil, the concentration of Fe was the maximum and variation in its concentration was several times higher than those reported by Kisku et al. [30].

Average concentration of metals during the dry season in the surface layer of the soil is higher than that in the wet season. The highest deposition of Fe (Figure 2) in soil might be due to its long-term use in the production of machine tools, paints, pigments, and alloying in various industries of the study area that may result in contamination of the soil and a change to the soil structure thus making it risky for use in cultivation [31].

TABLE 5. Different concentrations of metals in the agricultural soil of the study area over two seasons.

Metals	MAC in Agricultural Soil in China (mg/kg) [34]	Natural Background Soil in China (mg/kg) [34]	Safe Limit of India (mg/kg) [35]	DEPZ area (mg/kg) [37]	Present study					
					Dry season			Wet season		
					Mean (mg/kg)	SD	Range (Min-Max) (mg/kg)	Mean (mg/kg)	SD	Range (Min-Max) (mg/kg)
Fe			75–150	1715.8	30404	37.3	23016.4–38458	17103	12147	13.9–581
As		35			4073.1	1116	789.24–565.92	2326.2	3274	0–14307
Mn					339	142	149.94–22.77	305	131.6	107–582
Cu	100	35	135–200	39.14	60	16.5	19.92–728.25	90	66.9	20.4–281
Zn	300	100	300–600	115.4	209	193.6	75.33–859.95	194	120.4	0.15–474
Cr	200	90		53.7	49.66	34.7	22.77–170.83	34.2	26.5	0–89.78
Pb	350	35	250–500	49.7	27.6	7.9	9.79–41.08	23.83	11.3	0.511–45
Hg					486.6	229.3	132.7–5016.2	133.2	72.7	0–328.28
Ni	60	40		58.2	48.1	11.3	29–68.2	35.5	10.5	11.65–52
Cd	0.6	20	3–6	11.4	0.0072	0.02	0–0.09	1.04	2.03	0.25–8.8

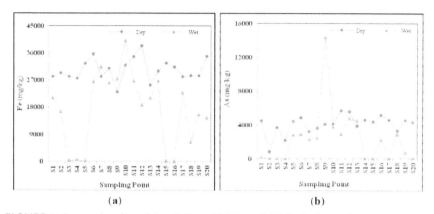

(a)   **(b)**

FIGURE 2: Seasonal and spatial variation of (a) Fe and (b) As in the study area.

High concentrations of As in the soils are due to significant anthro-pogenicity particularly industrial activities such as the metallurgical and chemical industries and the use of arsenical sprays [32]. Arsenic is a priority toxic element that can cause arsenicosis-related disease and internal cancers, even in trace amounts. The dry season showed about a two times higher concentration of arsenic in the soil than that of the wet season (Figure 2). The mean As content in the soil near industrial areas in Turkey was 9.53 mg/kg ranging from 1.50 to 65.60 mg/kg [32]. Arsenic concentration in uncontaminated

Mn is one of the commonly found elements in the lithosphere and its concentration in the dry season was about two times higher than in the wet season (Figure 3). Cu was distributed uniformly in the wet season, but in the dry some zigzag was found. This metal was about 1.5 times higher in the dry season compared to the wet (Figure 3). There was a sharp variation in average Cu concentration in the dry season at sampling point 8 due to point source contamination. More or less the average concentration of the metal was within the MAC of elements in agricultural soil found in China [34], and within the safe limit of India [35]. Cu content of soils in the Gebze region was between 7.87 and 725 mg/kg with an average of 95.88 mg/kg which was significantly greater than that in uncontaminated soils [32]. In some other works, Cu concentrations lower than in Gebze soils were recorded [33,36].

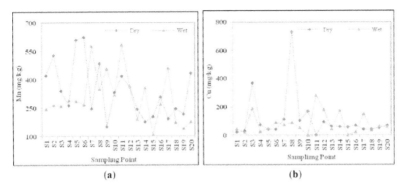

FIGURE 3: Seasonal and spatial variation of (a) Mn and (b) Cu in the study area.

Zn and Cr are heavy metals and their sources in industrial locations are usually anthropogenic [38]. The main anthropogenic sources of Zn are related to the non-ferric metal industry and agricultural practice [32,33]. Zinc is a very readily mobile element. High doses of Zn show toxic and carcinogenic effects and result in neurologic and hematological complications, hypertension, and kidney and liver function disorders [39].

The Cr content of topsoil is known to increase due to pollution from various sources of which the main ones are attributable to industrial wastes such as Cr pigment and tannery wastes, electroplating sludge, leather manufacturing wastes, and municipal sewage sludge etc. Cr behavior in soil is controlled by soil pH and redox potential, while long term exposure to Cr can cause liver and kidney damage [38]. The observed Zn and Cr concentrations in studied soil around DEPZ probably comes from construction materials in the form of alloys for protective coating for iron and steel. These metals are also used in the industries of DEPZ pigment and reducing agents; cotton processing, soldering and welding flux; rubber industry, glass, enamels, plastics, lubricants, cosmetics, pharmaceuticals, agents for burns and ointments [40]. Both metals were unevenly distributed in the study area (Figure 4). Both metals showed higher concentration in a similar study area to Ahmed and Gani [37].

Pb contamination in soils has been seriously emphasized in recent years since this metal is very toxic for humans and animals [32]. Pb enters human or animal metabolism via the food chain. Pb production

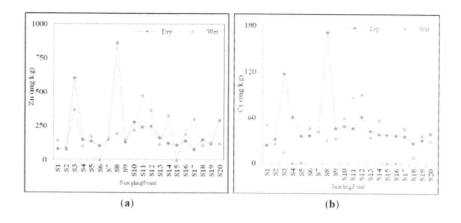

FIGURE 4: Seasonal and spatial variation of (a) Zn and (b) Cr in the study area.

and operation facilities without a waste-gas treatment system, battery production and scrap battery recovery facilities, thermal power plants, and iron–steel industries are the other lead sources. Moreover among the heavy metals, Pb is the most immobile element and its content in soil is closely associated with clay minerals, Mn-oxides, Al and Fe hydroxides, and organic material [32]. Although there was point wise variation in concentration of Pb (Figure 5), a very limited average variation was observed over two seasons. Pb concentrations in Gebze soils were between 17.07 and 8,469 mg/kg with an average of 246 mg/kg which is noticeably higher than values reported in the literature [32]. The average Pb concentration in the soils of the Thrace region was recorded as 33 mg/kg [41]. Long term exposure of Pb is risky. Bioaccumulation and bio-magnification can take place. Hence, regular consumption of vegetables from this area by residents could pose a serious neurological health problem from long term Pb exposure.

Hg is a toxic metal for environmental and human health. Base-metal processing and some chemical industrial activities are the main source for Hg contamination in soils. Mining activities, sewage wastes, and the use of fungicides also result in Hg pollution. The dry season contained about four times higher concentrations of this metal than that of the wet season (Figure 5). Seasonal rainfall may limit the concentration in soil during the

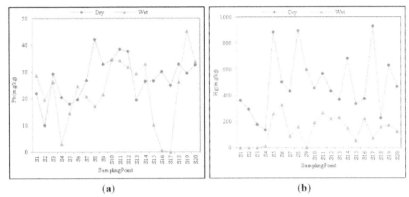

(a)   (b)

FIGURE 5: Seasonal and spatial variation of (a) Pb and (b) Hg in the study area.

wet season. Hg concentration in Gebze soils was between 9 μg/kg and 2,721 μg/kg with an average of 102 μg/kg [32], which is very much lower than present work.

Average concentration of Ni in the dry season was higher than the wet season (Figure 6). Major sources of Ni in the soil are poorly treated wastewater that is discharged from ceramics, steel and alloys and other metal processing industries. Long term exposure of Ni through the food chain may contribute to health problems like skin allergies, dermatitis, rhinitis, nasal sinusitis, lung injury and nasal mucosal injury [38].

There is a growing environmental concern about Cd being one of the most eco-toxic metals, exhibiting highly adverse effects on soil health, biological activity, plant metabolism, and the health of humans and animals [33]. The comparison of mean concentrations of heavy metals in the soil of the study area with the official Indian standard [35], and the MAC of elements in agricultural soil in China [34] showed that only the concentration of Cd was found to be 3-fold higher than the threshold level of India and 19 times higher than the Chinese standard (Figure 6). Higher concentration of Cd in the wet season may be due to more Cd containing waste water release from DEPZ compared to that of the dry season. The Cd concentrations in Gebze, Turkey soils varied from 0.05 to 176 mg/kg and the average was 4.41 mg/kg [32]. The average Cd concentration in northern

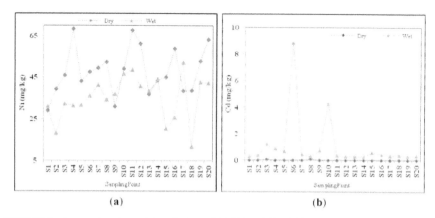

FIGURE 6: Seasonal and spatial variation of (a) Ni and (b) Cd in the study area.

Poland soils was 0.80 mg/kg [36] which is significantly lower than that in Gebze soils. Environmental levels are greatly enhanced by industrial operations as Cd is commonly used as a pigment in paint, plastics, ceramics and glass manufacture. Even at very low concentrations chronic exposure to this metal can lead to anemia, anosmia, cardiovascular diseases, renal problems and hypertension [15].

It is commonplace that the concentrations of the studied heavy metals were higher during the dry season, when the rainfall was comparatively low. During the wet season the values were in general low and fall within various standard levels. In the rainy season the pollution was lowest because of heavy rainfall, dilution and other runoff processes. Most of the suspended materials, which were not complexed and precipitated with soil, organic matter and other compounds, were flushed out through the canal into the adjoining vast flood zone. In the dry season rice, grasses and many other types of 'rabi' crops are grown in the contaminated soils which are irrigated with polluted water. Locally produced crops with attractive appearances might have high pollution content of heavy metals. Therefore long term intake of food grown in the area might create serious toxicological effects for the consumers.

TABLE 6: Average Igeo and contamination levels of soil in two seasons.

	Dry Season		Wet Season	
	I_{geo} Value	Contamination Level	I_{geo} Value	Contamination Level
Fe	−0.37	Uncontaminated	−1.09	Uncontaminated
As	3.23	Strongly contaminated	2.16	Moderately/strongly contaminated
Mn	0.61	Uncontaminated/moderately contaminated	−0.66	Uncontaminated
Cu	0.31	Uncontaminated/moderately contaminated	0.24	Uncontaminated/moderately contaminated
Zn	0.19	Uncontaminated/moderately contaminated	0.07	Uncontaminated/moderately contaminated
Cr	0.08	Uncontaminated/moderately contaminated	−0.45	Uncontaminated
Pb	0.06	Uncontaminated/moderately contaminated	−0.19	Uncontaminated
Hg	3.08	Strongly contaminated	2.09	Moderately/strongly contaminated
Ni	0.19	Uncontaminated/moderately contaminated	0.05	Uncontaminated/moderately contaminated
Cd	−0.03	Uncontaminated	0.53	Uncontaminated/moderately contaminated

9.3.2 INDEX OF GEOACCUMULATION

Average I_{geo} and contamination levels of different metals in soil are given in Table 6 while Figure 7 representsthe sampling point wise I_{geo} value in two seasons. I_{geo} is distinctly variable and suggests that soil around the DEPZ ranged from uncontaminated to strongly/extremely contaminated with respect to the analyzed metals. I_{geo} revealed that all the samples examined in both seasons in respect of Fe and Mn fell into class 0—uncontaminated. In the case of As in the dry season, 18 sampling points fell in class 4 and the average I_{geo} was 3.23 indicating strongly contaminated. During the wet season I_{geo} for As belong to moderately/strongly contaminated. This high index is caused mainly by the metallurgical industry; hence its content in the areas affected by industrial activity may be elevated. I_{geo} values for Cu in the dry season ranged from −2.74 to 1.28

with a mean value of 0.31 and most of the samples in both seasons fell into class 1 of uncontaminated to moderately contaminated. I_{geo} of Zn in the dry and wet season was 0.19 and 0.07, respectively and belongs to Igeo class 1. In the wet season, Cr showed an uncontaminated state, but uncontaminated/moderately contaminated in the dry season. A similar trend to Cr was also found for Pb, but different for Cd. Ni followed uncontaminated/moderately contaminated index over two seasons. I_{geo} of Hg in the dry season ranged from 2.7 to 3.98 and the value was 0 to 4.85 in the wet season. Figure 8 shows overall statistics of I_{geo} over two seasons.

9.3.3 CONTAMINATION FACTOR, DEGREE OF CONTAMINATION, MODIFIED DEGREE OF CONTAMINATION AND POLLUTION LOAD INDEX

The assessment of the overall contamination of the studied agricultural soil was based on C_f^i. In the dry season, the soil was classified as slightly contaminated with Fe, Mn and Cd, moderately contaminated with Zn, Cr, Pb and Ni, considerably contaminated with Cu and highly contaminated with As and Hg. In the wet season except for Cu and Cd, the contamination factor of all other metals decreased. However, there was a very limited change in the overall scenario. Cr was additionally added to the first category, Cd shifted from a slightly contamination to moderately contamination factor and Cu fell into the highly contaminated group. Overall the C_d values of the soil of the study area during the dry and wet seasons were 5,751.26 and 2,444.42, respectively. The maximum values of the contamination degree denoted very high contamination. The mC_d as proposed in the present study is based on integrating and averaging all the available analytical data for a set of soil samples. This modified method can therefore provide an integrated assessment of the overall enrichment and contamination impact of groups of pollutants in the soil. During the dry and wet seasons the mC_d varied as 575.13 and 244.44, respectively thus revealing an ultra high degree of contamination. Because of heavy

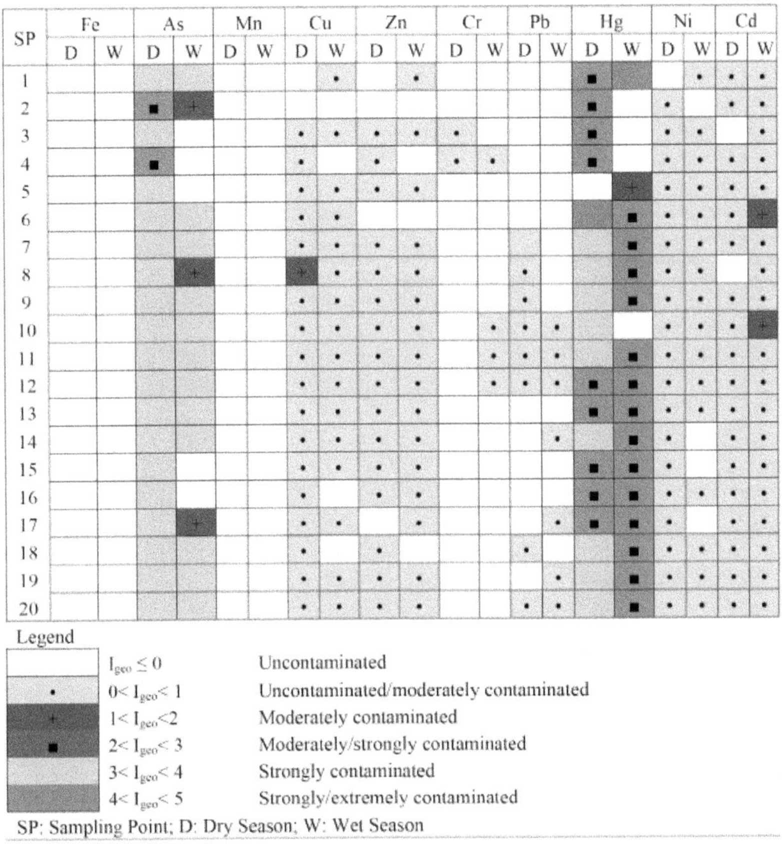

SP	Fe		As		Mn		Cu		Zn		Cr		Pb		Hg		Ni		Cd	
	D	W	D	W	D	W	D	W	D	W	D	W	D	W	D	W	D	W	D	W
1								•		•					■				•	•
2			■	+											■			•		•
3							•	•	•	•	•				■		•	•		•
4			■				•		•		•	•			■		•	•	•	•
5							•	•	•	•						+	•	•	•	•
6							•	•								■	•	•	•	
7				+			•	•	•	•						■	•	•	•	•
8				+			■	•	•	•				•		■	•	•		•
9							•	•	•	•				•		■	•	•		•
10							•	•	•	•		•	•	•			•	•	•	+
11							•	•	•	•		•	•	•		■	•	•		•
12							•	•	•	•		•	•	•	■	■	•	•		•
13							•	•	•	•					■	■	•	•		•
14							•	•	•	•				•	•	■	•			•
15							•	•	•	•					■	■	•	•		•
16							•		•	•					■	■	•	•	•	•
17				+			•	•	•	•				•	■	■	•		•	•
18							•		•					•		■	•	•	•	•
19							•	•	•	•				•		■	•	•	•	•
20							•	•	•	•			•	•		■	•	•	•	•

Legend

	$I_{geo} \leq 0$	Uncontaminated
•	$0 < I_{geo} < 1$	Uncontaminated/moderately contaminated
+	$1 < I_{geo} < 2$	Moderately contaminated
■	$2 < I_{geo} < 3$	Moderately/strongly contaminated
	$3 < I_{geo} < 4$	Strongly contaminated
	$4 < I_{geo} < 5$	Strongly/extremely contaminated

SP: Sampling Point; D: Dry Season; W: Wet Season

FIGURE 7: Representation of Igeo of metals at different sampling points and seasons.

rainfall, dilution and other run-off during thewet season, metals from the upper layer of the soil were flushed out to some extent through the canal into the adjoining vast flood zone and hence all the indices values were lower in this season compared to the dry season. The PLI values indicated immediate intervention to ameliorate pollution in both seasons. Average C_f^i, C_d and mC_d of the soil is given in Table 7.

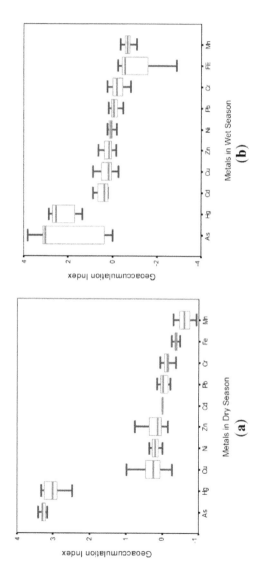

FIGURE 8: Statistics of Igeo of soil in (a) dry and (b) wet season.

TABLE 7: Average C_f^i, C_d, mC_d and pollution load index (PLI) of soil over two seasons.

	Dry Season	Wet Season
Fe	0.65	0.36
As	2715.36	1708.96
Mn	0.40	0.36
Cu	4.68	7.72
Zn	2.95	2.72
Cr	1.42	0.97
Pb	1.38	1.24
Hg	3021.98	717
Ni	2.40	1.77
Cd	0.02	3.49
C_d	5751.26	2444.42
mC_d	575.13	244.44
PLI	1801.3	50,047.5

9.3.4 CORRELATIONS MATRIX

Pearson correlation analysis [42] was performed between all the variables. The level of significance ($p \leq 0.05$ and $p \leq 0.01$) of multi-element correlation for soil samples was determined and the results are given in Table 8. The listed r values indicated the high degree of positive correlations and significant linear relation between various pairs of metals, reflecting their simultaneous release and identical source from the DEPZ zone, transport and accumulation in soil. The inter-metallic correlation coefficients in the soil samples with $p < 0.05$ during the dry season were: Fe-Ni, As-Pb, Cu-Zn, Cu-Cr, Cu-Cd, Zn-Pb, Zn-Cd and Cr-Cd. In the wet season the correlation trends were: Fe-As, Fe-Cr, Fe-Ni, As-Cr, Cu-Zn, Cu-Cr, Cu-Ni, Cr-Pb, Cr-Ni and Pb-Ni. The significant correlations indicate that they may have originated from common sources, presumably from other industrial (chemicals, paints) activities. The correlation of As with Cr and Fe indicate their common source from tannery industries. The strong association of Cd, Zn, and Cu indicates common sources, and these metals may have been derived from anthropogenic sources, especially the paint industry and municipal sewage system.

TABLE 8: Correlation coefficient matrix for the metals in soil around DEPZ during the dry and wet seasons.

	Fe	As	Mn	Cu	Zn	Cr	Pb	Hg	Ni	Cd
Dry Season										
Fe	1									
As	0.467*	1								
Mn	0.553 *	−0.048	1							
Cu	−0.004	−0.078	0.174	1						
Zn	0.112	−0.020	0.226	0.964**	1					
Cr	0.016	−0.096	0.164	0.972**	0.957**	1				
Pb	0.263	0.512*	−0.224	0.558 *	0.570**	0.469*	1			
Hg	0.185	0.079	0.281	−0.047	−0.129	−0.088	−0.165	1		
Ni	0.579**	0.193	0.142	0.161	0.248	0.213	0.393	−0.010	1	
Cd	−0.041	−0.122	0.217	0.978**	0.941**	0.960**	0.410	−0.064	0.060	1
Wet Season										
Fe	1									
As	0.753 **	1								
Mn	0.557 *	0.223	1							
Cu	0.315	0.417	0.535*	1						
Zn	0.110	0.173	0.511*	0.889 **	1					
Cr	0.750**	0.571**	0.555*	0.703 **	0.525*	1				
Pb	0.548 *	0.454 *	0.125	0.489 *	0.323	0.668**	1			
Hg	0.082	0.037	0.371	−0.158	−0.101	0.150	0.081	1		
Ni	0.623**	0.547 *	0.539*	0.599 **	0.473*	0.672**	0.622**	−0.092	1	
Cd	0.296	0.351	−0.069	0.003	−0.326	0.132	0.029	−0.091	0.099	1

*Correlation is significant at the 0.05 level (2-tailed); ** Correlation is significant at the 0.01 level (2-tailed).*

9.4 CONCLUSION

Soil is a great geochemical reservoir for contaminant as well as a natural buffer for transportation of chemical materials and elements in the atmosphere, hydrosphere, and biomass. For this, it is the most important component of the human biosphere. As soil is an important constituent of the human biosphere, any harmful change to this segment of the environment seriously affects the overall quality of human life. The most adverse effect of heavy metals is that they can be introduced into the food chain and threaten human health. Agricultural products growing on soils with high metal concentrations are represented by metal accumulations at levels harmful to human and animal health as well as to the bio-environment. The impact of anthropogenic heavy metal contamination on agriculture soil around DEPZ was evaluated in this study using seasonal variation and indices. All the indices more or less revealed that the study area was seriously affected by different metals. Dry seasons resulted in some multi-fold higher values of the overall indices. These metals with high concentrations in the studied soils may have been mixed with groundwater by leaching. High concentrations of heavy metals in soils around industrial facilities originate from an anthropogenic source which is associated with unrestrained solid release and untreated or poorly treated fluid wastes from these industrial facilities. To control soil contamination, legislative measures must be taken, legally binding the individual industries, forbidding discharge of untreated or poorly treated industrial effluents. Lowering the quality of soil health due to these industries can only be restricted if a zero discharge system ETP is implemented throughout the DEPZ. Immediate steps including regular monitoring of toxic metals in the agricultural soil is needed to check the environmental quality. Wastewater discharged from DEPZ could be recycled for the remediation of pollution in a sustainable and eco-specific way. Moreover different remediation measures should be taking promptly to remove excising metal contamination.

REFERENCES

1. USDA (United States Department of Agriculture Natural Resources Conservation Services Soils Quality Institute). Urban Technical Note, 2001.
2. Thuy, H.T.T.; Tobschall, H.J.; An, P.V. Distribution of heavy metals in urban soils—A case study of Danang-Hoian area (vietnam). Environ. Geogr. 2000, 39, 603–610.
3. Mingorance, M.D.; Valdes, B.; Oliva-Rossini, S. Strategies of heavy metal uptake by plants growing under industrial emissions. Environ. Int. 2007, 33, 514–520.
4. Alam, M.G.M.; Snow, E.T.; Tanaka, A. Arsenic and heavy metal concentration of vegetables grown in Samta village, Bangladesh. Sci. Total Environ. 2003, 111, 811–815.
5. Arora, M.; Kiran, B.; Rani, A.; Rani, S.; Kaur, B.; Mittal, M. Heavy metal accumulation in vegetables irrigated with water from different sources. Food Chem. 2008, 111, 811–815.
6. Bhuiyan, M.A.H.; Suruvi, N.I.; Dampare, S.B.; Islam, M.A.; Quraishi, S.B.; Ganyaglo, S.; Suzuki, S. Investigation of the possible sources of heavy metal contamina-

tion in lagoon and canal water in the tannery industrial area in Dhaka, Bangladesh. Environ. Monit. Assess. 2011, 175, 633–649.

7. Ahmed, G.; Uddin, M.K.; Khan, G.M.; Rahman, M.S.; Chowdhury, D.A. Distribution of trace metal pollutants in surface water system connected to effluent disposal points of Dhaka Export Processing Zone (DEPZ), Bangladesh: A statistical approach. J. Nat. Sci. Sus. Technol. 2009, 3, 293–304.

8. Mortula, M.M.; Rahman, M.S. Study on waste disposal at DEPZ. Bangladesh Environ. (BAPA). 2002, 2, 807–817.

9. Yadav, R.K.; Goyal, B.; Sharma, R.K.; Dubey, S.K.; Minhas, P.S. Post irrigation impact of domestic sewage effluent on composition of soils, crops and ground water—A case study. Environ. Int. 2002, 28, 481–486.

10. Singh, K.P.; Mohan, D.; Sinha, S.; Dalwani, R. Impact assessment of treated/untreated wastewater toxicants discharged by sewage treatment plants on health, agricultural, and environmental quality in the wastewater disposal area. Chemosphere 2004, 55, 227–255. Appl. Sci. 2012, 2 600

11. Chen, Y.; Wang, C.; Wang, Z. Residues and sources identification of persistent organic pollutants in farmland soils irrigated by effluents from biological treatment plants. Environ. Int. 2005, 31, 778–783.

12. Mapanda, F.; Mangwayana, E.N.; Nyamangara, J.; Giller, K.E. The effect of long-term irrigation using wastewater on heavy metal contents of soils under vegetables in Harare, Zimbabwe. Agric. Ecosys. Environ. 2005, 107, 151–165.

13. Luo, X.; Yu, S.; Zhu, Y.; Li, X. Trace metal contamination in urban soils of China. Sci. Total Environ. 2012, 441–442, 17–30.

14. Muchuweti, M.; Birkett, J.W.; Chinyanga, E.; Zvauya, R.; Scrimshaw, M.D.; Lester, J. Heavy metal content of vegetables irrigated with mixture of wastewater and sewage sludge in Zimbabwe: Implications for human health. Agric. Ecosys. Environ. 2006, 112, 41–48.

15. Sharma, R.K.; Agrawal, M.; Marshall, F. Heavy metal contamination in vegetables grown in wastewater irrigated areas of Varanasi, India. Bull. Environ. Contam. Toxicol. 2006, 77, 312–318.

16. Sharma, R.K.; Agrawal, M.; Marshall, F. Heavy metal (Cu, Zn, Cd and Pb) contamination of vegetables in urban India: A case study in Varanasi. Environ. Poll. 2008, 154, 254–263.

17. Khan, M.K.; Alam, A.M.; Islam, M.S.; Hassan, M.Q.; Al-Mansur, M.A. Environmental pollution around Dhaka EPZ and its impact on surface and groundwater. Bangladesh J. Sci. Ind. Res. 2011, 46, 153–162.

18. Khanam, D.; Rahman, S.H.; Islam, M.S.; Ahsan, M.A.; Shaha, B.; Akbor, M.A.; Beg, R.U.; Adyel, T.M. Seasonal implication of heavy metal contamination of surface water around Dhaka Export Processing Zone (DEPZ), savar, bangladesh. Jahangirnagar Univ. J. Sci. 2011, 34, 21–35.

19. Mahfuz, M.A.; Ahmad, J.U.; Sultana, M.S.; Rahman, M.M.; Goni, M.A.; Rahman, M.S. Status of physicochemical properties of wastewater in Bangladesh: A cash studies in Dhalai Beel of DEPZ. Bangladesh J. Environ. Res. 2004, 2, 9–15.

20. Rahman, S.H.; Neelormi, S.; Tareq, S.M. Environmental impact assessments of textile and dyeing industries on ecosystem of Karnopara canal at Savar, Bangladesh. Jahangirnagar Univ. J. Sci. 2008, 31, 19–32.

21. Muller, G. Index of Geoaccumulation in sediments of the Rhine River. Geo. J. 1969, 2, 108–118

22. Taylor, S.R.; Mclennan, S.M. The geochemical evolution of the continental crust. Rev. Geophys. 1995, 33, 241–265

23. Muller, G. Die Schwermetallbelstung der sediment des Neckars und seiner Nebenflusse: Eine Bestandsaufnahme. Chem. Zeitung. 1981, 105, 156–164.

24. Hakanson, L. An ecological risk index for aquatic pollution control, a sedimentological approach. Water Res. 1980, 14, 975–1001.

25. Krzysztof, L.; Wiechula, D.; Korns, I. Metal contamination of farming soils affected by industry. Environ. Int. 2004, 30, 159–165.

26. Abrahim, G.M.S.; Parker, R.J. Assessment of heavy metal enrichment factors and the degree of contamination in marine sediments from Tamaki Estuary, Auckland, New Zealand. Environ. Monit. Assess. 2008, 136, 227–238.

27. Tomlinson, D.L.; Wilson, J.G.; Harris, C.R.; Jeffrey, D.W. Problems in the assessment of heavy metal levels in estuaries and the formation of a pollution index. Helgolaender Meeresunter. 1980, 33, 566–575.

28. Turekian, K.K.; Wedepohl, K.H. Distribution of the elements in some major units of the earth's crust. Bull. Geo. Soc. Am. 1961, 72, 175–192.

29. Angula, E. The Tomlinson Pollution Index applied to heavy metal, Mussel–Watch data: A useful index to assess coastal pollution. Sci. Total Environ. 1996, 187, 19–56.

30. Kisku, G.C.; Barman, S.C.; Bhargava, S.K. Contamination of soil and plants with potentially toxic elements irrigated with mixed industrial effluent and its impact on the environment. Water Air Soil Poll. 2000, 120, 121–137.

31. Islam, M.M.; Halim, M.A.; Safiullah, S.; Hoque, S.A.M.W.; Islam, M.S. Heavy metal (Pb, Cd, Zn, Cu, Cr and Mn) content in textile sludge in Gazipur, Bangladesh. Res. J. Environ. Sci. 2009, 3, 311–315.

32. Yaylali-Abanuz, G. Heavy metal contamination of surface soil around Gebze industrial area, Turkey. Microchem. J. 2011, 99, 82–92.

33. Kabata-Pendias, A. Trace Element in Soils and Plants; CRC Press: Boca Raton, FL, USA, 2000; p. 413.

34. SEPA. The Limits of Pollutants in Food; GB2762-2005; State Environmental Protection Administration: Beijing, China, 2005.

35. Awashthi, S.K. Prevention of Food Adulteration Act No. 37 of 1954. Central and State Rules as Amended for 1999; Ashoka Law House: New Delhi, India, 2000; p. 2000.

36. Loska, K.; Wiechula, D.; Korus, I. Metal contamination of farming soils affected by industry. Environ. Int. 2004, 30, 159–165.

37. Ahmed, J.U.; Gani, M.A. Heavy metal contaminations in water, soil and vegetation of the industrial areas in dhaka, Bangladesh. Environ. Monit. Assess. 2010, 166, 347–357.

38. Mondol, M.N.; Chamon, A.S.; Faiz, B.; Elahi, S.F. Seasonal variation of heavy metal concentrations in Water and plant samples around Tejgaon industrial Area of Bangladesh. J. Bangladesh Acad. Sci. 2011, 35, 19–41.

39. Roa, M.S.; Gopalkrishnan, R.; Venkatesh, B.R. Medical geology—An emerging field in environmental science. In National Symposium on Role of Earth Sciences, Integrated and Related Societal Issues, Lucknow, India, 2–4 November 2001, GSI Spl. Pub. No. 65, 2001; pp. 213–222.

40. Ahmed, G.; Miah, M.A.; Anawar, H.M.; Chowdhury, D.A.; Ahmad, J.U. Influence of multi-industrial activities on trace metal contamination: An approach towards surface water body in the vicinity of Dhaka Export Processing Zone (DEPZ). Environ. Monit. Assess. 2011; doi:10.1007/s10661-011-2254-9.

41. Coşkun, M.; Steinnes, E.; Frontasyeva, V.F.; Sjobakk, T.E.; Demnika, S. Heavy metal pollution of surface soil in the Thrace region, Turkey. Environ. Monit. Assess. 2006, 119, 545–556.

42. Edwards, A.L. The Correlation Coefficient. In Introduction to Linear Regression and Correlation; W. H. Freeman and Company: San Francisco, CA, USA, 1976; Chapter 4, pp. 33–46.

PART IV

REMEDIATION OF HEAVY METAL CONTAMINATION

CHAPTER 10

PHYTOREMEDIATION OF HEAVY METALS: A GREEN TECHNOLOGY

P. AHMADPOUR, F. AHMADPOUR, T. M. M. MAHMUD, ARIFIN ABDU, M. SOLEIMANI, AND F. HOSSEINI TAYEFEH

10.1 INTRODUCTION

Since the beginning of the industrial revolution, heavy metal contamination of the biosphere has increased considerably and became a serious environmental concern. Contamination by heavy metals can be considered as one of the most critical threats to soil and water resources as well as to human health (Yoon et al., 2006). During the past decades, the annual widespread release of heavy metals reached 22000 t (metric ton) for Cd, 939000 t for Cu, 1350000 t for Zn, and 738000 t for Pb (Singh et al., 2003).

Sources of metal contamination include anthropogenic and geological activities. Industrial pollutants, smelting, mining, military activities, fuel production and agricultural chemicals are some of the anthropogenic activities that cause metal contamination (Jadia and Fulekar, 2009). The application of phosphate fertilizers to the agricultural soil has led to increase in Cd, Cu, Zn and As (Zarcinas et al., 2004). Indeed, the increasing demand for agricultural products has led to extensive cultivation in agri-

This chapter was originally published under the Creative Commons Attribution License. Ahmadpour P, Ahmadpour F, Mahmud TMM, Abdu A, Soleimani M, and Tayefeh FH. Phytoremediation of Heavy Metals: A Green Technology. African Journal of Biotechnology 11,76 (2012), doi: 10.5897/ AJB12.459.

cultural lands. Applying fertilizers, pesticides and herbicides is necessary to protect the quality and quantities of these products. However, the excessive use of these agro-chemicals creates environmental problems, such as accumulation of these chemical substances in the soil and plant uptake (Sahibin et al., 2002). Unlike organic matter, these metals cannot be altered by micro-organisms. The toxicity of heavy metals is a very serious issue, because they have a long persistence in the environment. The half-life of these toxic elements is more than 20 years (Ruiz et al., 2009). According to the United States Environmental Action Group (USEAG), this environmental problem has threatened the health of more than 10 million people in many countries (Environ-mental News Service, 2006). Heavy metal pollution has spread throughout the world. 53 elements are classified as heavy metals. Their densities exceed 5 g cm-3, and they are known as universal pollutants in industrial areas (Sarma, 2011).

10.2 ENVIRONMENTAL POLLUTION AND SOURCES OF CONTAMINATION

Environmental pollution is the present of chemicals at poisonous levels in land, water and air. Pollution can be defined as an accidental or deliberate contamination of the environment with waste generated by human activities. Our environment has been contaminated with organic and inorganic pollutants, because pollutants are released into the environment through many different ways. Soil, water and air have been contaminated as a result of industrial activities and the unmanageable growth of large cities. Metals such as Cd, Cu, Cr, Ni, Zn and Pb are known to be serious environmental pollutants.

10.2.1 SOIL CONTAMINATION BY HEAVY METALS

The two different ways that heavy metals enter the environment are from natural and anthropogenic sources. Natural sources of heavy metals contamination usually result from the weathering of mines, which are themselves created anthropogenically (Wei et al., 2008). Heavy metal is defined as any element with metallic characteristics, such as density, conductivity, stability as cations, and an atomic number greater than 20 (Raskin et al., 1994). Heavy metal pollution is a crucial environ-mental concern throughout the world. It occurs in the soil, in water, in living organisms, and at the bottom of the sediments. Environmental contami-nation by heavy metals as a result of industrial and mining activities became widespread in the late 19th and early 20th centuries (Benavides et al., 2005). Heavy metals, including Cd, Cu, Cr, Zn, Ni and Pb as critical pollutants, have an adverse effect on the environment, specifi-cally at high concentrations in areas with severe anthro-pogenic activi-ties (United States Protection Agency, 1997). Although they are natural components of the earth's crust, heavy metals' biochemical equivalence and geochemical cycles have changed noticeably due to human activi-ties (Baccio et al., 2003). These metals are just being transformed from one form to another, because of their inability to degrade naturally. The heavy metals namely Cu, Fe, Zn, Mo and Mn are micronutrients and are considered to be essential to maintaining life in biological systems. However, at higher concentrations, these metals become highly toxic and threaten the health of animals and humans by influencing the quality of crops, water and atmosphere. The heavy metals Cd, Cu, Ni and Hg create greater phytotoxicity than Zn and Pb (Raskin et al., 1994).

The pollution of soil is a crucial matter that has attracted considerable public attention over the past few decades. A large proportion of land has become hazardous and non-arable for humans and animals, because of ex-tensive pollution. It is unusual to have soils without at least traces of heavy metals, and the levels of these elements become more toxic due to anthro-pogenic or natural activities that are harmful for living systems (Turan and Esringu, 2007). Organic pollutants are anthropogenic and degrade in the soil compared to heavy metals, which are non-degradable and occur naturally in the environment (Garbisu and Alkorta, 2001). Heavy metals can be developed by industrial activities, volcanic operations and parent material. Generally, depending on the type of element and its location, the concentration of metals in the soil ranges from traces levels to as high as 100000 mg kg^{-1} (Blaylock and Huang, 2000).

10.2.2 TOXICITY OF HEAVY METALS IN PLANTS

Heavy metals can be poisonous for macro- and micro-organisms through direct influence on the biochemical and physiological procedures, reducing growth, deterio-rating cell organelles, and preventing photosynthesis. Regarding the transportation of metals from roots to the aerial parts of the plants, some metals (especially Pb) tend to be accumulated in roots more than in aerial parts, because of some barriers that prevent their movement. However, other metals, such as Cd, moves easily in plants (Garbisu and Alkorta, 2001). Generally, all plants are able to accumulate essential elements, such as Cu, Fe, Zn, Ca, K, Mg and Na, from soil solutions for growth and development. However, during this process, plants also accumulate some non-essential elements, such as Cd, As, Cr, Al and Pb that have no biological activity.

10.3 REMEDIATION OF HEAVY METALS

Current conventional methods to remediate heavy metal-contaminated soil and water, such as ex situ excavation, landfill of the top contaminated soils (Zhou and Song, 2004), detoxification (Ghosh and Singh, 2005), and physico-chemical remediation, are expensive (Danh et al., 2009), time consuming, labor exhaustive and increase the mobilization of contaminants, and destroy the biotic and structure of the soil. Therefore, these remediation techniques are not technically or financially suitable for large contaminated areas (Baccio et al., 2003). Bioremediation was developed as a technology to degrade pollutants into a low toxic level by using microorganisms. However, the use of this technology to remediate contaminated areas by applying living organism was less successful for extensive metal and organic pollutants. Plants are able to metabolize sub-stances produced in natural ecosystems (Vidali, 2001). Phytoremediation is an approach in which plants are applied to detoxify contaminated areas (Garbisu and Alkorta, 2001; Mangkoedihardjo and Surahmaida, 2008).

10.4 DEFINITION AND GENERAL TYPES OF PHYTOREMEDIATION

Phytoremediation is a promising new technology that uses plants to clean up contaminated areas. It is a low cost, long term, environmentally and aesthetically friendly method of immobilizing/stabilizing, degrading, trans-ferring, removing, or detoxifying contaminants, including metals, pesticides, hydrocarbons, and chlorinated solvents (Susarla et al., 2002; Jadia and Fulekar, 2008; Zhang et al., 2010). Over the past 2 decades, it has become a highly accepted means of detoxifying conta-minated water and soil (U.S.EPA, 2001). Historically, phytoremediation has been considered a natural process, first identified and proved more than 300 years ago (Lasat, 2000). The specific plant and wild species that are used in this technique are effective at accumulating increasing amounts of toxic heavy metals (Ghosh and Singh, 2005; Brunet et al., 2008). These plants are known as accumulators. They accumulate heavy metals at higher concentrations ($\geq$ 100 times) above ground than do non-hyperaccumulators growing in the same conditions, without showing any observable symptoms in their tissues (Barceló and Poschenrieder, 2003). Phytoremediation can be applied to detoxify areas with trivial pollution of metal, nutrients, organic matter, or contaminants. Nagaraju and Karimulla (2002) described that some species, including *Jatropha curcas* (from *Euphorbiaceae*), *Dodonaea viscose* (from *Sapindaceae*), and *Cassia auriculata* (from *Fabaceae*), had potential for remediation of soils polluted with different kinds of trace and major elements. Phytoremediation can be classified into different applications, such as phytofiltration or rhizofiltration, phytostabilization, phytovolatilization, phytodegradation (Long et al., 2002), and phyto-extraction (Jadia and Fulekar, 2009).

10.4.1 PHYTOFILTRATION OR RHIZOFILTRATION

Phytofiltration or rhizofiltration is the removal by plant roots of contaminants in waste water, surface water, or extracted ground water (Pivetz,

2001). Abhilashet al. (2009) investigated the potential of *Limnocharis flava* (L.) Buchenau, grown for phytofiltration of Cd in polluted water with low concentrations of Cd in a hydroponic experiment. They spiked 45-day-old seedlings of *L. flava* with different concentrations of Cd (0.5, 1, 2 and 4 mg L^{-1}). The concentration of Cd in different parts of the plant was highest in the roots followed by leaves and peduncle. This suggested that *L. flava* was a suitable species for phytofiltration of low concentrations of Cd in water.

10.4.2 PHYTOSTABILIZATION

Phytostabilization is a simple, cost-effective, and less environmental invasive approach to stabilize and reduce the bioavailability of contaminants by using plants. In fact, this approach uses plant roots to restrict the mobility and bioavailability of contaminants in the soil (Jadia and Fulekar, 2009). Plants can reduce the future adverse effects of pollutants in the environment by keeping them from entering the ground water or spreading in the air. This method is applicable when there is no prompt action to detoxify contaminated areas (for example, if a responsible company only exists for a short time, or if an area is not of high concern on a remediation agenda) (Garbisu and Alkorta, 2001). In this approach, the chemical and biological characteristics of polluted soils are amended by increasing the organic matter content, cation exchange capacity (CEC), nutrient level, and biological actions (Alvarenga et al., 2008). In phyto-stabilization, plants are responsible for reducing the percolation of water within the soil matrix, which may create a hazardous leachate, inhibiting direct contact with polluted soil by acting as barrier, and interfering with soil erosion, which results in the spread of toxic metals to the other sites (Raskin and Ensley, 2000). Phyto-stabilization is a suitable technique to remediate Cd, Cu, As, Zn and Cr. Alvarenga et al. (2009) investigated the effect of three organic residues, sewage sludge, municipal solid waste compost, and garden waste compost, on the phytostabilization of an extremely acidic metal-contaminated soil. The plant species used in this experiment was perennial ryegrass (*Loliumperenne L.*). The organic residues were used at 25, 50 and 100 Mg ha^{-1} (dry weight basis). These reagents immobilized and decreased the mobile fraction of Cu, Pb and Zn. It was inferred that ryegrass had the potential to be used in phytostabilization for mine-polluted soil and municipal solid waste compost, and to a lesser extent, sewage sludge, used at 50 Mg ha^{-1}, and that it is efficient in the in situ immobilization of metals, developing the chemical properties of the soil, and greatly enhancing the plant biomass.

10.4.3 PHYTOVOLATILIZATION

Phytovolatilization is the use of green plants to extract volatile contaminants, such as Hg and Se, from polluted soils and to ascend them into the air from their foliage (Karami and Shamsuddin, 2010). Bañuelos (2000) perceived that some plants were able to transform Se in the form of dimethylselenide and dimethyldiselenide in high-selenium media.

10.4.4 PHYTODEGRADATION

Phytodegradation is the use of plants and micro-organisms to uptake, metabolize, and degrade the organic contaminant. In this approach, plant roots are used in association with microorganisms to detoxify soil contaminated with organic compounds (Garbisu and Alkorta, 2001). It is also known as phytotransformation. Some plants are able to decontaminate soil, sludge, sediment, and ground and surface water by producing enzymes. This approach involves organic compounds, including herbicides, insecticides, chlorinated solvents, and inorganic contaminants (Pivetz, 2001).

10.4.5 PHYTOEXTRACTION

Phytoextration is a phytoremediation technique that uses plants to remove heavy metals, such as Cd, from water, soil, and sediments (Yanai et al., 2006; Van Nevel et al., 2007). It is an ideal method for removing pollutants from soil without adversely affecting the soil's properties. Furthermore, in this approach, metals accumulated in harvestable parts of the plant can be simply restored from the ash that is produced after drying, ashing, and composting these harvestable parts (Garbisu and Alkorta, 2001). Phyto-extraction has also been called phytomining or biomining (Pivetz, 2001). This technology is a more advanced form of phytoremediation, in which high-biomass crops grown in the contaminated soil are used to bioharvest and recover heavy metals. It can be applied in mineral industry to com-mercially produce metals by cropping (Sheoran et al., 2009).

The ability of plants to transport and uptake heavy metals from the soil into their above-ground shoots and the harvestable parts of their under-ground roots is the key to successful phytoextraction (Garbisu and Alkor-ta, 2001; Chen et al., 2003). Robinson et al. (2006) stated that a few field experiments and commercial exercises have been done in the past decade to investigate successful phytoextraction. Moreover, for phytoextraction to be considered successful, the contaminated areas need to be detoxi-fied to a level specified by environ-mental rules and for a lower cost than conventional techniques (Kos and Le tan, 2003). Nascimento and Xing (2006) expressed that phytoextraction may be considered as a commercial technology in the future.

Several literatures have described the potential of various species for the phytoextraction of heavy metals contaminated areas (Grispen et al., 2006; Daghan et al., 2008; Neugschwandtner et al., 2008; Zadeh et al., 2008; Liu et al., 2010; Nwaichi and Onyeike, 2010). Mangkoedihardjo and Surahmaida (2008) examined the potential of *J. curcas L.* for decon-tamination of Cd- and Pb-contaminated soil. The garden soil was artifi-cially contaminated by $Pb(NO_3)_2$ and $Cd(NO_3)_2$. Plants were treated with different levels of these elements in a separate and mixed compound for a period of one month. The result showed that the primary concentration (50 mg kg^{-1}) of these two elements had no harmful effects on the plants. The researchers found that *J. curcas L.* has a potential for phytoremediation of Cd and Pb in contaminated areas. Ang et al. (2003) studied the removal of Cd, Pb, As and Hg from slime tailings at Forest Research Institute, Ma-laysia. Various timber plants, including *Hopea odorata, Acacia mangium, Swietenia macrophylla,* and *Intsia palembanica*, were planted in slime tailing to determine their potential for bioaccumulation of Cd, Hg, Pb and As. The results of this study suggested that *H. odorata* and *I. palembanica* had the potential for Cd removal in a short period of time compared with others, whereas *A. mangium* was suitable for removal of As. Jiang et al. (2004) determined the growth performance and ability for Cu phytoextraction of Elsholtzia splendens. In a greenhouse study, $CuSO_4.5H_2O$ was applied in various concentrations such as 100, 200, 400, 600, 800, 1000 and 1200 mg kg^{-1}. The plant species showed a high ability to tolerate Cu toxicity and performed its usual growth when exposed to Cu up to 80 mg kg^{-1} of available Cu. Li et al. (2009) studied the phytoextraction potential of carambola (*Averrhoa carambola*) as a tree species with high biomass. After 170 days growing on low Cd-contaminated soil, this species produced more biomass of shoots (18.6 t ha^{-1}) and accumulated 213 g Cd per hectare. The researchers suggested that carambola is a suitable choice for low Cd-contaminated soils. The phytoextraction potential of a hybrid poplar (*Populus deltoids* × *Populus nigra*) in two purple and alluvial soils contaminated with Cd was evaluated by Wu et al. (2009). It was observed that the accumulation of Cd in plant parts increased with the increase of this element in the two soils. The accu-mulation of Cd in plant roots was higher than in shoots and then in leaves in purple soil, but the reverse was true in alluvial soil.

Murakami et al. (2007) also examined the ability of soybeans (*Glycine max L.* Merr., cv. Enrei and Suzuyutaka), maize (*Zea mays L.*, cv. Gold Dent), and rice (*Oryza sativa L.*, cv. Nipponbare and Milyang 23) for phytoextraction of Cd-polluted areas. The species were planted on one andosol and two fluvisols contaminated with low concentrations of Cd (0.83 to 4.29 mg Cd kg^{-1}) for 60 days. The results indicated that the accumulation of Cd in shoots of Milyang 23 rice was 1 to 15% of the total Cd in the soil. It was inferred that these species had the potential to remediate paddy soil with low Cd contents.

10.4.6 SELECTION OF HYPER-ACCUMULATOR OF PLANTS

Successful phytoremediation requires recognition of suitable plant species to accumulate metals in toxic levels as well as creating high biomass (Clemens et al., 2002; Odoemelam and Ukpe, 2008). Generally, the ideal plants for phytoextraction should have high capacity to accumulate toxic levels of metals in their aerial parts (shoots), high growth rates, and tolerance to high salinity and high pH. Moreover, these plants must produce high dry biomass, simply grown and completely harvestable, and must uptake and translocate metals to aerial parts efficiently (Alkorta et al., 2004; Sarma, 2011). Overall, it is recommended to use the native plant species that grow locally near the site. These species are less competitive under local conditions and will reduce the metal concentration to an acceptable level for normal plant growth (Rajakaruna et al., 2006).

10.4.7 ADVANTAGES OF PHYTOREMEDIATION

Phytoremediation is a low-cost and effective strategy to clean up contaminated soils without requiring high-cost tools and expert human resources (Environmental protection agency, 2000; Ghosh and Singh, 2005). As a green technology, it is applicable for different kinds of organic and inorganic pollutants and provides aesthetic benefits to the environment by using trees and creating green areas, which is socially and psychologically beneficial for all (Ghosh and Singh, 2005; Lewis, 2006). This green technology is suitable for large areas in which other approaches would be expensive and ineffective (Vidali, 2001; Prasad and Freitas, 2003). In addition, as a practical approach to decontaminating soil and water, residues can be reused with minimal harm to the environment (Schnoor, 2002). Furthermore, the expan-sion of contaminants to air and water is reduced by preventing leaching and soil erosion that may result from wind and water activity (Pivetz, 2001; Ghosh and Singh, 2005).

10.4.8 DISADVANTAGES OF PHYTOREMEDIATION

Time is the most serious limitation of phytoremediation, because this approach may require several years for effective remediation (Vidali, 2001; Rajakaruna et al., 2006). Moreover, preserving the vegetation in extensively contaminated areas is complicated (Vidali, 2001), and human health could also be threatened by entering the pollutant into the food chain through animals feeding on the contaminated plants (Pivetz, 2001). This technology is not impressive when just a small part of the contami-nant is bio-available for plants in the soil (Rajakaruna et al., 2006). Beside that, it is limited to the low or mildly contaminated areas enclosed by the plant root district (Ghosh and Singh, 2005).

10.5 CRITERIA FOR METAL ACCUMULATION IN PLANTS

All plant species have the ability to uptake metals; however, some can accumulate greater amounts of metals (100 times more than the average plant in the same condition without showing any adverse effect). The woody or herbaceous plants that accumulate and tolerate heavy metals in an amount greater than the toxic levels in their tissue are known as hyperaccumulators (Baker et al., 2000; Barceló and Poschenrieder, 2003; Zhou and Song, 2004). In recent years, the use of hyperaccumulators for remediation of contaminated sites due to their capacity to take up heavy metals from polluted soil and accumulate them in their shoots has been receiving a great deal of attention from researchers (Sun et al., 2007, 2009). The main criteria for hyper-accumulators are; (i) accumulating capability, (ii) tolerance capability, (iii) removal efficiency (RE) based on plant biomass, (iv) bio-concentration factor (BCF) and (v) transfer factor (TF).

10.5.1 ACCUMULATING CAPABILITY

Accumulating capability is the natural capacity of plants to accumulate metals in their above-ground parts (the threshold concentration) in amounts greater than 100 mg kg^{-1} for Cd (Zhou and Song, 2004), 1000 mg kg^{-1} for Cu, Cr, Pb, and Co, 10 mg kg^{-1} for Hg (Baker et al., 2000) and 10000 mg kg^{-1} dry weight of shoots for Ni and Zn (Lasat, 2002).

10.5.2 TOLERANCE CAPABILITY

Tolerance capability is the ability of plants to grow in heavy metal-contaminated sites and to have consi-derable tolerance to heavy metals without showing any reverse effects, such as chlorosis, necrosis, whitish-brown color, or reduction in the above-ground biomass (or at least not a significant reduction) (Sun et al., 2009).

10.5.3 REMOVAL EFFICIENCY

Removal efficiency based on plant biomass is the total concentrations of metals and dry biomass of plants to the total loaded metals in the growth media (Soleimani et al., 2010).

TABLE 1: Some hyperaccumulator species of metals.

Species	Metal	Reference
Clerodendrum infortunatum	Cu	Rajakaruna and Böhm (2002)
Croton bonplandianus	Cu	Rajakaruna and Böhm (2002)
Thordisa villosa	Cu	Rajakaruna and Böhm (2002)
Pityrogramma calomelanos	As	Dembitsky and Rezanka (2003)
Pistia stratiotes	Zn, Pb, Ni, Hg, Cu, Cd, Ag, and Cr	Odjegba and Fasidi (2004)
Alyssum lesbiacum	Ni	Cluis (2004)
Helicotylenchus indicus	Pb	Sekara et al. (2005)
Bidens pilosa	Cd	Sun et al. (2009)
Thlaspi caerluescens	Cd, Zn, and Pb	Cluis (2004) and Banasova et al. (2008)
Lonicera japonica	Cd	Liu et al. (2009)
Solanum nigrum L.	Cd	Sun et al. (2008)
Sedum alferedii	Cd	Sun et al. (2007)
Brassica junceae	Ni and Cr	Saraswat and Rai (2009)

10.5.4 BCF

BCF index is the ratio of heavy metal concentration in plant roots to that in the soil (Yoon et al., 2006). Cluis (2004) reported that the BCF for hyperaccumulators is > 1, and in some cases can increase up to 100.

10.5.5 TF

TF is the capability of plants to take up heavy metals in their roots and to translocate them from the roots to their above-ground parts (shoots). Therefore, it is the ratio of heavy metal concentration in aerial parts of the plant to that in its roots (Mattina et al., 2003; Liu et al., 2010). This specific criterion for hyperaccumulators should reach > 1 to indicate that the concentration of heavy metals above ground is greater than that below ground (roots). Therefore, it can be concluded that this criteria is more crucial in phytoextraction, where harvesting the aerial parts of the plant is the most important objective (Wei and Zhou, 2004; Karami and Shamsuddin, 2010). Baker and Whiting (2002) reported that excluders can be identified by a TF < 1, whereas accumulators are characterized by a TF > 1. BCF, TF and RE are calculated by the following equations:

McGrath and Zhao (2003) reported that BCFs and TFs are > 1 in hyperaccumulators. More than 400 species from 45 families all over the world have been classified as hyperaccumulators (Sun et al., 2009). Sarma (2011) reported the latest number of metal hyperaccumulators. According to his report, more than 500 plant species consisting of 101 families are classified as metal hyperaccumulators, including Euphorbiaceae, Violaceae, Poaceae, Lamiaceae, Flacourtiaceace, Cunouniaceae, Asateraceae, Brassicaceae, Caryophyllace, and Cyperaceae. Zhou and Song (2004) reported that the hyperaccumulation of Cd and As occurs rarely in the plant families. They found that because hyper-accumulators produce low shoot biomass with long periods of maturity and long growing seasons, there are only a few plants with high metal accumulation ability and high biomass. However, Baker et al. (2000) found many species that can be classified as hyperaccumulators based on their capacity to tolerate toxic concentrations of metals, such as Cd, Cu, As, Co, Mn, Zn, Ni, Pb and Se (Table 1).

10.6 CONCLUSION

Heavy metals are one of the most critical threats to the soil and water resources, as well as to human health. These metals are released into the environment through mining, smelting of metal ores, industrial emissions, and the application of pesticides, herbicides and fertilizers. Metals, such as Cd, Cu, Pb, Zn , and metalloids (e.g. As), are considered to be metallic pollutants. Conven-tional remediation technologies are expensive, time consuming and environmentally divesting. Therefore, it is inevitable to use a low cost and environmentally friendly technology to remediate polluted soils with heavy metals, specifically in developing countries. Phytoremediation of metals is the most effective plant-based method to remove pollutants from contaminated areas. This green technology can be applied to remediate the polluted soils without creating any destructive effect of soil structure. Some specific plants, such as herbs and woody species, have been proven to have noticeable potential to absorb toxic metals. These plants are known as hyper-accumulators. Researchers are trying to find new plant species that are suitable to be used in removing heavy metals from contaminated soils.

REFERENCES

1. Abhilash P, Pandey VC, Srivastava P, Rakesh PS, Chandran S, Singh N, Thomas AP (2009). Phytofiltration of cadmium from water by Limnocharis flava (L.) Buchenau grown in free-floating culture system. J. Hazard. Mater. 170(2-3):791-797.
2. Alkorta I, Hernández-Allica J, Becerril J, Amezaga I, Albizu I, Garbisu C (2004). Recent findings on the phytoremediation of soils contaminated with environmentally toxic heavy metals and metalloids such as zinc, cadmium, lead, and arsenic. Rev. Environ. Sci. Biotechnol. 3(1):71-90.
3. Alvarenga P, Goncalves AP, Fernandes RM, De Varennes A, Vallini G, Duarte E, Cunha-Queda AC (2008). Evaluation of composts and liming materials in the phytostabilization of a mine soil using perennial ryegrass. Sci. Total Environ. 406(1-2):43-56.
4. Alvarenga P, Gonçalves AP, Fernandes RM, de Varennes A, Vallini G, Duarte E, Cunha-Queda AC (2009). Organic residues as immobilizing agents in aided phytostabilization: (I) Effects on soil chemical characteristics. Chemosphere 74(10):1292-1300.
5. Ang LH, Tang LK, Hui TF, Ho WM, Theisera GW (2003). Bioaccumulation of heavy metals by Acacia mangium, Hopea odorata, Intsia palembanica and Swietenia macrophylla grown on slime tailings. Project No. 05-03-10-SF0038, Forest Research Institute Malaysia (FRIM), Malaysia pp.22-26.
6. Baccio D, Tognetti R, Sebastiani L, Vitagliano C (2003). Responses of Populus deltoides× Populus nigra (Populus× euramericana) clone I-214 to high zinc concentrations. New Phytologist. 159(2):443-452. DOI: 410.1046/j.1469-8137.2003.00818.x.

7. Baker AJM, McGrath SP, Reeves RD, Smith JAC (2000). Metal hyperaccumulator plants: A review of the ecology and physiology of a biological resource for phytoremediation of metal-polluted soils. In N. Terry and G. Banuelos (Ed.): Lewis Publishers, Boca Raton, FL pp. 5-107.

8. Baker AJM, Whiting SN (2002). In search of the Holy Grail–a further step in understanding metal hyperaccumulation? New Phytologist 155(1):1-4.

9. Banasova V, Horak O, Nadubinska M, Ciamporova M (2008). Heavy metal content in Thlaspi caerulescens J. et C. Presl growing on metalliferous and non-metalliferous soils in Central Slovakia. Int. J. Environ. Pollut. 33(2):133-145.

10. Bañuelos GS (2000). Phytoextraction of selenium from soils irrigated with selenium-laden effluent. Plant Soil 224(2):251-258.

11. Barceló J, Poschenrieder C (2003). Phytoremediation: principles and perspectives. Contrib. Sci. 2(3):333-344.

12. Benavides MP, Gallego SM, Tomaro ML (2005). Cadmium toxicity in plants. Braz. J. Plant Physiol. 17:21-34.

13. Blaylock MJ, Huang JW (2000). Phytoextraction of metals. Phytoremediation of toxic metals: using plants to clean-up the environment. New York, John Wiley and Sons, Inc. pp.53-70.

14. Brunet J, Repellin A, Varrault G, Terryn N, Zuily-Fodil Y (2008). Lead accumulation in the roots of grass pea (Lathyrus sativus L.): a novel plant for phytoremediation systems? Comptes Rendus Biologies 331(11):859-864.

15. Chen YX, Lin Q, Luo YM, He YF, Zhen SJ, Yu YL, Tian GM, Wong MH (2003). The role of citric acid on the phytoremediation of heavy metal contaminated soil. Chemosphere 50(6):807-811.

16. Clemens S, Palmgren MG, Krämer U (2002). A long way ahead: understanding and engineering plant metal accumulation. Trends Plant Sci. 7(7):309-315.

17. Cluis C (2004). Junk-greedy Greens: phytoremediation as a new option for soil decontamination. BioTech J. 2:61-67.

18. Daghan H, Schaeffer A, Fischer R, Commandeur U (2008). Phytoextraction of cadmium from contaminated soil using transgenic Tobacco Plants. Int. J. Environ. Appl. Sci. 3(5):336-345.

19. Danh LT, Truong P, Mammucari R, Tran T, Foster N (2009). Vétiver grass, Vetiveria zizanioides: A choice plant for phytoremediation of heavy metals and organic wastes. Int. J. Phytoremediation 11(8):664-691.

20. Dembitsky VM, Rezanka T (2003). Natural occurrence of arseno compounds in plants, lichens, fungi, algal species, and microorganisms. Plant Sci. 165(6):1177-1192.

21. Environmental protection agency (2000). Selecting and using phytoremediation for site clean up. Environmental protection agency (EPA). July 2001.office of solid waste and emergency response.brownfileds technology primer, Retrieved 17 July, 2010, from http://www.brownfieldstsc.org/pdfs/phytoremprimer.pdf

22. Environmetal News Service (2006). Environmetal News Service. New York.

23. Garbisu C, Alkorta I (2001). Phytoextraction: A cost-effective plant-based technology for the removal of metals from the environment. Bioresour. Technol. 77(3):229-236.

24. Ghosh M, Singh SP (2005). A review on phytoremediation of heavy metals and utilization of it's by products. Appl. Ecol. Environ. Res. 3(1):1-18.

25. Grispen VMJ, Nelissen HJM, Verkleij JAC (2006). Phytoextraction with Brassica napus L.: A tool for sustainable management of heavy metal contaminated soils. Environ. Pollut. 144(1):77-83.

26. Jadia CD, Fulekar MH (2008). Phytoremediation: The application of vermicompost to remove zinc, cadmium, copper, nickel and lead by sunflower plant. Environ. Eng. Manag. J. 7(5):547-558.

27. Jadia CD, Fulekar MH (2009). Phytoremediation of heavy metals: Recent techniques. Afr. J. Biotechnol. 8(6):921-928.

28. Jiang LY, Yang XE, He ZL (2004). Growth response and phytoextraction of copper at different levels in soils by Elsholtzia splendens. Chemosphere 55(9):1179-1187.

29. Karami A, Shamsuddin ZH (2010). Phytoremediation of heavy metals with several efficiency enhancer methods. Afr. J. Biotechnol. 9(25):3689-3698.

30. Kos B, Le tan D (2003). Influence of a biodegradable ([S,S]-EDDS) and nondegradable (EDTA) chelate and hydrogel modified soil water sorption capacity on Pb phytoextraction and leaching. Plant Soil 253(2):403-411.

31. Lasat MM (2000). Phytoextraction of metals from contaminated soil: a review of plant/soil/metal interaction and assessment of pertinent agronomic issues. J. Hazard. Substance Res. 2(5):1-25.

32. Lasat MM (2002). Phytoextraction of toxic metals: A review of biological mechanisms. J. Environ. Qual. 31:109-120.

33. Lewis AC (2006). Assessment and comparison of two phytoremediation systems treating slow-moving groundwater plumes of TCE. Master thesis. Ohio University p.158.

34. Li JT, Liao B, Dai Z, Zhu R, Shu WS (2009). Phytoextraction of Cd-contaminated soil by carambola (Averrhoa carambola) in field trials. Chemosphere 76(9):1233-1239.

35. Liu W, Zhou Q, An J Sun Y, Liu R (2010). Variations in cadmium accumulation among Chinese cabbage cultivars and screening for Cd-safe cultivars. J. Hazard. Mater. 173(1-3):737-743.

36. Liu Z, He X, Chen W, Yuan F, Yan K, Tao D (2009). Accumulation and tolerance characteristics of cadmium in a potential hyperaccumulator-Lonicera japonica Thunb. J. Hazard. Mater. 169(1-3):170-175.

37. Long X, Yang X, Ni W (2002). Current situation and prospect on the remediation of soils contaminated by heavy metals. Ying yong sheng tai xue bao= The journal of applied ecology/Zhongguo sheng tai xue xue hui, Zhongguo ke xue yuan Shenyang ying yong sheng tai yan jiu suo zhu ban 13(6):757-762.

38. Mangkoedihardjo S, Surahmaida A (2008). Jatropha curcas L. for phytoremediation of lead and cadmium polluted soil. World Appl. Sci. J. 4(4):519-522.

39. Mattina MJI, Lannucci-Berger W, Musante C, White JC (2003). Concurrent plant uptake of heavy metals and persistent organic pollutants from soil. Environ. Pollut. 124(3):375-378.

40. McGrath SP, Zhao FJ (2003). Phytoextraction of metals and metalloids from contaminated soils. Curr. Opin. Biotechnol. 14(3):277-282.

41. Murakami M, Ae N, Ishikawa S (2007). Phytoextraction of cadmium by rice (Oryza sativa L.), soybean (Glycine max (L.) Merr.), and maize (Zea mays L.). Environ. Pollut. 145(1):96-103.

42. Nagaraju A, Karimulla S (2002). Accumulation of elements in plants and soils in and around Nellore mica belt, Andhra Pradesh, India: a biogeochemical study. Environ. Geol. 41(7):852-860.

43. Nascimento CWA, Xing B (2006). Phytoextraction: A review on enhanced metal availability and plant accumulation. Scientia Agricola 63:299-311.

44. Neugschwandtner RW, Tlustos P, Komárek M, Száková J (2008). Phytoextraction of Pb and Cd from a contaminated agricultural soil using different EDTA application regimes: laboratory versus field scale measures of efficiency. Geoderma 144(3-4):446-454.

45. Nwaichi EO, Onyeike EN (2010). Cu tolerance and accumulation by Centrosema Pubescen Benth and Mucuna Pruriens Var Pruriens. Arch. Appl. Sci. Res. 2(3):238-247.

46. Odjegba VJ, Fasidi IO (2004). Accumulation of trace elements by Pistia stratiotes: implications for phytoremediation. Ecotoxicology 13(7):637-646.

47. Odoemelam SA, Ukpe RA (2008). Heavy meal decontamination of polluted soils using Bryophyllum pinnatum. Afr. J. Biotechnol. 7(23):4301-4303.

48. Pivetz BE (2001). Ground Water Issue: Phytoremediation of contaminated soil and ground water at hazardous waste sites pp.1-36.

49. Prasad MNV, Freitas H (2003). Metal hyperaccumulation in plants-Biodiversity prospecting for phytoremediation technology. Electron. J. Biotechnol. 6(3):285-321.

50. Rajakaruna N, Böhm BA (2002). Serpentine and its vegetation: A preliminary study from Sri Lanka. J. Appl. Botany-Angewandte Botanik 76:20-28.

51. Rajakaruna N, Tompkins KM, Pavicevic PG (2006). Phytoremediation: An affordable green technology for the clean-up of metal-contaminated sites in Sri Lanka. Ceylon J. Sci. (Biological Sciences) 35:25-39.

52. Raskin I, Ensley BD (2000). Phytoremediation of toxic metals:Using plants to clean up the environment.: John Wiley & Sons, Inc. Publishing, New York. p. 304.

53. Raskin I, Kumar PBA, Dushenkov S, Salt DE (1994). Bioconcentration of heavy metals by plants. Curr. Opin. Biotechnol. 5(3):285-290.

54. Robinson B, Schulin R, Nowack B, Roulier S, Menon M, Clothier B, Green S, Mills T (2006). Phytoremediation for the management of metal flux in contaminated sites. Forest, Snow Landscape Res. 80(2):221-234.

55. Ruiz JM, Blasco B, Ríos JJ, Cervilla LM, Rosales MA, Rubio-Wilhelmi M M, Sánchez-Rodríguez E, Castellano R, Romero L (2009). Distribution and efficiency of the phytoextraction of cadmium by different organic chelates. Terra LatinoAmericana 27(4):296-301.

56. Sahibin AR, Zulfahmi AR, Lai KM, Errol P, Talib ML (2002). Heavy metals content of soil under vegetables cultivation in Cameron highland In: Proceedings of the regional symposium on environment and natural resources 10-11th April 2002, Kuala Lumpur, Malaysia 1:660-667.

57. Saraswat S, Rai JPN (2009). Phytoextraction potential of six plant species grown in multimetal contaminated soil. Chem. Ecol. 25(1):1-11.

58. Sarma H (2011). Metal hyperaccumulation in plants: A review focusing on phytoremediation technology. J. Environ. Sci. Technol. 4:118-138.

59. Schnoor JL (2002). Phytoremediation of soil and groundwater. Technol. Eval. Report TE-02 1:252-630.

60. Sekara A, Poniedzialeek M, Ciura J, Jedrszczyk E (2005). Cadmium and lead accumulation and distribution in the organs of nine crops: Implications for phytoremediation. Polish J. Environ. Stud. 14(4):509-516.

61. Sheoran V, Sheoran AS, Poonia P (2009). Phytomining: A review. Miner. Eng. 22(12):1007-1019.

62. Singh OV, Labana S, Pandey G, Budhiraja R, Jain RK (2003). Phytoremediation: an overview of metallic ion decontamination from soil. Appl. Microbiol. Biotechnol. 61(5):405-412.

63. Ahmadpour et al. 14043

64. Soleimani M, Hajabbasi MA, Afyuni M, Mirlohi A, Borggaard OK, Holm PE (2010). Effect of endophytic fungi on cadmium tolerance and bioaccumulation by Festuca arundinaceaand Festuca pratensis. Int. J. Phytoremediation 12(6):535-549.

65. Sun Q, Ye ZH, Wang XR, Wong MH (2007). Cadmium hyperaccumulation leads to an increase of glutathione rather than phytochelatins in the cadmium hyperaccumulator Sedum alfredii. J. Plant Physiol. 164(11):1489-1498.

66. Sun Y, Zhou Q, Diao C (2008). Effects of cadmium and arsenic on growth and metal accumulation of Cd-hyperaccumulator Solanum nigrum L. Bioresour. Technol. 99(5):1103-1110.

67. Sun Y, Zhou Q, Wang L, Liu W (2009). Cadmium tolerance and accumulation characteristics of Bidens pilosa L. as a potential Cd-hyperaccumulator. J. Hazard. Mater. 161(2-3): 808-814.

68. Susarla S, Medina VF, McCutcheon SC (2002). Phytoremediation: an ecological solution to organic chemical contamination. Ecol. Eng. 18(5):647-658.

69. Turan M, Esringu A (2007). Phytoremediation based on canola (Brassica napus L.) and Indian mustard (Brassica juncea L.) planted on spiked soil by aliquot amount of Cd, Cu, Pb, and Zn. Plant Soil Environ. 53(1):7-15.

70. United States Environmental Protection Agency (2001). Citizen's guide to phytoremediation. Retrieved 4 January, 2011, from http://www.clu-in.org/download/citizens/citphyto.pdf

71. United States Protection Agency (1997). Introduction to Phytoremediation, EPA 600/R-99/107. Environmental Protection Agency (EPA), Office of Research and Development, Cincinnati, OH,US.

72. Van Nevel L, Mertens J, Oorts K, Verheyen K (2007). Phytoextraction of metals from soils: How far from practice? Environ. Pollut. 150(1):34-40.

73. Vidali M (2001). Bioremediation. An overview. Pure Appl. Chem. 73(7):1163-1172.

74. Wei S, Teixeira da Silva JA, Zhou Q (2008). Agro-improving method of phytoextracting heavy metal contaminated soil. J. Hazard. Mater. 150(3):662-668.

75. Wei SH, Zhou QX (2004). Discussion on basic principles and strengthening measures for phytoremediation of soils contaminated by heavy metals. Chin. J. Ecol. 23(1):65-72.

76. Wu F, YangW, Zhang J, Zhou L (2009). Cadmium accumulation and growth responses of a poplar (Populus deltoids× Populus nigra) in cadmium contaminated purple soil and alluvial soil. J. Hazard. Mater. pp.1-6.

77. Yanai J, Zhao FJ, McGrath SP, Kosaki T (2006). Effect of soil characteristics on Cd uptake by the hyperaccumulator Thlaspi caerulescens. Environ. Pollut. 139(1):167-175.

78. Yoon J, CaoX, Zhou Q, Ma LQ (2006). Accumulation of Pb, Cu, and Zn in native plants growing on a contaminated Florida site. Sci. Total Environ. 368(2-3):456-464.
79. Zadeh BM, Savaghebi-Firozabadi GR, Alikhani HA, Hosseini HM (2008). Effect of Sunflower and Amaranthus culture and application of inoculants on phytoremediation of the soils contaminated with cadmium. American-Eurasian J. Agric. Environ. Sci. 4(1):93-103.
80. Zarcinas BA, Ishak CF, McLaughlin MJ, Cozens G(2004). Heavy metals in soils and crops in Southeast Asia. Environ. Geochem. Health 26(3):343-357.
81. Zhang BY, Zheng JS, Sharp RG (2010). Phytoremediation in engineered wetlands: Mechanisms and applications. Procedia Environ. Sci. 2:1315-1325.
82. Zhou QX, Song YF (2004). Principles and methods of contaminated soil remediation: Science Press, Beijing p.568.

CHAPTER 11

ASSESSMENT OF THE EFFICACY OF CHELATE-ASSISTED PHYTOEXTRACTION OF LEAD BY COFFEEWEED (*Sesbania exaltata* Raf.)

GLORIA MILLER, GREGORIO BEGONIA, MARIA BEGONIA, JENNIFER NTONI, AND OSCAR HUNDLEY

11.1 INTRODUCTION

Metals differ from other toxic substances in that they are neither created nor destroyed by humans. Nevertheless, utilization by humans influences the potential for health effects in at least two major ways: first by environmental transport such as anthropogenic contributions to air, water, food and soil; and second, by altering the speciation or biochemical form of the element. Lead, in particular, depending upon the reactant surface, pH, redox potential, and other factors can bind tightly to the soil [1-3] with a retention time of 150 to 5000 years [1, 4].

In spite of the ever-growing number of toxic metalcontaminated sites, the most commonly used methods of dealing with heavy metal pollution are either the extremely costly process of excavation and reburial or simply isolation of the contaminated sites. Remediation costs in the U.S. have

*This chapter was originally published under the Creative Commons Attribution License. Miller G, Begonia G, Begonia M, Ntoni J, and Hundley O. Assessment of the Efficacy of Chelate-Assisted Phytoextraction of Lead by Coffeeweed (*Sesbania exaltata Raf.). International Journal of Environmental Research and Public Health 5,5 (2008), 428-435. doi:10.3390/ijerph5050428.*

been estimated at \$7 billion to \$8 billion per year, approximately 35% of which involves remediation of metals [5].

Recently, heavy metal phytoextraction has emerged as a promising, cost-effective alternative to the conventional methods of remediation [6-10]. The objective of phytoextraction is to reduce heavy metal levels below regulatory limits within a reasonable time frame [7]. To achieve this objective, plants must accumulate high levels of heavy metals and produce high amounts of biomass [8-11]. Soil-metal interaction by sorption, precipitation, and complexation processes, and differences between plant species in metal uptake efficiency, transport, and susceptibility make a general prediction of soil metal bioavailability and risks of plant metal toxicity difficult [12]. Moreover, the tight binding characteristic of Pb to soils and plant materials make Pb especially challenging for phytoremediation [13]. Previous hydroponic studies revealed that uptake and translocation of heavy metals in plants are enhanced by increasing heavy metal concentration in the nutrient solution [14]. The bioavailability of heavy metals in the soil is therefore, of paramount importance for successful phytoextraction. Lead has limited solubility in soils, and its availability for plant uptake is minimal due to complexation or binding with organic and inorganic soil colloids, and precipitation as carbonates, hydroxides, and phosphates [15-18].

Studies have shown that chelates can not only desorb heavy metals from the soil matrix into soil solution [19], but can enhance the bioavailability of Pb for uptake by the plant. Also, it has been reported that chelate amendments may aid in moving the Pb that is sequestered in the xylem cell wall of the roots upwards and into the shoots [3, 20- 24]. Specifically, this study was conducted to determine whether amendments of chelates such as EDTA, EGTA or HAc can enhance the solubility of Pb and make it more bioavailable for root uptake. Also we wanted to assess the efficacy of chelates in facilitating translocation of the metal into the shoots of coffeeweed. Our aim in this study was not only to utilize a Pb-specific chelate, but to also apply it in a time-efficient manner so that we could harvest the plant during its peak phytoextraction period. In a practical field application, this could reduce the risk of water pollution due to chelates and/or chelate-metal complexes migrating from the soil. Results of this study will be used to help establish an optimal time frame for harvesting *Sesbania exaltata* after chelate amendment, thereby limiting the likelihood of exposure to grazing animals.

11.2 MATERIALS AND METHODS

In order to minimize discrepancies in the experimental results that could arise from heterogeneous soil samples, a laboratory contaminated soil sample was used throughout this experiment so that we could create a test sample with consistent lead concentration and speciation, soil composition, contamination process, and contamination period.

Delta top soil and humus peat were air-dried to approximately 1 – 3% moisture content for 3 – 4 days under greenhouse conditions. Top soil and peat were cleaned of debris using a 1-cm sieve. Soil was prepared by mixing sieved soil, and peat in a 2:1 volumetric proportions. Representative samples of the prepared soil mixture were sent to Mississippi State University Soil Testing Laboratory, Mississippi State, MS to determine its physical and chemical characteristics.

11.2.1 SOLUBILITY TEST AND CHELATE SELECTION

To prepare lead-contaminated soil for the leadsolubility test, approximately 550 g of the dry, sieved delta topsoil, peat mixture (2:1 v/v) were placed in a plastic Ziploc bag and amended with 1000 mg Pb/kg dry soil mixture using lead nitrate. Deionized distilled water was added to each bag of soil mixture to adjust the soil moisture content to approximately 30% field capacity. The bags of soil were left to equilibrate (age) on a laboratory bench in the greenhouse for six weeks. The bags were occasionally turned and mixed during the incubation period to ensure thorough mixing. After incubation, 2 g of Pb-contaminated soil were placed into each 15 mL BD Falcon tube. The appropriate chelate (ethylenediaminetetraacetic acid [EDTA], ethylene glycol tetraacetic acid [EGTA], or acetic acid [HAc] in a 1:1 ratio with the metal) was added to each tube for a final volume of 10 mL (1:5 soil to chelate ratio). Deionized distilled water was used as a control. The soilchelate mixture was then agitated in a Classic C-1 platform

shaker for 0, 1, 2, 5, 6, and 7 days at room temperature. To ensure homogeneity, the mixture was shaken for 10 minutes for time 0. At the end
of each time period, the suspension was centrifuged at 5000 rpm (Fisher
Scientific Centrific® Centrifuge Model 225) with a 5-minute acceleration
and 30-minute deceleration. The supernatant was then filtered through a
Whatman 0.45 μm filter paper and Pb contents of each filtrate were quantified using inductively coupled plasma-optical emission spectrometry
(ICP-OES; Perkin Elmer Optima 3300 DV).

11.2.2 GREENHOUSE EXPERIMENT

Plants were grown in the Jackson State University greenhouse to evaluate the effectiveness of soil-applied chelating agents in enhancing metal
uptake by *Sesbania* exaltata. Lead-spiked soil for the greenhouse experiment was prepared similarly as previously described for the solubility test.
Approximately 550 g of the dry, sieved delta topsoil, peat mixture (2:1
v/v) were placed in a plastic Ziploc bag and amended with either 0, 1000,
or 2000 mg Pb/kg dry soil mixture using lead nitrate. The Pb concentrations were chosen to simulate a moderately contaminated soil. Deionized
distilled water was added to the bags to adjust the soil moisture content
to approximately 30% field capacity. The bags of soil were left to age on
a laboratory bench in the greenhouse for six weeks. The bags were occasionally turned and mixed during the aging period to ensure thorough
mixing. Sowing of seeds: In order to soften the seed coat, *Sesbania* seeds
were placed in a beaker that had been filled with deionized distilled water and the water was heated to 40°C. The heat was then turned off, and
the seeds were left to soak in the water for 24 hours. To prepare planting
tubes, brown Wipe-All paper towels were folded and pushed to the bottom
of each 656 mL Deepot tube (Stuewe and Sons, Inc., Corvallis, OR). The
holes on the sides and at the bottom of the tubes were then wrapped with
parafilm to prevent water from leaching from the bottom of the tube and
to allow aeration at the root zone. Each Deepot tube was then filled with
550 g of the appropriate Pb-spiked soil mixture (0, 1000, 2000 mg Pb/kg
dry soil) and 6 pre-soaked *Sesbania* seeds were planted and watered with
20 mL deionized distilled water.

11.2.2.1 PLANT ESTABLISHMENT AND MAINTENANCE

Plants were irrigated every 2 days with 20 mL of either deionized distilled water or with a modified [23] Hoagland's nutrient solution (pH 6.5) which contained the following nutrients in mM: NH_4NO_3, 5; K_2SO_4, 1.25; $CaCl_2.2H_2O$, 2.0; $MgSO_4.7H_2O$, 0.5; K_2HPO_4, 0.15; $CaSO_4.2H_2O$, 6.0; the following in µM: H_3BO_3, 2.3; $MnSO_4.H_2O$, 0.46; $ZnSO_4.7H_2O$, 0.6; $CuSO_4.5H_2O$, 0.15; $NaMoO_4.2H_2O$, 0.10; $2.6H_2O$, 10.0; and 20 mg/L Fe sequestrene. A 250- mL plastic cup was placed under each tube for leachate collection and to prevent cross contamination among treatments. Any leachate collected in each plastic cup was poured back into its corresponding tube. Periodically, these cups were rinsed with deionized distilled water and the resulting washing solutions were poured back into the respective tubes. The volume of water and/or nutrient solution ensured that soil moisture content was maintained at field capacity. Plants were maintained in a naturally-lit greenhouse throughout the experimental period. Emerged seedlings were thinned out to a desired population density of 2 plants per tube at 5 days after emergence.

Ethylenediaminetetraacetic acid (1:1 ratio with the Pb) was applied as 100 mL aqueous solution 6 weeks after emergence. One hundred mL aqueous HAc solutions were also added to some treatments immediately after EDTA applications. This experiment consisted of twelve treatments, arranged in a randomized complete block design (RCBD) with 2 plants per tube replicated 4 times for each harvesting period (0, 6, and 7 days after chelate amendment).

During harvest, shoots and roots were separated, and roots were washed with distilled water to remove any adhering debris, then oven-dried in a Thelco convection oven at approximately 75°C for at least 48 hours.

11.2.3 METAL EXTRACTION AND ANALYSES

Dried samples were weighed and ground in a Wiley mill equipped with a 425 µm (40-mesh) screen. Lead contents were extracted using EPA test method 3040B with slight modifications. Briefly, 40 ml of 50% aqueous nitric acid were added to a 250 mL Erlenmeyer flask containing a representative sample of ground plant tissue. The acidified sample was heated to 35°C, refluxed for 15 minutes without boiling, and then allowed to cool. Another 10 mL of 50% aqueous nitric acid were added and the sample was again heated and refluxed without boiling. To initiate the peroxide reaction, 2 mL of deionized distilled water and 3 mL of 30% hydrogen peroxide were added to the concentrated digestate and heated until effervescence subsided. Another 7 mL of 30% hydrogen peroxide were added continuously in 1 mL aliquots as the digestate was again heated. The digestate was heated until effervescence was minimal and its volume reduced to approximately 5 mL. After cooling, the final digestate was diluted to approximately 15 mL with deionized water. The digestate was filtered through a filter paper (Whatman No. 1) and the final volume was adjusted to 25 mL with deionized, distilled water. Lead concentrations were quantified using Inductively Coupled Plasma-Optical Emission Spectrometry [ICP-OES; Perkin Elmer Optima 3300 DV] and expressed as μg Pb/g dry weight of plant tissue.

11.3 STATISTICAL ANALYSES

The solubility test and chelate selection experiment consisted of 4 treatments replicated 3 times for each time period (0, 1, 3, 5, 6, and 7 days). The greenhouse experiment consisted of 2 plants per tube, replicated 4 times for each of 3 harvesting periods (0, 6, and 7 days after chelate amendment). Data were analyzed using Statistical Analysis System (SAS V9). Treatment comparisons were done using Fisher's Protected Least Significant Difference (LSD) test. A probability of less than 5% ($p < 0.05$) was considered to be statistically significant.

11.4 RESULTS AND DISCUSSION

For soil remediation initiatives, it is important to characterize both the chemical and physical parameters of the soil. Composition (e.g., nutrients, organic and inorganic materials), and soil mixture may all influence how the contaminant will behave. In general, in can be concluded that the chemistry of metal interaction with soil matrix is central to the phytoremediation concept. While the soil used in this study was high in phosphorus, potassium, and magnesium, these were nonetheless essential plant nutrients. Results of the soil analysis by Mississippi State University Soil Testing Laboratory are summarized in Table 1, and showed that the parameters of the soil used in this study were well within limits for our objectives.

TABLE 1: Physical and chemical characteristics of the soil

Characteristic	Extractable Levels (lbs/acre)
Soil Acidity (pH)	6.3
Phosphorus	130*
Potassium	301*
Calcium	4537
Magnesium	726**
Zinc	4.2*
Sodium	161
CEC	17.6
% Clay	7.50
% Silt	80.08
% Sand	12.4

High* Very High**

11.4.1 SOLUBILITY TEST AND CHELATE SELECTION

Results of ICP-OES analyses revealed that among the chelates tested (EDTA, EGTA, and HAc), EDTA was the most effective in solubilizing soil-bound Pb (Fig. 1). Lead concentrations in soil solution increased with extraction time and remained constant 6 to 7 days after chelate amendment.

After 6 days, Pb concentration in soil solution of the EDTA-treated soil was 10.53 µg Pb/mL, as compared to 1.057, 0.047, and 0.048 µg Pb/mL for EGTA, HAc, and H_2O, respectively. These findings are consistent with those reported by Shen and his colleagues [25], who found that

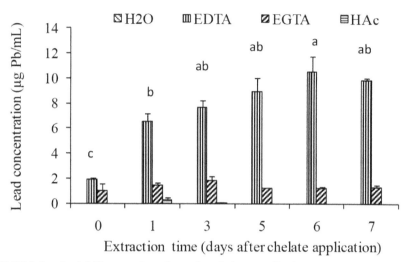

FIGURE 1: Lead solubility at various times after chelates application. Values and error bars represent means and standard errors of 3 replicates. Treatments with common letters do not differ significantly (Fisher's LSD, $p < 0.05$). Values not shown for a given extraction time indicates that Pb was not detected.

the Pb-concentration in soil solution of the EDTAtreated soil was 42-fold higher than that of the control soil, and that citric acid application to the soil produced only a small increase in the Pb concentration of soil solution and used by itself was much less effective than other chelates used.

We hypothesized that the efficacy of EDTA in solubilizing Pb from the soil may be related to the high binding capacity of EDTA for Pb as shown in previous studies [3, 20]. It may also be presumed that EGTA and HAc are more rapidly degraded than EDTA. Our aim in the solubility study was not only to utilize a Pb-specific chelate, but to also apply it in a time-efficient manner so that we could harvest the plant during its peak phytoextraction period. In a practical field application, this would reduce the likelihood of herbivores eating the contaminated plants, as well as limit the risk of water pollution due to chelates and/or chelate-metal complexes migrating from the soil.

11.4.2 GREENHOUSE EXPERIMENT

Lead-tissue concentration: Lead taken up by plants is usually increased with Pb concentrations in soil [26, 27]. Lead concentrations in root tissues were highest at 2000 mg Pb/kg, with EDTA alone or in combination with HAc

Also, Pb concentrations were higher for day 6 than for day 7. When no chelates were applied, shoot Pb concentrations slightly increased with increasing levels of soil-applied Pb. This could be due to Pb binding to ion exchangeable sites on the cell wall and extracellular deposition mainly in the form of Pb carbonates deposited on the cell wall as previously demonstrated [28]. Lead being a soft Lewis acid, forms a strong covalent bond

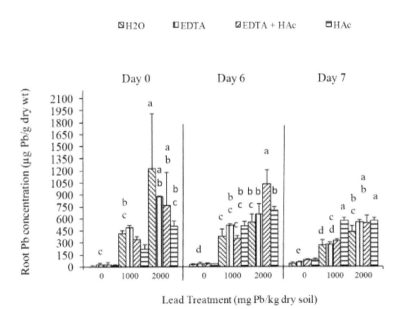

FIGURE 2: Effects of lead and chelates on root Pb concentrations (μg Pb/g dry wt) of Sesbania exaltata Raf. at different days (0,6, 7) after chelate application. Chelates were applied at 6 weeks after seedling emergence. Values and error bars represent means and standard errors of 4 replicates. Treatments with common letters do not differ significantly from other treatments within the same time period (Fisher's LSD p < 0.05).

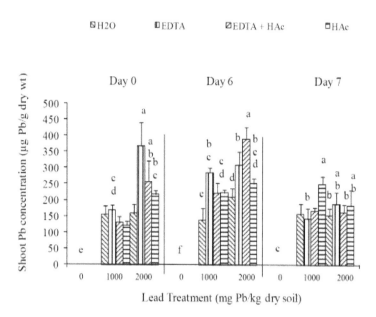

FIGURE 3: Effects of lead and chelates on shoot Pb concentrations (µg Pb/g dry wt.) of Sesbania exaltata Raf. at different days (0,6, 7) after chelate application. Chelates were applied at 6 weeks after seedling emergence. Values and error bars represent means and standard errors of 4 replicates. Treatments with common letters do not differ significantly from other treatments within the same time period (Fisher's LSD $p < 0.05$).

not only with the soil, but with plant tissues as well [20]. It is believed that since the xylem cell walls have a high cation exchange capacity, the upward movement of metal cations are severely retarded [10].

In the absence of chelates Pb concentration in shoots of coffeeweed plants grown at 1000 and 2000 mg Pb/kg were minimal (Fig. 3). With the addition of chelates alone or in combination with HAc, translocation significantly increased in day 0 and day 6. With the addition of chelates, it appeared that not only did the roots absorb more lead, but the metal was translocated to the shoots, facilitating some of the desirable characteristics of a hyperaccumulator, such as a high metal uptake by the roots, and translocation of the metal from the root to the above ground shoots. By day 7, however, there was not only a decline in the translocation of Pb to the shoots, but the decline was significantly lower in the 2000 mg Pb/kg treatments as compared to 1000 mg Pb/kg treatments.

TABLE 2: Effects of lead and chelates on root dry biomass of *Sesbania* exaltata Raf. at 0, 6, and 7 days after chelate application

Treatment		Root Dry Biomass (mg/plant)										
		Day 0				Day 6				Day 7			
Pb (mg/ kg)	EDTA	Mean	**	±	SEM	Mean	**	±	SEM	Mean	**	±	SEM
0	0	15.12	c	±	5.2	34.37	c	±	1.7	34.00	ab	±	1.9
0	1000	20.25	abc	±	6.9	37.50	bc	±	6.4	29.62	ab	±	3.2
0	1000*	19.87	abc	±	4.7	36.62	bc	±	7.1	31.37	ab	±	6.7
0	HAc	15.75	abc	±	5.4	34.50	c	±	5.8	27.75	b	±	7.8
1000	0	16.25	abc	±	2.0	36.12	bc	±	5.0	38.37	ab	±	6.7
1000	1000	17.87	abc	±	3.9	41.52	abc	±	8.5	26.50	ab	±	3.5
1000	1000*	24.12	abc	±	4.4	47.00	abc	±	1.9	40.50	ab	±	2.1
1000	HAc	23.62	abc	±	3.4	49.87	ab	±	3.3	41.25	ab	±	8.8
2000	0	29.75	a	±	3.6	51.00	ab	±	6.4	48.12	a	±	5.3
2000	2000	24.25	abc	±	3.7	42.37	abc	±	3.6	47.12	a	±	7.2
2000	2000*	21.87	abc	±	6.4	53.50	a	±	5.7	47.37	a	±	8.7
2000	HAc	28.87	a	±	3.1	45.87	abc	±	3.6	40.83	ab	±	11.1

*Aqueous solutions of EDTA and HAc were applied alone or in combination in a 1:1 ratio with [Pb(NO3)2] six weeks after seedling emergence. Plants were harvested at 0, 6, and 7 days after chelate application. *Indicates that HAc was added following EDTA amendment. SEM = standard error of the mean of 4 replications. **Means followed by a common letter are not significantly different from other treatments within the same harvesting period (p < 0.05).*

Our results are comparable with experiments by other investigators who have reported that bringing the Pb into solution with a chelating agent, not only makes more Pb bioavailable for root uptake [10, 20, 29] , but also moves the Pb that is sequestered in the xylem cell wall upwards and into the shoots. Blaylock [3] demonstrated with Indian mustard (*Brassica juncea* (L.), and Huang et al. [20] demonstrated with peas (*Pisum sativum L.*) and corn (*Zea mays L.*) that the addition of EDTA to Pbcontaminated soil increased the shoot Pb concentrations by 300-fold, 111-fold, and 57-fold, respectively. Transport across root cellular membrane is an important process which initiates metal absorption into plant tissues. Several studies

have shown that sequestration in root vacuole may prevent the translocation of some metals from root to shoot [30] whereas in hyperaccumulating
plants, the mechanism of vacuolar sequestration may be disabled, allowing metal translocation and hyperaccumulation in leaves [30]. Therefore,
it is generally agreed that the ability of plants to move the Pb upwards into
the shoots varies much more than their ability to accumulate metals in the
roots [29, 31].

11.4.2.1 BIOMASS

The capacity of plants to remove contaminants from the soil is a function
of biomass per unit area and concentration of the contaminant in the plants
[32]. Lead is not considered to be an essential element for plant growth
and development, rather Pb inhibits growth, reduces photosynthesis (by
inhibiting enzymes unique to photosynthesis), interferes with cell division
and respiration, reduces water absorption and transpiration, accelerates
abscission or defoliation and pigmentation, and reduces chlorophyll and
adenosine triphosphate (ATP) synthesis [33]. At maturity, metalenriched
aboveground biomass is harvested and a fraction of soil metal contamination removed [29, 34, 35]. Our results revealed that root biomass was
not significantly different across the treatments. However, roots that were
grown in 0 mg Pb/kg soil had the lowest biomass, followed by treatments
of 1000 mg Pb/kg soil. The highest root biomass was seen in treatments
of 2000 mg Pb/kg soil (Table 2). Shoot biomass followed the trend of root
biomass. For each harvesting period (0, 6, and 7 days) shoots that were
grown in the highest lead treatments (2000 mg Pb/kg soil) alone or in
combination with chelates had the highest biomass. It is known that metal
phytotoxicity causes stress to the plant resulting in a reduction in biomass
and eventual death (in some cases). However, we observed no discern-
ible phytotoxic symptoms in neither roots nor shoots. We concluded from
this study and from earlier studies [24] that *Sesbania* may be tolerant to
Pb-EDTA complex. Vassil et al. [21] demonstrated with Indian mustard
that EDTA appears to chelate Pb outside of the plant, and then the soluble
Pb-EDTA complex is transported through the plant, via the xylem, and
accumulates in the leaves. Further, they found that toxicity symptoms in

Indian mustard exposed to Pb and EDTA were strongly correlated with the presence of free protonated EDTA in solution.

Another explanation for the apparent metal tolerance seen in *Sesbania* may be the presence of natural metalbinding peptides. Cunningham and Ow [2] described the presence of specific high-affinity ligands as one of the metal-resistant mechanisms existing in some plants.

These ligands, known as phytochelatins and metallothioneins, are reported to make the metal less toxic to the plant [36, 37]. We are not certain whether this resistance mechanism may exist in *Sesbania*, however, corollary studies regarding the relationship between Pb uptake and phytochelatin synthesis in coffeeweed are being investigated in our laboratory. Results of this study indicated that EDTA can be applied to selected Pb-contaminated soils in a timeefficient manner so that plants can be harvested during their peak phytoextractive period, thereby limiting the likelihood of exposure to herbivores as well as reducing the risk of water pollution due to chelates and /or chelatemetal complexes migrating from the soil.

REFERENCES

1. Kumar, P. B. A. N.; Dushenkov, V.; Motto, H.; Raskin, I.: Phytoextraction: The use of plants to remove heavy metals from soils. Environ. Sci. Technol. 1995, 29, (5), 1232-1238.
2. Cunningham, S. D.; Ow, W. D. Promises and prospects of phytoremediation. Plant Physiology, 1996, 110, 715-719.
3. Blaylock, M. J.; Salt, D. E.; Dushenkov, S.; Zakharova, O.; Gussman, C.; Kapulnik, Y.; Ensley, B. D.; Raskin, I. Enhanced accumulation of Pb in Indian mustard by soil-applied chelating agents. Environ. Sci. Technol. 1997, 31, (3), 860-865.
4. Friedland, A. J.: Heavy metal tolerance in plants: Evolutionary aspects. In Shaw, A. J., Ed. CRC Press: Boca Raton, FL, 1990.
5. Glass, D. J.: Economic potential of phytoremediation. In Phytoremediation of toxic metals: Using plants to clean up the environment. Raskin, I.; Ensley, D. D., Eds. John Wiley & Sons, Inc.: New York, 2000; pp 1-14.
6. McGrath, S. P.; Sidoli, C. M. D.; Baker, A. J. M.; Reeves, R. D.: The potential for the use of metalaccumulating plants for the in situ decontamination of metal-polluted soils. In Integrated soil sediment research: A basis for proper protection, Eijascker, H. J. P.; Hamers, T., Eds. Academic Publ: Dordrecht, Netherlands, 1993; pp 673-677.
7. Raskin, I.; Kumar, P. B. A. N.; Dushenkov, V.; Salt, D. E.: Bioconcentration of heavy metals by plants. Current Opinions in Biotechnology 1994, 5, 285-290.

8. Chaney, R. L.; Malik, M.; Li, Y. M.; Brown, S. L.; Brewer, E. P.; Angle, J. S.; Baker, A. J. M.: Phytoremediation of soil metals. Curr. Opin. Biotechnol. 1997, 8, (279-284).
9. Salt, D. E.; Blaylock, M.; Kumar, P. B. A. N.; Sushenkov, V.; Ensley, B. D.; Chet, I.; Raskin, I.: Phytoremediation: A novel strategy for the removal of toxic metals from the environment using plants. Biotechnology 1995, 13, 468-475.
10. Salt, D. E.; Smith, R. D.; Raskin, I. Phytoremediation. Annu. Rev. Plant Physiol. Plant Mol. Biol. 1998, 49, 643-668.
11. Shen, Z. G.; Zhao, E. J.; McGrath, S. P. Uptake and transport of zinc in the hyperaccumulator Thlaspi caerulescens and the nonhyperaccumulator Thlaspi ochroleucum. Plant Cell Environment 1997, 20, 898- 906.
12. Keltjens, W. G.; van Beusichem, M. L. Phytochelatins as biomarkers for heavy metal toxicity in maize: Single metal effects of copper and cadmium. J. Plant Nutr. 1998, 21, (4), 635 - 648.
13. Lim, T. T.; Tay, J. H.; Wang, J. Y.: Chelating-agentenhanced heavy metal extraction from a contaminated acidic soil. J. Environ. Eng. 2004, 130, 59 - 66.
14. Ghosh, S.; Rhyne, C.: Influence of EDTA on Pb uptake in two weed species, *Sesbania* and Ipomoea, in hydroponic culture. J. Miss. Acad. Sci. 1999, 44, (1), 11.
15. McBride, M. B.: Environmental chemistry of soils. Oxford University Press: 1994.
16. Kinniburgh, D. G.; Jackson, M. L.; Syers, J. K. Adsorption of alkaline-earth, transition, and heavymetal cations by hydrous oxide gels of iron and aluminum. Soil Science Society American Journal 1976, 40, 796-799.
17. Ruby, M. V.; Schoof, R.; Brattin, W.; Goldade, M.; Post, G.; Harnois, M.; Mosby, D. E.; Casteel, S. W.; Berti, W.; Carpenter, M.; Edwards, D.; Cragin, D.; Chappell, W.: Advances in evaluating the oral bioavailability of inorganics in soil for use in human health risk assessment. Environ. Sci. Technol. 1999, 32, 3697-3705.
18. Grcman, H.; Vodnik, D.; Velikonja-Bolta, S.; Lestan, D.: Heavy metals in the environment: Ethylenediaminedissuccinate as a new chelate for environmentlaly safe enhanced lead phytoextraction. J. Environ. Qual. 2003, 32, 500-506.
19. Jorgensen, S. E.: Removal of heavy metals from compost and soil by ecotechnological methods. Ecol. Eng. 1993, 2, 89-100.
20. Huang, J. W.; Chen, J.; Berti, W. R.; Cunningham, S. D.: Phytoremediation of lead-contaminated soils: Role of synthetic chelates in lead phytoextraction. Environ. Sci. Technol. 1997, 31, (3), 800-805.
21. Vassil, A. D.; Kapulnik, Y.; Raskin, I.; Salt, D. E. The role of EDTA in lead transport and accumulation Int. J. Environ. Res. Public Health 2008, 5(5) 435 by Indian mustard. Plant Physiology 1998, 117, 447 - 453.
22. Wu, J.; Hsu, F. C.; Cunningham, S. D. Chelateassisted Pb phytoextraction: Pb availability uptake, and translocation constraints. Environ. Sci. Technol. 1999, 33, 1898-1905.
23. Begonia, G. B.; Begonia, M. F. T.; Miller, G. S.; Kambhampati, M.: Phytoremediation of metalcontaminated soils: Jackson State University research initiatives. In Metal ions in biology and medicine, Centeno, J. A.; Collery, P.; Vernet, G.; Finkelman, R. B.; Gibb, H.; Etienne, J. C., Eds. 2000; Vol. 6, pp 682-684.

24. Begonia, G. B.; Miller, G. S.; Begonia, M. F. T.; Burks, C. Chelate-enhanced phyto-extraction of leadcontaminated soils using coffeeweed (*Sesbania* exaltata raf.). Bull. Environ. Contam. Toxicol. 2002, 69, (5), 624-654. 37.

25. Shen, Z.G.; Li, X.D.; Wang, C.C.; Chen, H.M.; Chua, H.: Lead phytoextraction from contaminated soil with high-biomass plant species. J. Environ. Qual. 2002, 31, 1893-1900.

26. Turpeinen, R.; Salminen, J.; Kairesalo, T. Mobility and bioavailability of lead in contaminated boreal forest soil. Environ. Sci. Technol. 2000, 34, 5152- 5156.

27. Cao, X.; Ma, L. Q.; Chen, M.; Donald W. Hardison, J.; Harris, W. G. Weathering of lead bullets and their environmental effects at outdoor shooting ranges. J. Environ. Qual. 2003, 32, 526-534.

28. Dushenkov, V.; Kumar, P. B. A. N.; Motto, H.; Raskin, I.: Rhizofiltration: The use of plants to remove heavy metals from aqueous streams. Environ. Sci. Technol. 1995, 29, 1239 - 1245.

29. Lasat, M. M.: Phytoextraction of toxic metals: A review of biological mechanisms. J Environ Qual. 2002, 31, 109-120.

30. Lasat, M. M.; Baker, A. J. M.; Kochian, L. V.: Altered Zn compartmentation in the root symplasm and simulated Zn absorption into the leaf as mechanisms involved in Zn hyperaccumulation in Thlaspi caerulescens. Plant Physiology 1998, 118, 875 - 883.

31. Kayser, A.; Wenger, K.; Keller, A.; Attinger, W.; Felix, H. R.; Gupta, S. K.; Schulin, R. Enhancement of phytoextraction of Zn, Cd, and Cu from calcareous soil: The use of NTA and sulfur amendments. Environ. Sci. Technol. 2000, 34, 1778 - 1783.

32. Fuhrmann, M.; Lasat, M. M.; Ebbs, S. D.; Kochian, L. V.; Cornish, J. Plant and envi-ronmental interactions: Uptake of cesium-137 and strontium-90 from contaminated soil by three plant species; application to phytoremediation. J. Environ. Qual. 2002, 31, 904-909.

33. Toxnet Hazardous substances data bank. http://toxnet.nlm.nih.gov. Available 08-15-06.

34. Ebbs, S. D.; Lasat, M. M.; Brady, D. J.; Cornish, J.; Gordon, R.; Kochian, L. V.: Phytoextraction of cadmium and zinc from a contaminated soil. J. Environ. Qual. 1997, 26, 1424-1430.

35. Lombi, E.; Zhao, F. J.; Dunham, D. J.; McGrath, S. P. Phytoremediation of heavy metal-contaminated soils: Natural hyperaccumulation versus chemically enhanced phytoextraction. J. Environ. Qual. 2001, 30, 1919-1926.

36. Grill, E.; Winnacker, E. L.; Zenk, M. H.: Phytochelatins: The principal heavy-metal complexing peptides of higher plants. Science (Washington, D. C., 1883) 1985, 230, 674-676.

37. Grill, E.; Winnacker, E. L.; Zenk, M. H. Phytochelatins, a class of heavy-metal-binding peptides from plants, are functionally analogous to metallothioneins. Proc. Natl. Acad. Sci. USA. 1987, 84, 439 - 443.

CHAPTER 12

SUSTAINABLE SOURCES OF BIOMASS FOR BIOREMEDIATION OF HEAVY METALS IN WASTE WATER DERIVED FROM COAL-FIRED POWER GENERATION

RICHARD J. SAUNDERS, NICHOLAS A. PAUL, YI HU, AND ROCKY DE NYS

12.1 INTRODUCTION

The use of algae to remove pollutants from water, algal bioremediation, has been well studied over the past 40 years [1], [2], [3], [4]. Since the 1980s considerable research effort has been devoted to the development of algal biosorbents to remediate pollutants, particularly heavy metals [5]. At the laboratory scale these preparations have proven spectacularly successful at sorbing pollutants, especially heavy metals [5], [6]. However, uptake of the concept has been lack-lustre, evidenced by the lack of successful commercialisation (e.g. AlgaSORB circa 1991). This is likely because available algal (seaweed) biomass that is produced has established markets as food and as food ingredients (see Chopin and Sawhney [7] for market details). Furthermore, amongst the most successful preparations

This chapter was originally published under the Creative Commons Attribution License. Saunders RJ, Paul NA, Hu Y, and de Nys R. Sustainable Sources of Biomass for Bioremediation of Heavy Metals in Waste Water Derived from Coal-Fired Power Generation. PLoS ONE 7,5 (2012), doi:10.1371/journal.pone.0036470.

developed are those from brown macroalgae [8] which already have particularly well established markets and command a high price. A cheaper, reliable and locally derived source of biomass is critical [4], and remains a bottleneck for commercial applications of algae in bioremediation.

TABLE 1: Elemental composition (mg.L^{-1}) of the different water sources used in the growth and elemental uptake experiment. < indicates that concentration was below the limits of detection.

Element	Ash Dam	Ash Dam +f/2	Town Supply +f/2
Aluminum	0.08	0.06	<0.01
Arsenic	0.0175	0.017	<0.001
Boron	2.26	2.28	<0.05
Cadmium	0.0004	0.00035	<0.0001
Calcium	197.0	189.5	10.0
Chromium	<0.001	0.001	<0.001
Copper	0.004	0.0185	0.0215
Iron	0.275	1.55	0.7
Lead	<0.001	<0.001	<0.001
Magnesium	69.5	60.5	2.0
Manganese	0.002	0.0775	0.103
Mercury	<0.0001	<0.0001	<0.0001
Molybdenum	0.8595	0.9345	0.017
Nickel	0.016	0.026	<0.001
Phosphorous	<1.0	1.0	1.0
Potassium	30	31.5	5.5
Selenium	0.06	0.02	<0.01
Sodium	335.5	332	34.5
Strontium	1.365	1.43	0.05
Vanadium	0.565	0.6	<0.01
Zinc	0.231	0.3585	0.16

Bioconcentration, defined as the accumulation of a substance from the environment by the live algal biomass, offers an alternative approach to biosorption, defined as adsorption of metal ions on dead biomass [6]. We have used the term bioconcentration rather than bioaccumulation which is often associated with the process of trophic level transfer, and thus can be confused with biomagnification of pollutants [9], [10]. While there has been substantial research into algal biosorption, there has been remarkably little research devoted to algal bioconcentration for heavy metal bioremediation, however see Sternberg and Dorn [11]. The common justifications for researching algal biosorption are that the biomass is inexpensive [5] and has greater binding capacity than live biomass [12], [13]. However, biosorption approaches rely largely on specific binding of elements to active sites on cell walls [6] whereas bioconcentration may occur in numerous cellular structures or compartments, e.g. vacuolar accumulation of heavy metals [14] and can occur simultaneously for metals in different ionic states, cf. anionic or cationic arsenic: Ghimire et al. [15]. The key factors for bioconcentration to be successful are the ability of the algae to target numerous heavy metals [16], [17] and the capacity to grow and survive in the waste water stream. Thus, bioremediation with living biomass is a combination of both bioconcentration and biomass productivity, as high growth rates will provide new cellular material to bind and capture metals. The process is complicated when different growth states or age of algal tissue influence the selectivity and concentrations of specific metals [17]. In these cases factors that affect growth may also impact capacity for bioconcentration, making it essential to simultaneously quantify bioconcentration and algal growth in the relevant waste water stream.

Living macroalgae play an important role in pollution management [18]. The earliest uses of algae in pollution control were developed for sewage waste water where their uptake of N and P was harnessed [1]. The capacity for some algae for luxury uptake or bioconcentration of N and P has also been utilised for bioremediation of waste water from aquaculture (integrated aquaculture) [19]. More recently, the role of carbon in human induced climate change has resulted in a considerable focus on carbon capture and storage (CCS). One approach that has attracted considerable commercial interest is the culture of algal biomass to sequester or recycle carbon (BioCCSR) [20]. The success of these applications of algal bioremediation processes lies in the fact that the biomass is cultured in situ and does not require wild harvested biomass. In our study, we have applied this concept and considered the culture of algae in polluted water as a remediation strategy, circumventing the need to source biomass for remediation and

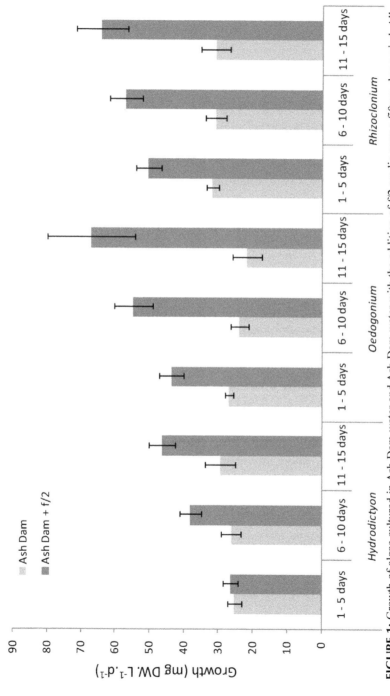

FIGURE 1: Growth of algae cultured in Ash Dam water and Ash Dam water with the addition of f/2 media over a fifteen day period. All three species of algae had higher growth rates with the addition of f/2 and there was also a significant influence of time on growth for all species. Error bars are standard error.

providing a continuous management strategy for heavy metal extraction. The use of macroalgae rather than microalgae also negates the difficulties associated with harvesting biomass. The culture of live biomass also has the additional value-adding potential for BioCCSR as well as by-product development [21]. Such by-products depend entirely on the waste water streams that are to be remediated.

Coal fired power stations produce large volumes of polluted waste water when the ash collected in the flue, and that remaining in the furnace after the combustion of coal, is washed out. The contaminants in this water vary depending on the source of the coal but commonly include high concentrations of As, V, Mo and Se [22], [23], [24]. This presents a significant problem for industry as the water is often contaminated to such a degree that it must be stored and/or treated, at considerable cost. Storage dams containing large volumes of contaminated water are often associated with coal fired power stations and these ash storages come with environmental and human health risks [24]. Previous research on the use of algae for bioremediation has usually focussed on only one or two problematic elements, particularly Cd [6], [25]. However, most waste water streams including water contained in Ash Dams are complex and contain numerous hazardous elements [23], [24]. Therefore, a broad approach that considers uptake across a great number of elements is optimal in terms of bioremediation potential. In this study we investigated the potential of freshwater green algae to bioconcentrate a wide variety of elements in water sourced from an Ash Dam associated with the Tarong coal fired power station in south-eastern Queensland. Furthermore, we develop baseline data to establish a model for the bioremediation of Ash Dam water, and metals removal and recovery.

12.2 METHODS

12.2.1 GENERAL

Three species of freshwater green macroalgae were utilised to investigate growth performance and to determine the bioconcentration factors for

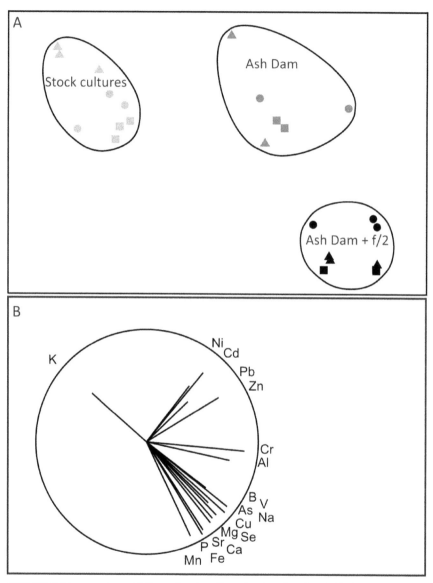

FIGURE 2. Non metric multidimensional scaling plot showing the similarity between algal species and treatments based on elemental composition. (A) nMDS plot (Stress = 0.05) with the groups from the cluster analysis superimposed. Triangles represent Rhizoclonium, circles represent Oedongonium and squares represent Hydrodictyon. (B) The same nMDS as A, with vectors superimposed, the length and direction of which indicates the strength of the correlation and direction of change between the two nMDS axes. Only elements with a correlation coefficient of 0.5 or greater are shown.

a variety of elements when grown in Ash Dam water sourced from the Tarong coal fired power station in south-eastern Queensland. The Tarong Power Station has a total generating capacity of 1400 megawatts and is amongst the largest power stations in Queensland, Australia.

TABLE 2: Elements grouped according to relative concentrations across all treatments and control (above).

Relative concentration across species between treatments and control	Elements
Stock culture < Ash Dam < Ash Dam + f/2	As, B, Ca, Cu, Mg, Se, Sr, V
(Stock culture ≈ Ash Dam +f/2) < Ash Dam	Cd, Ni, Pb, Zn
(Stock Culture ≈ Ash Dam) < Ash Dam + f/2	Fe, P
(Stock Culture ≈ Asch DAm) > Ash Dam + f/2	K
Stock culture < (Ash Dam ≈ Ash Dam + f/2)	Cr
Variation in pattern between species and treatment	Al, Mo, Na, Mn
Relative bioconcentration factors across species between treatments	Elements
Ash Dam < Ash Dam + f/2	As, B, Ca, Mg, P, Se, Sr, V
Ash Dam + f/2 < Ash Dam	Cd, Mn, Ni, Zn
Variation in pattern between species and treatment	Al, Cr, Cu, Fe, K, Mo

12.2.2 ALGAE COLLECTIONS

The three species of green macroalgae were collected from aquaculture ponds and irrigation channels in Queensland. These were *Hydrodictyon sp., Oedogonium sp.,* and *Rhizoclonium sp.* Species were identified to genus level using taxonomic keys [26] as each lacked defining characteristics to allow for identification to species level. The three freshwater algae are cosmopolitan genera from freshwater systems and are therefore representative of the macroalgae available in many freshwater environments. Furthermore, all can have rapid growth, particularly under eutrophic conditions, and are pest species in these environments [27], [28], [29], [30], [31], [32]. These algae range from the unbranched filamentous *Oedogonium* (cell diameter >2 μm) and *Rhizoclonium* (cell diameter >10–50 μm)

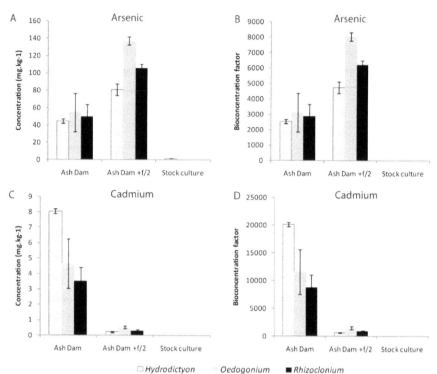

FIGURE 3. Examples of patterns of concentration and bioconcentration of metals in algae when cultured in Ash Dam water, Ash Dam water with f/2 and the stock cultures. (A) Concentration of arsenic. (B). Bioconcentration of arsenic. (C) Concentration of cadmium. (D) Bioconcentration of cadmium.

to the net-forming *Hydrodictyon* (water net). Stock cultures of all algae were maintained in standard f/2 media [33]in the aquaculture facility at James Cook University prior to the experimental testing of growth and elemental uptake.

12.2.3 CULTURE METHODS

The three algal species, *Hydrodictyon sp., Oedogonium sp.* and *Rhizoclonium sp.*, were cultured in two treatments. The first was neat Ash Dam water, to determine growth potential without nutrient supplementation. The second treatment was an f/2 medium [33] in which freshwater was substituted for Ash Dam water (Ash Dam +f/2). This treatment was to determine growth and remediation potential of the algae in Ash Dam water under conditions where N, P and essential trace elements were not limiting. Ash dam water, with and without f/2 nutrients, had an initial pH of 7.0. The algae were cultured in 1.0 L Schott bottles at a stocking density of 1.0 g.L−1. The Schott bottles were placed randomly in a Sanyo Versatile Environmental Test Chamber (MLR-351). Mean light levels within the cabinet were 88 µmol photons. m^{-2}.s^{-1} with a photoperiod of 12L: 12D and temperature was maintained at a constant 24°C. A complete water change was done at 10:00 am each day and the Schott bottles were rotated within the cabinet each day to avoid light bias. Every five days for a total of 15 days the replicates were dried using paper towel and weighed to the nearest 0.1 g. A stocking density of 1.0 g.L^{-1} was re-established at this time by increasing the volume of water. At the end of the 15 day culture period, the algae were harvested, patted dry on paper towel and then dried in a dehydrator for 48 hours at 45°C. The dry biomass was then weighed to the nearest 0.01 g. Dry matter content (DM) was calculated for each species and treatment at the end of the experiment (DM = dry weight/wet weight). All the dried biomass was stored in snap lock plastic bags at 4.0°C until it was analysed for elemental composition.

12.2.4 GROWTH

Growth rates for each combination of treatment (Ash Dam and Ash Dam +f/2) and species were calculated using the fresh weight determined for three consecutive growth periods at day 5, 10 and 15 of the experiment. Growth rate (GR) was calculated using the equation GR = (Mf−Mi)/15*(DM) where, Mf = fresh mass at day 15 and Mi = initial mass. Mean growth rate (mg DW.L^{-1}.d^{-1}) are presented for each species x treatment combination (±1 standard error, from the three growth periods).

12.2.5 ELEMENTAL ANALYSIS (ALGAE)

The concentrations of 21 different elements, listed in Table 1, were determined for the algae grown in the two treatments (Ash Dam water and Ash

Dam water with f/2) and unexposed biomass from the original stock that was maintained in dechlorinated water with f/2 media, henceforth referred to as stock cultures. All biomass was prepared for the analysis by drying in a dehydrator for 48 hours at 45°C. A minimum of 100 mg dry weight of algae was required for accurate determination of the elemental composition (see below). Three replicates were available in the majority of cases but in a few samples were pooled to provide duplicates.

For the elemental analyses, 100 mg samples of the dried algae were placed into digestion vessels with 2.5 mL SupraPure (Merck Germany) double distilled HNO_3 and 1.0 mL AR Grade H_2O_2. The mixture was left to stand in the fume-hood for two hours to allow the reaction to complete. The vessels were then heated to 180°C in a microwave oven (Milestone Starter D) and maintained at this temperature for ten minutes. After cooling to room temperature, the digested samples were diluted to 100 mL with Milli-Q water in a volumetric flask. No further dilution was needed before elemental analysis.

Sample analysis was carried out using two instruments. Major elements (Al, Ca, K, Na and P) were measured using a Varian Liberty Series II Inductively Coupled Plasma Optical Emission Spectrometer (Melbourne, Australia). The remaining elements were measured using a Varian 820-MS Inductively Coupled Plasma Mass Spectrometer (ICP-MS) (Melbourne, Australia). External calibration strategy was used for both instruments with a series of multi-element standard solution containing all the elements of interest and the results were reported after subtracting the procedure blanks. Algae may be subject to Cl- polyatomic ion interference, thus elements such as V, As, Se are susceptible to false positives. To assess this, one algal sample was spiked with 1 ppb As, Se and V and measured three times for quality control with recovery between 102 and 108% indicating no significant interferences. These analyses were done by the Advanced Analytical Centre (AAC) at James Cook University (JCU).

12.2.6 ELEMENTAL ANALYSIS (WATER)

The three water sources in which the algae were cultured were analysed for the concentration of the same 21 elements as the algae (Table 1). These were Ash Dam water, the Ash Dam water with the addition of f/2 nutrients and stock culture water. The stock culture water was Townsville city supply that had been dechlorinated using a charcoal filter and supplemented with f/2 nutrients. Two replicate water samples of 200 mL each were taken from these three separate water sources. The samples were collected using a 200 mL syringe and passed through a Minisart 0.45 μm filter to remove particulates. The elemental measurements were done according to the USEPA 6020 ICP-MS standard following an acid digest. These measurements were done by the Australian Centre for Tropical Freshwater Research (ACTFR).

12.2.7 BIOCONCENTRATION FACTOR

Bioconcentration factor (BCF) is the ratio of the chemical concentration in the organism to the water [34]. BCF was calculated using the equation $BCF = Cb/Cw$ where, Cb = concentration of elements in the dry algal biomass $(mg.kg{-}1)$ and Cw = concentration of elements in the water $(mg. L{-}1)$.

12.2.8 STATISTICAL ANALYSIS

Growth rates were compared between treatment (Ash Dam and Ash Dam +f/2) and species (n = 3) by analysis of variance (ANOVA) with time (three, five day growth periods) as a blocked (random) factor in the model. Growth rate was log transformed to meet the assumptions of ANOVA. Mean Square (MS) error terms in the mixed model ANOVA were adjusted for calculation of F-ratios for treatment (MS treatment x time), species (MS species x time) and treatment x species (MS treatment x species x time). Multivariate statistics were used to determine if there were differences in elemental composition between species and algal biomass cultured in neat Ash Dam water, Ash Dam water +f/2 and stock culture biomass that was never exposed to ash dam water. A similarity matrix was calculated from the 4th root transformed concentrations of all the different elements and a hierarchical agglomerative cluster analysis was done and superimposed on an nMDS. The 2D plots of the similarity matrix illustrate the clustering of the different treatments and show the direction and strength of change in the elemental composition of the algae.

12.3 RESULTS AND DISCUSSION

12.3.1 WATER ANALYSIS

The elemental composition of the Ash Dam water was complex and contained several heavy metals, such as V and As, at high concentrations (Table 1). The addition of f/2 nutrients to the ash water marginally increased the concentration of the essential elements Cu, Mn, Mo, Sr and Zn. The addition of f/2 nutrients increased Fe concentration by nearly 1.0 mg.L^{-1}. The majority of heavy metals were undetectable in the stock culture water (dechlorinated town supply with f/2), although the f/2 nutrients provided low concentrations of some essential elements such as Fe, Cu, Sr, Mo, Mn and Zn (Table 1). The growth medium f/2 also supplied some nutrient detectable by the ICP- MS analysis as P.

12.3.2 GROWTH AND METAL UPTAKE

All three algae grew in the neat Ash Dam water and in Ash Dam water +f/2 nutrients. *Oedogonium* and *Rhizoclonium* had substantially higher growth rates with the addition of f/2 whereas *Hydrodictyon* grew marginally more with the addition of f/2 (Figure 1; ANOVA, species x treatment $F_{2,4}$ = 20.04, p = 0.008). There was a subtle but significant influence of time on growth for all species with and without nutrient-addition, with a trend for increased growth in Ash Dam with f/2 over the three consecutive growth periods (Figure 1; ANOVA, treatment x time $F_{2,54}$ = 7.91, p = 0.001). Growth was stable in the Ash Dam water from the first week 27.98 mg DW.L^{-1}.d^{-1} (±1.16 SE) through to the third week 27.37 mg DW.L^{-1}.d^{-1} (±2.29 SE). By contrast each species increased growth rate over the same period by ~50% with the addition of f/2 to the Ash Dam water, on average from 40.12 mg $DW.L^{-1}.d^{-1}$ (±3.36 SE) to 59.14 mg $DW.L^{-1}.d^{-1}$ (± 4.84 SE). The addition of f/2 nutrients aided in the acclimatisation of the algae, evidenced by their increasing growth rate over the three week period of the experiment compared to the neat Ash Dam water, in which growth was stable over the period (Figure 1). Growth rates in the treatments were broadly comparable to those for all three species across a range of environments [27], [28], [29], [30], [31], [35].

The stock cultures had the lowest concentration of all metals. The pattern of concentration of elements was more similar between algae, than across treatments, indicating that the water in which the algae were cultured significantly influenced their elemental composition. The multivariate analysis indicated that the stock cultures were uniformly low in heavy metal concentration compared to the Ash Dam and Ash Dam +f/2 treatments and that the addition of f/2 had a strong impact on the elemental composition of the algae cultured in the Ash Dam water (Figure 2).

Many elements showed consistent patterns of concentration across species and treatments. These could be categorised into several groups (Table 2). The most common pattern among treatments and species was that the concentration in the algal biomass of the element was Stock cultures<Ash Dam<Ash Dam+f/2. This pattern held true for the majority of elements including As, B, Ca, Cu, Mg, Se, Sr and V (see example of As, Figure 3a,b). The addition of f/2 media provided better conditions for uptake of these elements, presumably through increased metabolic rate. This group was identified when the vectors indicating the strength and direction of correlation on the nMDS plot were considered. These vectors indicated a greater concentration of this group of elements in the biomass cultured in the Ash Dam+f/2 water over that cultured in neat Ash Dam water (Figure 2a,b).

The second most common pattern, that occurred was [Stock cultures≈ Ash Dam +f2] <Ash Dam for Cd, Ni, Pb and Zn (see example of Cd, Figure 3c,d). These elements appear to be excluded when growth is increased by the addition of f/2 media. This group of elements was also defined clearly in the nMDS where the vectors for Cd, Ni, Pb and Zn indicated a similar strength and direction of change (Figure 2b). Thus, to maximise bioremediation of these elements it would be important to choose a compromise between algal growth promoted with f/2 media, and

bioconcentration, which is substantially higher under sub-optimal growth conditions. Elements that were at very low concentrations in the water remained so in the algae. For example, Hg was undetectable in the water (Table 1) and was also undetectable in the algae (Table 2). However, Pb, which was undetectable in any of the water sources was found, albeit at low concentrations (<2 mg.kg^{-1}), in the algae.

As bioconcentration factor is derived from the elemental composition of the algae it follows similar patterns. Bioconcentration was exhibited for all elements in at least one of the treatments and species with the exception of Na, which was always near equilibrium between the algal biomass and the water. BCF could not be calculated for Al, As, B, Cd, Ni, Se and V in the stock cultures because the concentration in the town water with f/2 nutrients was below detection limits (Table 1). However, none of these elements were concentrated to any significant extent in the stock cultures. BCF could not be calculated for any algae for Hg or Pb because these were undetectable in any of the water sources. Given that the detection limits for Pb in the water were at least 0.001 mg.L^{-1} and concentration in the algae was over 1.0 mg.kg^{-1}, a BCF of at least 1000 can be inferred. Extremely high bioconcentration factors of over 10,000 were observed for many elements including Al, Cu, Mn, Ni, P and Zn. BCF for each alga was highest in the stock cultures treatments for Sr, Mg and K. The same was true for Mo for all algae except Oedogonium which appeared to exclude, or at least limit uptake of Mo substantially. Considerable variation in BCF was exhibited between species and treatments for the elements Al, Fe, Cu and Ca.

BCF was consistently highest for As, B, P, Se and V in the algae grown in Ash Dam water with f/2 media (Table 2; example for As Figure 3a,b). All algae also exhibited the fastest growth in this treatment, indicating that the rate of uptake of these elements is increased when growth rate increases. BCF was highest for Cd, Mn, Ni and Zn in all the algae grown in Ash Dam water without f/2 media (example for Cd Figure 3c,d).

In this study we have not distinguished between adsorption and absorption but rather concentrated on the total metal concentration of the biomass as we were primarily concerned with total remediation potential of the biomass. Overall, growth of algae, either with, or without additional nutrients (f/2), demonstrated bioremediation of a very complex waste water stream. While the f/2 media enhanced algal growth it also resulted in markedly different but consistent patterns of bioconcentration across groups of elements. Two key elements in the Ash Dam water, As and V, were bioconcentrated to a higher level when additional nutrients (f/2) were provided. For these elements, the optimised bioremediation strategy would be to target optimised biomass productivities, e.g. through the supply of limiting nutrients. However, increased growth resulted in a concomitant reduction in the bioconcentration of Zn, Ni and Cd. This provides opportunities to tailor remediation strategies within targeted waste streams by controlling growth, through the provision of nitrogen in particular. Importantly, there is a trade off between growth and bioconcentration. If growth is sufficiently high, then there will be more biomass available for bioconcentration and thus the reduced rate of elemental uptake of Zn, Ni and Cd will have a negligible impact on overall bioremediation. These opposing patterns of bioconcentration could result from reactions with the f/2 media that force these specific elements into a form that is not readily bound to the algae, similar to other changes in water medium that influence ionic state and uptake [36]. These have not been measured. Alternatively, the altered metabolic state of the algae could impact the binding mechanisms on and within the cells [17]. Some of the observed differences may then be explained by changes in bioavailability [37], [38]. Future research should consider ways to provide nutrient that maximises metal bioavailability as well as growth.

12.3.3 COMPARISON TO BIOSORPTION

The majority of algal biosorption studies report metal binding capacity in the range of 10 to 100 mg.g^{-1} [5], [6], although some report even higher levels of biosorption, achieved by pretreatment of the biomass and manipulation of pH of the waste water [5]. In our study, individual metal concentration in the biomass rarely exceeded 5 mg.g^{-1} but total heavy metal load was comparable to the majority of biosorption studies, reaching over 60 mg.g^{-1} total heavy metals for all species cultured with f/2 medium. The high concentrations achieved in our study may be the result of the live algae bioconcentrating within the cell vacuole as well as binding to the cell walls. This provides additional sites for elemental storage in live algae

[14], indeed, metals can even be bound in free sugars e.g. arseno-sugars for arsenic [39]. This, coupled with the ability of live biomass to grow and provide new substrate continuously, may make up for any short-fall in uptake capacity. Bioremediation may also be enhanced by manipulation of growing conditions of the algae. Integrating culture with other waste streams, such as municipal waste to provide N and P, and control of pH through the addition of CO_2, from flue gas from the power station, could yield higher rates of bioremediation through increased biomass productivity in the supply of the growth limiting nutrient carbon e.g. Israel et al. [40]. Furthermore, the control of waste water pH (by manipulating the dissolved CO_2 from coal-fired power station) offers additional opportunities to target bioconcentration of specific elements [36] or ionic states [41].

The justification for biosorption using algal biomass for heavy metal remediation often relies on costs, notably without comprehensive life cycle analyses that incorporate costs of processing and/or transport of biomass to and from polluted sites [5]. Using the elemental concentrations of problematic elements in the algae in this study (*Oedogonium* with V concentration at 1543 mg.kg^{-1} and *Rhizoclonium* with As concentration at 105 mg.kg^{-1}) and realistic algal biomass yields of 20 g DW.m^{-2}.d^{-1}, which equates to 73 tonne.ha^{-1}.annum^{-1} [42], a 100 ha culture area could remove one tonne of As per annum, providing significant removal of an environmentally sensitive element. In addition, such a culture would remove nearly 11 tonnes of V per annum as well as substantial removal of most other metals at the same time. However, estimating the costs of large scale algal culture is difficult [43] and estimates will vary widely by region. While algal culture may not be a panacea for heavy metal pollution, consideration of complementary remediation strategies where regional alliances between organic waste producers (N and P) and chemical waste producers (metals) are developed [5] will be an important step in developing practical, cost effective algal bioremediation.

REFERENCES

1.   Ryther JH, Tenore KR, Dunstan WM, Huguenin JE (1972) Controlled eutrophication-increasing food production from sea by recycling human wastes. Bioscience 22: 144–&.

2. Kuyucak N, Volesky B (1988) Biosorbents for recovery of metals from industrial solutions. Biotechnology Letters 10: 137–142. doi: 10.1007/BF01024641.

3. Romero-Gonzalez ME, Williams CJ, Gardiner PHE (2001) Study of the mechanisms of cadmium biosorption by dealginated seaweed waste. Environmental Science & Technology 35: 3025–3030. doi: 10.1021/es991133r.

4. Fu F, Wang Q (2011) Removal of heavy metal ions from wastewaters: A review. Journal of Environmental Management 92: 407–418. doi: 10.1016/j.jenvman.2010.11.011.

5. Hubbe MA, Hasan SH, Ducoste JJ (2011) Cellulosic substrates for removal of pollutants from aqueous systems: A review. 1. Metals Bioresources 6: 2161-U2914.

6. Mehta SK, Gaur JP (2005) Use of algae for removing heavy metal ions from wastewater: Progress and prospects. Critical Reviews in Biotechnology 25: 113–152. doi: 10.1080/07388550500248571.

7. Chopin T, Sawhney M (2009) Seaweeds and their mariculture. In: Steele JH, Thorpe SA, Turekian KK, editors. The Encyclopedia of Ocean Sciences. Oxford: Elsevier. pp. 4477–4487.

8. Davis TA, Volesky B, Mucci A (2003) A review of the biochemistry of heavy metal biosorption by brown algae. Water Research 37: 4311–4330. doi: 10.1016/S0043-1354(03)00293-8.

9. Senn DB, Chesney EJ, Blum JD, Bank MS, Maage A, et al. (2010) Stable Isotope (N, C, Hg) Study of Methylmercury Sources and Trophic Transfer in the Northern Gulf of Mexico. Environmental Science & Technology 44: 1630–1637. doi: 10.1021/es902361j.

10. Mason RP, Reinfelder JR, Morel FMM (1995) Bioaccumulation of mercury and methylmercury. Water, Air, & Soil Pollution 80: 915–921. doi: 10.1007/BF01189744.

11. Sternberg SPK, Dorn RW (2002) Cadmium removal using Cladophora in batch, semi-batch and flow reactors. Bioresource Technology 81: 249–255. doi: 10.1016/S0960-8524(01)00131-6.

12. Abu Al-Rub FA, El-Naas MH, Benyahia F, Ashour I (2004) Biosorption of nickel on blank alginate beads, free and immobilized algal cells. Process Biochemistry 39: 1767–1773. doi: 10.1016/j.procbio.2003.08.002.

13. Volesky B, Holan ZR (1995) Biosorption of heavy-metals. Biotechnology Progress 11: 235–250. doi: 10.1021/bp00033a001.

14. Volland S, Andosch A, Milla M, Stöger B, Lütz C, et al. (2011) Intracellular metal compartmentalization in the green algal model system Microsterias denticulata (Streptophyta) measured by transmission electron microscopy-coupled electron energy loss spectroscopy. Journal of Phycology 47: 565–579. doi: 10.1111/j.1529-8817.2011.00988.x.

15. Ghimire KN, Inoue K, Ohto K, Hayashida T (2008) Adsorption study of metal ions onto crosslinked seaweed Laminaria japonica. Bioresource Technology 99: 32–37. doi: 10.1016/j.biortech.2006.11.057.

16. Hedouin L, Bustamante P, Fichez R, Warnau M (2008) The tropical brown alga Lobophora variegata as a bioindicator of mining contamination in the New Caledonia lagoon: A field transplantation study. Marine Environmental Research 66: 438–444. doi: 10.1016/j.marenvres.2008.07.005.

17. Stengel DB, McGrath H, Morrison LJ (2005) Tissue Cu, Fe and Mn concentrations in different-aged and different functional thallus regions of three brown algae from western Ireland. Estuarine, Coastal and Shelf Science 65: 687–696. doi: 10.1016/j.ecss.2005.07.003.

18. Fei XG (2004) Solving the coastal eutrophication problem by large scale seaweed cultivation. Hydrobiologia 512: 145–151. doi: 10.1023/B:HYDR.0000020320.68331.ce.

19. Chopin T, Robinson S (2004) Defining the appropriate regulatory and policy framework for the development of integrated multi-tropihic aquaculture practices: introduction to the workshop and positioning of the issues. Bulletin of the Aquaculture Association of Canada 104: 4–10.

20. Tang Q, Zhang J, Fang J (2011) Shellfish and seaweed mariculture increase atmospheric CO2 absorption by coastal ecosystems. Marine Ecology Progress Series 424: doi: 10.3354/meps08979.

21. Bird MI, Wurster CM, de Paula Silva PH, Bass AM, de Nys R (2011) Algal biochar-production and properties. Bioresource Technology In Press, Corrected Proof.

22. Alberts JJ, Newman MC, Evans DW (1985) Seasonal-variations of trace-elements in dissolved and suspended loads for coal ash ponds and pond effluents. Water Air and Soil Pollution 26: 111–128. doi: 10.1007/BF00292062.

23. Cornelis G, Johnson CA, Van Gerven T, Vandecasteele C (2008) Leaching mechanisms of oxyanionic metalloid and metal species in alkaline solid wastes: A review. Applied Geochemistry 23: 955–976. doi: 10.1016/j.apgeochem.2008.02.001.

24. Ruhl L, Vengosh A, Dwyer GS, Hsu-Kim H, Deonarine A, et al. (2009) Survey of the Potential Environmental and Health Impacts in the Immediate Aftermath of the Coal Ash Spill in Kingston, Tennessee. Environmental Science & Technology 43: 6326–6333. doi: 10.1021/es900714p.

25. Ahluwalia SS, Goyal D (2007) Microbial and plant derived biomass for removal of heavy metals from wastewater. Bioresource Technology 98: 2243–2257. doi: 10.1016/j.biortech.2005.12.006.

26. Yee N, Entwisle TJ (2011) ALGAKEY - Interactive Identification of Australian Freshwater Algae, Version # 1:

27. Davis LS, Hoffmann JP, Cook PW (1990) Seasonal succession of algal periphyton from a waste-water treatment facility. Journal of Phycology 26: 611–617. doi: 10.1111/j.0022-3646.1990.00611.x.

28. Francke JA, Denoude PJ (1983) Growth of Stigeoclonium and Oedogonium species in artificial ammonium-N and phosphate-P gradients Aquatic Botany 15: 375–380. doi: 10.1016/0304-3770(83)90005-0.

29. Hawes I, Smith R (1993) Influence of environmental factors on the growth in culture of a New Zealand strain of the fast spreading alga Hydrodictyon reticulatum (water-net). Journal of Applied Phycology 5: 437–445. doi: 10.1007/BF02182736.

30. Kebede-Westhead E, Pizarro C, Mulbry WW (2006) Treatment of swine manure effluent using freshwater algae: Production, nutrient recovery, and elemental composition of algal biomass at four effluent loading rates. Journal of Applied Phycology 18: 41–46. doi: 10.1007/s10811-005-9012-8.

31. Mulbry WW, Wilkie AC (2001) Growth of benthic freshwater algae on dairy manures. Journal of Applied Phycology 13: 301–306. doi: 10.1023/a:1017545116317.

32. Wells RDS, Hall JA, Clayton JS, Champion PD, Payne GW, et al. (1999) The rise and fall of water net (Hydrodictyon reticulatum) in New Zealand. Journal of Aquatic Plant Management 37: 49–55.

33. Guillard RRL, Ryther JH (1962) Studies of marine plank-tonic diatoms. I. Cyclotella nana Hustedt, and Detonula confervacea (Cleve) Gran. Canadian Journal of Microbiology 8: 229–239. doi: 10.1139/m62-029.

34. DeForest DK, Brix KV, Adams WJ (2007) Assessing metal bioaccumulation in aquatic environments: The inverse relationship between bioaccumulation factors, trophic transfer factors and exposure concentration. Aquatic Toxicology 84: 236–246. doi: 10.1016/j.aquatox.2007.02.022.

35. Jasonsmith JF, Maher W, Roach AC, Krikowa F (2008) Selenium bioaccumulation and biomagnification in Lake Wallace, New South Wales, Australia. Marine and Freshwater Research 59: 1048–1060. doi: 10.1071/MF08197.

36. Boullemant A, Lavoie M, Fortin C, Campbell PGC (2009) Uptake of Hydrophobic Metal Complexes by Three Freshwater Algae: Unexpected Influence of pH. Environmental Science & Technology 43: 3308–3314. doi: 10.1021/es802832u.

37. Batley GE, Apte SC, Stauber JL (2004) Speciation and bioavailability of trace metals in water: Progress since 1982. Australian Journal of Chemistry 57: 903–919. doi: 10.1071/CH04095.

38. Slaveykova VI, Wilkinson KJ (2005) Predicting the bioavailability of metals and metal complexes: Critical review of the biotic ligand model. Environmental Chemistry 2: 9–24. doi: 10.1071/EN04076.

39. Geng W, Komine R, Ohta T, Nakajima T, Takanashi H, et al. (2009) Arsenic speciation in marine product samples: Comparison of extraction-HPLC method and digestion-cryogenic trap method. Talanta 79: 369–375. doi: 10.1016/j.talanta.2009.03.067.

40. Israel A, Gavrieli J, Glazer A, Friedlander M (2005) Utilization of flue gas from a power plant for tank cultivation of the red seaweed Gracilaria cornea. Aquaculture 249: 311–316. doi: 10.1016/j.aquaculture.2005.04.058.

41. Murphy V, Hughes H, McLoughlin P (2008) Comparative study of chromium biosorption by red, green and brown seaweed biomass. Chemosphere 70: 1128–1134. doi: 10.1016/j.chemosphere.2007.08.015.

42. de Silva PH, McBride S, de Nys R, Paul NA (2008) Integrating filamentous 'green tide' algae into tropical pond-based aquaculture. Aquaculture 284: 74–80. doi: 10.1016/j.aquaculture.2008.07.035.

43. Stephens E, Ross IL, King Z, Mussgnug JH, Kruse O, et al. (2010) An economic and technical evaluation of microalgal biofuels. Nature Biotechnology 28: 126–128. doi: 10.1038/nbt0210-126.

There is a table missing from this version of the article. To see it, as well as other supplemental information, please visit the original version of the article as cited on the first page of this chapter.

CHAPTER 13

CHARACTERIZATION OF THE METABOLICALLY MODIFIED HEAVY METAL-RESISTANT *Cupriavidus metallidurans* STRAIN MSR33 GENERATED FOR MERCURY BIOREMEDIATION

LUIS A. ROJAS, CAROLINA YÁCEZ, MYRIAM GONZÁLEZ, SOLEDAD LOBOS, KORNELIA SMALLA, AND MICHAEL SEEGER

13.1 INTRODUCTION

Environmental decontamination of polluted sites is one of the main challenges for sustainable development. Bioremediation is an attractive technology for the clean-up of polluted waters and soils [1]–[6]. Mercury is one of the most toxic elements in the environment [7], [8]. Metal mining, fossil combustion and the chloralkali and acetaldehyde industries have raised mercury levels in water bodies and soils. Mercury enters from industrial sources mainly as Hg^{2+} into the environment [9]–[11]. Physicochemical and biological processes have been applied for mercury removal from contaminated environments. Physicochemical processes for heavy

This chapter was originally published under the Creative Commons Attribution License. Rojas LA, Yáñez C, González M, Lobos S, Smalla K, and Seeger M. Characterization of the Metabolically Modified Heavy Metal-Resistant Cupriavidus metallidurans *Strain MSR33 Generated for Mercury Bioremediation.* PLoS ONE *6,3 (2011), doi:10.1371/journal.pone.0017555.*

metal removal such as ion exchange and precipitation treatment procedures result in large volumes of mercury-contaminated sludge and are of high cost [12]–[13]. As an alternative to physicochemical processes, bacteria have been applied for the remediation of mercury pollution [1], [3], [10], [14]. The biological processes for mercury removal are of low cost, simple and environmentally friendly [10]. Mercury-polluted sites are often contaminated with other heavy metals [15]. Therefore, bacteria with resistance to several heavy metals may be useful for remediation.

In bacteria, mercury resistance mer genes are generally organized in operons located on plasmids and transposons [16]–[17]. The narrow-spectrum mercury resistance merRTPADE operon confers resistance to inorganic mercury and the broad-spectrum mercury resistance merRTPAGBDE operon confers resistance to inorganic and organic mercury species (Fig. 1). The bacterial mechanism of mercury resistance includes the uptake and transport of Hg^{2+} by the periplasmic protein MerP and the inner membrane protein MerT. MerE is a membrane protein that probably acts as a broad mercury transporter mediating the transport of both methylmercury and Hg^{2+} [18]. The cytosolic mercuric reductase MerA reduces Hg^{2+} to less toxic elemental mercury [9]. The gene merB encodes an organomercurial lyase that catalyses the protonolytic cleavage of carbon-mercury bonds [19]–[20]. The merG gene product is involved in the reduction of cellular permeability to organomercurial compounds [21]. MerD probably acts as a distal co-repressor of transcriptional activation [9], [22]. MerR is the activator or repressor of the transcription of mer genes in presence or absence of mercury ions, respectively [23], [24]. At mercury stress condition the transcriptional activator MerR triggers the expression of the structural mer genes [25]. Sequencing of the native IncP-1β plasmid pTP6 that was originally isolated from mercury-polluted river sediment in a triparental mating showed that all these genes are part of transposon Tn50580 [26].

MerA, mercuric reductase; MerB, organomercurial lyase; MerP, periplasmic mercury-binding protein; MerT, membrane mercury transport protein; MerG, periplasmic protein involved in cell permeability to phenylmercury; MerE membrane protein that probably acts as a broad mercury transporter; MerR, transcriptional activator or repressor of the transcription of mer genes (black pentagon); MerD, co-represor of transcriptional activation; Pt, promoter of merT and merB2 genes.

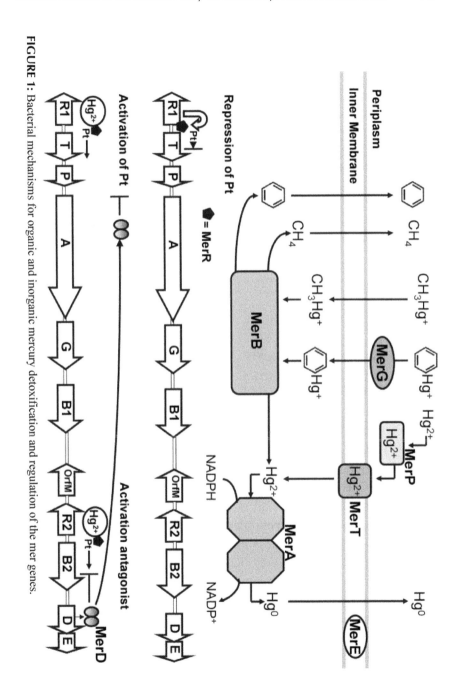

FIGURE 1: Bacterial mechanisms for organic and inorganic mercury detoxification and regulation of the mer genes.

The heavy metal-resistant model bacterium *Cupriavidus metallidurans* strain CH34 harbors two large plasmids, pMOL28 and pMOL30, which carry genetic determinants for heavy metal resistance [27]. Each plasmid contains a merRTPADE operon that confers a narrow-spectrum mercury resistance. To improve inorganic and organic mercury resistance of strain *C. metallidurans* CH34, the IncP-1β plasmid pTP6 was introduced into strain CH34 in this study. The transconjugant strain *Cupriavidus metallidurans* MSR33 showed a broad-spectrum mercury resistance and was able to efficiently remove mercury from polluted water.

13.2 MATERIALS AND METHODS

13.2.1 CHEMICALS

$HgCl_2$ (analytical grade), $CuSO_4 \cdot 5H_2O$, K_2CrO_4, $NaBH_4$, NaOH, HCl (Suprapur) and standard Titrisol solution were obtained from Merck (Darmstadt, Germany). CH_3HgCl (analytical grade) were obtained from Sigma Aldrich (Saint Louis, MO, USA). Stock solutions of Cu^{2+} (5,000 µg ml^{-1}); CrO_4^{2-} (2,500 µg ml^{-1}); Hg^{2+} (1,000 µg ml^{-1}) and CH_3Hg^+ (100 µg ml^{-1}) were prepared. $NaBH_4$ solution (0.25%) was prepared in NaOH (0.4%) solution. High purity hydrochloric acid was used for mercury dilutions before quantification by inductively coupled plasma optical emission spectrometer (ICP-OES). Sodium succinate and salts for media preparation were obtained from Merck (Darmstadt, Germany). Taq DNA polymerase and bovine serum albumin for PCR were obtained from Invitrogen (Carlsbad, CA, USA). RNA was extracted using an RNeasy Protect Bacteria Mini kit from Qiagen (Hilden, Germany). For RNA quantification the Quant-iT™ RNA Assay kit from Invitrogen (Carlsbad, CA, USA) was used. RT-PCR was performed using SuperScriptTM III One-Step RT-PCR System and Taq DNA Polymerase Platinum® from Invitrogen (Carlsbad, CA, USA).

13.2.2 BACTERIA, PLASMIDS, AND CULTURE CONDITIONS

Cupriavidus metallidurans CH34 [28], [29], *Escherichia coli* JM109 (pTP6) [26] and *Cupriavidus necator* JMP134 (pJP4) [30] were grown in Luria Bertani (LB) medium (contained per l: 10 g tryptone, 5 g yeast extract and 10 g NaCl) [31] in absence or presence of Hg^{2+}. *Cupriavidus metallidurans* strains CH34 and MSR33 were cultivated also in low-phosphate Tris-buffered mineral salts (LPTMS) medium. The LPTMS medium contained (per l): 6.06 g Tris, 4.68 g NaCl, 1.49 g KCl, 1.07 g NH_4Cl, 0.43 g Na_2SO_4, 0.2 g $MgCl_2 \cdot 6H_2O$, 0.03 g $CaCl_2 \cdot 2H_2O$, 0.23 g $Na_2HPO_4 \cdot 12H_2O$, 0.005 g $Fe(III)(NH_4)$ citrate, and 1 ml of the trace element solution SL 7 of Biebl and Pfennig [28]. Succinate (0.3%) was used as sole carbon source. To study the effect of inorganic mercury on growth, Hg^{2+} (0.04 mM) was added at time 0 or at exponential phase. Bacterial growth was determined by measuring turbidity at 600 nm.

The IncP-1β plasmid pTP6 that was originally isolated in a triparental mating from mercury-polluted river sediments contains merR1TPAGB1 and merR2B2D2E gene clusters conferring a broad-range mercury resistance [26].

13.2.3 GENERATION OF TRANSCONJUGANT BACTERIAL STRAINS

Plasmid pTP6 was transferred from *E. coli* JM109 to *C. metallidurans* CH34 by biparental mating. Donor and recipient cells were mixed (1:4) and placed in a sterile filter onto plate count agar (PCA) (Merck, Darmstadt, Germany). After overnight incubation at 28°C, transconjugants were selected on LPTMS agar plates containing succinate (0.3%) as sole carbon source and supplemented with Hg^{2+} (0.04 mM). The transconjugants were further tested for growth in liquid LPTMS medium containing succinate (0.3%) as sole carbon source and supplemented with Hg^{2+} (0.09 mM).

Genomic DNA purification was performed by standard procedures [31]. Plasmid extraction from *C. metallidurans* transconjugants, *C. metallidurans* CH34, *C. necator* JMP134 (pJP4) and *E. coli* JM109 (pTP6) was performed using the method of Kado and Liu with minor modifications [32], [33]. The presence of heavy metal resistance genes in genomic and plasmid DNA was analyzed by polymerase chain reaction (PCR) with specific primers. Primers used for merB gene amplification were the forward 5′-TCGCCCCATATATTTTAGAAC-3′ and the reverse 5′-GTCGGGA-CAGATGCAAAGAAA-3′ [34]. Primers used for amplification of chrB gene were the forward 5′-GTCGTTAGCTTGCCAACATC-3′ and the reverse 5′-CGGAAAGCAAGATGTCGAATCG-3′ [35], [36]. Primers used for copA gene amplification were the degenerated forward Coprun F2, 5′-GGSASBTACTGGTRBCAC-3′ and the degenerated reverse Coprun R1, 5′-TGNGHCATCATSGTRTCRTT-3′ [37]. Plasmids and PCR products were visualized in agarose gel electrophoresis (0.7%) in TAE buffer (0.04 M Tris, 0.04 M acetate, 0.001 M EDTA, pH 8.0).

The stability of the pTP6 plasmid in strain *C. metallidurans* MSR33 was evaluated through repeated cultivation in LB medium under non-selective conditions. A single colony of *C. metallidurans* MSR33 was used to inoculate 25 ml LB broth. After growth during 24 h at 28°C, serial dilutions were plated on PCA and 25 µl was used to inoculate fresh LB broth. The procedure was repeated six times. Forty-eight colonies randomly selected were picked from PCA plates of appropriate dilutions obtained when grown in LB broth for 1, 2, 3, 4, 5, 6 and 7 d and streaked on PCA supplemented with Hg^{2+} (0.5 mM). Plasmids were isolated and analyzed from 5 randomly selected mercury-resistant colonies from cultures grown during more than 70 generations under non-selective conditions.

13.2.4 SEQUENCE ANALYSES OF MERA AND MERB

The amino acid sequences of MerA and MerB proteins were obtained from GenPept database (http://www.ncbi.nlm.nih.gov). The multiple MerA and MerB sequences were aligned using Clustal W software [38].

13.2.5 DETERMINATION OF MINIMUM INHIBITORY CONCENTRATION (MIC) VALUES

For testing the bacterial resistance to Hg^{2+}, CH_3Hg^+, Cu^{2+} and CrO_4^{2-}, LPT-MS agar plates were used. Ten µl of cultures grown overnight in LPTMS were placed onto the agar plates supplemented with different concentrations of each heavy metal. LPTMS was supplemented with mercury (II) (chloride) in the concentration range from 0.01 to 0.16 mM (in increasing concentration of 0.01 mM steps), copper (sulphate) in the concentration range from 3.0 to 4.2 mM (in increasing concentration of 0.1 mM steps), and (potassium) chromate in the concentration range from 0.5 to 1.0 mM (in increasing concentration of 0.1 mM steps). The plates were incubated at 28°C for 5 d. The lowest concentration of metal salts that prevented growth was recorded as the MIC. MIC for methylmercury in the concentration range from 0.005 to 0.1 mM (in increasing concentration of 0.005 mM steps) was studied on paper discs on plates. The methylmercury in solution (10 μl) was added to sterile paper discs (diameter 6 mm) and the impregnated paper discs were placed on the bacterial culture in an agar plate. Methylmercury diffused in the area surrounding each paper disc. MIC for methylmercury was recorded as the lowest concentration at which growth inhibition was observed. MIC analyses were done in triplicate.

13.2.6 TRANSMISSION ELECTRON MICROSCOPY

To evaluate the morphology of cells exposed to Hg^{2+}, cells were observed by transmission electron microscopy. Strain MSR33 and strain CH34 cells were grown in LPTMS medium and further incubated at exponential phase in presence of Hg^{2+} (0.04 mM) for 2 h. Cells were harvested by centrifugation and washed with 50 mM phosphate buffer (pH 7.0). Cells were fixed and treated for transmission electron microscopy as described [39]. Briefly, the cells were fixed with Karnowsky solution (2.5% glutaraldehyde, 3.0% formaldehyde) in 0.2 M cacodylate buffer, post fixed with 2% osmium tetroxide and dehydrated in ethanol and acetone series. Finally, cells were embedded in an epoxy resin (Eponate 812). Thin sections (500 nm) were obtained with diamond knife in an Ultracut E ultramicrotome (Reichert). Sections were contrasted with uranyl acetate and lead citrate and observed with a Zeiss EM900 electron microscope. Micrographs were obtained and enlarged for analysis.

13.2.7 SODIUM DODECYL SULFATE-POLYACRYLAMIDE GEL ELECTROPHORESIS (SDS-PAGE)

SDS-PAGE was performed as previously reported [40]. Cells grown in LB medium supplemented with Hg^{2+} (0.075 mM) were harvested at exponential phase (Turbidity600 nm = 0.7) and washed twice with 50 mM phosphate buffer (pH 7.0). Cells were treated as previously described [40]. Briefly, Laemmli sample buffer was added (100 μl per 5 mg of wet weight of bacteria), cells were disrupted by boiling for 5 min and cell debris were removed by centrifugation at 10,733 g for 10 min at 4°C. Proteins were quantified using a Qubit fluorometer (Invitrogen). Proteins were separated by SDS-PAGE using 7 to 15% linear polyacrylamide gradient and visualized by staining with Coomassie brilliant blue R-250.

13.2.8 BIOREMEDIATION EXPERIMENTS IN BIOREACTORS AND MERCURY QUANTIFICATION

A bioreactor was designed for mercury bioremediation assays. Each bioreactor contained 50 ml of Hg^{2+} (0.10 mM and 0.15 mM) spiked aqueous solution (50 mM phosphate buffer, pH 7.0) in a 250 ml sterile flask and was aerated by a blower at 300 ml min^{-1}. Volatilized mercury (Hg0) was trapped on an external HNO_3 (10%) solution in a 15 ml sterile tube. Bioreactors were bioaugmented by inoculating MSR33 or CH34 cells grown in LPTMS medium supplemented with Hg^{2+} (0.01 mM) for the induction of mer genes (cells were added to reach a final concentration of 2×108 cells ml^{-1}). In control bioreactors, cells were not added. Thioglycolate (5 mM) was added to selected bioreactors. The bioreactors were incubated at 30°C for 5 h. All treatments were performed in triplicate. Samples (1 ml) were taken and centrifuged (10,733 g for 10 min) to remove the cells. The supernatants were used for mercury quantification. To quantify total mercury, samples were oxidized by adding 5 ml of HNO_3 (analytical grade), heated at 90°C and then diluted to a final volume of 10 ml with HCl 10% (v/v). Samples were diluted to concentrations <100 μg of Hg per liter. Total mercury was quantified by cold vapor emission spectroscopy using a flow injection system Perkin Elmer (model FIAS 200) linked to an inductively coupled plasma optical emission spectrometer (ICP-OES) Perkin Elmer (model Optima 2000). Ionic mercury (II) from samples was reduced in the FIAS system (1.5 ml min⁻¹) with $NaBH_4$ (0.25% m/v) in NaOH (0.4% m/v) (1.4 ml min⁻¹) to metallic mercury, which was volatilized using argon that carried atomic mercury to the plasma and detected at 253.652 nm. The standard Hg (Titrisol, Merck) concentration in the calibration solutions ranged from 10 to 80 µg l⁻¹. Values were calculated as the mean ±SD of three independent experiments.

13.3 RESULTS

13.3.1 GENERATION OF A BROAD RANGE MERCURY-RESISTANT C. METALLIDURANS STRAIN

To generate a heavy metal-resistant strain with broad-range mercury species resistance, novel and additional mer genes were introduced into heavy metal-resistant *C. metallidurans* CH34. Plasmid pTP6 carrying a complex set of mer genes encoding a broad-spectrum mercury resistance was selected to transfer mer genes into strain CH34. Plasmid pTP6 is an IncP-1β plasmid that only carries mercury resistance genes including merB and merA genes and no additional resistance or catabolic genes [26]. MerB1 of pTP6 is closely related to MerB1 of Tn5058 (*Pseudomonas sp.* strain ED23–33) and MerB from pMR62 (*Pseudomonas sp.* strain K-62), whereas MerB2 of pTP6 is closely to MerB2 of Tn5058 and MerB of the isolated plasmid pQBR103. MerA from pTP6 is closely related to MerA of Tn5058 (Pseudomonas sp. strain ED23–33) and pMR26 (*Pseudomonas sp.* strain K-62), whereas MerA of the plasmids pMOL28 and pMOL30 are closely related to MerA of the catabolic plasmid pJP4 (*C. necator* strain JMP134) and the virulence plasmid pWR501 (*S. flexneri* strain 5a), respectively.

Plasmid pTP6 was transferred from *E. coli* JM109 (pTP6) to *C. metallidurans* strain CH34 by biparental conjugation. Transconjugants were selected by growth in LPTMS agar medium in presence of Hg^{2+} (0.04 mM).

More than 100 mercury-resistant transconjugants were observed in the plate. Parental strain CH34 and *E. coli* strain JM109 (pTP6) did not grow in presence of Hg^{2+} (0.04 mM). All forty-four transconjugants that formed a large colony in presence of Hg^{2+} (0.04 mM) and that were picked were capable to grow in liquid LPTMS medium containing succinate (0.3%) as sole carbon source and Hg^{2+} (0.09 mM). One of these transconjugants, *C. metallidurans* strain MSR33, was selected for further characterization.

13.3.2 GENETIC CHARACTERIZATION OF THE TRANSCONJUGANT STRAIN MSR33

Strain MSR33 was genetically characterized. The presence of plasmid pTP6 in strain MSR33 was confirmed by plasmid isolation and visualization by agarose gel electrophoresis. The presence of merB gene in the isolated plasmid, pTP6, and in genomic DNA of strain MSR33 was confirmed by PCR (Fig. 2). Additionally, the *C. metallidurans* chromate-resistance chrB gene and the copper resistance copA gene were detected in the genome of transconjugant strain MSR33.

The genetic stability of a modified microorganism is critical to its release into the environment [6]. Therefore, the stability of the plasmid pTP6 in the transconjugant strain MSR33 was studied under non-selective conditions. All colonies of strain MSR33 grown under non-selective pressure after 70 generations, maintained their mercury resistance. All the colonies analyzed with improved mercury resistance after 70 generations, possessed the plasmid pTP6. These results indicated that additional mercury resistance mer genes were stable in strain MSR33. It has been reported that the IncP-1β plasmid pTP6 was stably maintained in different proteobacterial hosts under no-selective conditions [26].

13.3.3 SYNTHESIS OF MERB AND MERA IN C. METALLIDURANS MSR33

Plasmid pTP6 provided to the strain MSR33 novel genes that encode two organomercurial lyases MerB and an additional gene that encodes

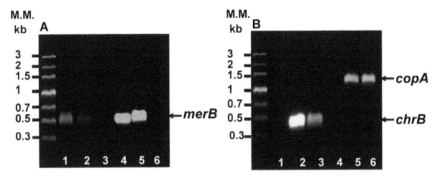

FIGURE 2: Detection by PCR of heavy metal resistance genes in C. metallidurans MSR33. A, detection of merB gene in plasmid (lanes 1–3) and genomic DNA (lanes 4–6). B, detection of chrB (lanes 1–3) and copA (lanes 4–6) genes in genomic DNA. Lanes: 1 and 4, E. coli JM109 (pTP6); 2 and 5, C. metallidurans strain MSR33; 3 and 6, C. metallidurans strain CH34.

the mercuric reductase MerA. The synthesis of MerB and MerA in strain MSR33 grown in presence of Hg^{2+} was studied. The protein pattern of strain MSR33 in presence of Hg^{2+} was analyzed by SDS-PAGE (Fig. 3). The induction of proteins of approximately 23 kDa that probably are MerB enzymes was observed only in the transconjugant strain MSR33. Additionally, the induction of a protein of 50 KDa that probably is the mercuric reductase MerA was observed in strains MSR33 and CH34 during growth in presence of Hg^{2+}.

13.3.4 HEAVY METAL RESISTANCE OF STRAIN MSR33

In order to study the inorganic mercury and organic mercury resistances of strain MSR33, cells were grown on LPTMS agar plates in presence of Hg^{2+} and CH_3Hg^+. Strain MSR33 showed a high resistance to Hg^{2+} (0.12 mM) whereas strain CH34 has a lower resistance to Hg^{2+} (0.05 mM). Therefore, the presence of plasmid pTP6 increased 2.4 fold Hg^{2+} resistance of strain MSR33. Noteworthy, strain MSR33 possesses a high resistance to methylmercury (0.08 mM). In contrast, strain CH34 was sensitive to

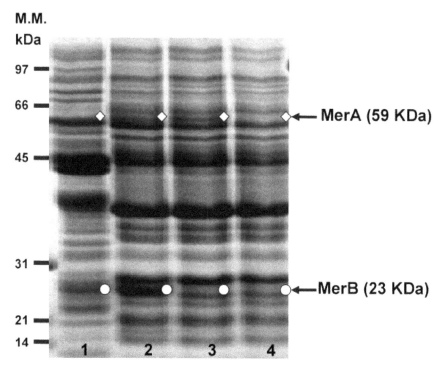

FIGURE 3: Synthesis of MerB and MerA proteins in C. metallidurans strain MSR33. Proteins of cells grown in LPTMS medium in presence (lanes 1–3) and absence of Hg2+ (lane 4) were separated by SDS-PAGE and stained with Coomassie blue. Lanes: 1, E. coli JM109 (pTP6) grown in presence of Hg2+; 2, C. metallidurans strain MSR33 grown in presence of Hg2+; 3, C. metallidurans strain CH34 grown in presence of Hg2+; 4, C. metallidurans strain CH34 grown in absence of mercury. White diamonds and circles indicate MerA and MerB proteins, respectively. Molecular mass of protein standards in kD are shown on the left side of gel.

methylmercury. Strain MSR33 maintained the Cu^{2+} and CrO_4^{2-} resistances of the parental strain CH34 (Table 1).

13.3.5 EFFECT OF HG^{2+} ON THE GROWTH AND CELL MORPHOLOGY

The effect of exposure to Hg^{2+} on the growth of strain MSR33 was studied (Fig. 4). Cells were grown in LPTMS medium with succinate as the sole carbon source in the absence or presence of Hg^{2+} (0.04 mM). Hg^{2+} did not affect the growth of MSR33 ($\mu = 0.051$ h^{-1}), whereas the growth of strain CH34 decreased slightly in presence of Hg^{2+} ($\mu = 0.044$ h^{-1}) (Fig. 4). Strains MSR33 and CH34 showed a similar growth rate in absence of Hg^{2+}. In addition, the effect of Hg^{2+} (0.04 mM) exposure at exponential phase of growth was studied. Interestingly, strain MSR33 growth was not affected after addition of Hg^{2+} at exponential phase (Fig. 4). In contrast, no further growth of strain CH34 was observed when 0.04 mM of Hg^{2+} were added at exponential phase.

TABLE 1: Minimum inhibitory concentrations of heavy metals for transconjugant C. metallidurans strain MSR33.

Metal (mM)	Strain MSR33	Strain CH34	Increased resistance (fold)
Hg^{2+}	0.12	0.05	2.4
CH_3HG	0.08	<0.005	>16
Cu^{2+}	3.80	3.80	0
CrO_4^{2-}	0.70	0.70	0

To evaluate changes in cell morphology by exposure to mercury, cells of MSR33 grown on LPTMS medium with succinate as sole carbon source and exposed to Hg^{2+} were analyzed by transmission electron microscopy (Fig. 5). Strain MSR33 cells exposed to Hg^{2+} showed no changes in cell morphology. In contrast, CH34 cells exposed to Hg^{2+} exhibited a fuzzy outer membrane. Electron dense granules in the cytoplasm of MSR33 and CH34 strains exposed to Hg^{2+} were also observed (Fig. 5).

13.3.6 BIOREMEDIATION OF MERCURY-POLLUTED WATER

In this study, the effectiveness of *C. metallidurans* strain MSR33 as a bio-catalyst for the bioremediation of two Hg^{2+}-polluted aqueous solutions in a bioreactor was assessed. Bioaugmentation was studied in aqueous solutions containing Hg^{2+} (0.10 mM and 0.15 mM). The effect of addition of thioglycolate, a compound that provides an excess of exogenous

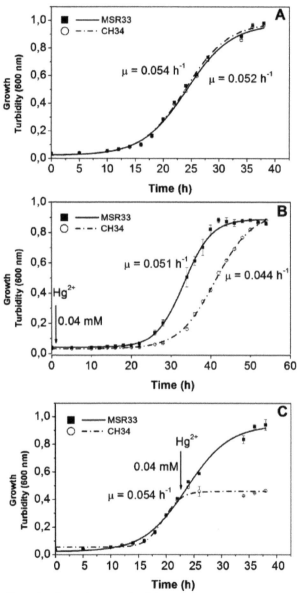

FIGURE 4: Effect of Hg^{2+} on the growth of *C. metallidurans* strains MSR33 and CH34. Cells were grown in LPTMS medium in absence of Hg^{2+} A, presence of Hg^{2+} (0.04 mM) B, or Hg^{2+} (0.04 mM) were added to LPTMS medium at exponential phase C. The exposure to Hg^{2+} started at the point indicated with an arrow. Each value is the mean ± SD of three independent assays.

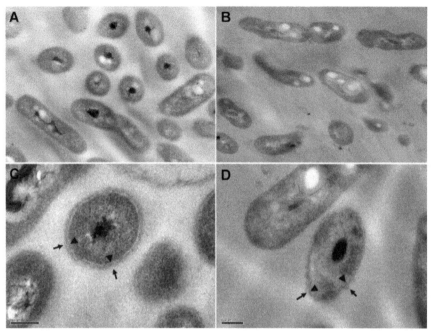

FIGURE 5: Ultrathin section of C. metallidurans strain MSR33 exposed to Hg2+. Cells grown in LPTMS medium with succinate until exponential phase were further incubated in presence of Hg2+ (0.04 mM) for 1 h. A and C, MSR33 cells; B and D, CH34 cells. Arrows and arrowheads indicate the outer and the inner membrane, respectively. The bars represent 0.2 μm.

thiol groups [41], [42] was evaluated on mercury bioremediation by strain MSR33. The effect of bioaugmentation by strain MSR33 on Hg^{2+} removal in two polluted waters is shown in Figure 6. In presence of thioglycolate, *C. metallidurans* strain MSR33 completely removed mercury (0.10 mM and 0.15 mM) from the two polluted waters after 2 h. In absence of thioglycolate, strain MSR33 removed 88% of Hg^{2+} (0.10 mM) after 3 h. In contrast, wild type *C. metallidurans* strain CH34 was not capable to remove mercury either in presence or absence of thioglycolate. No mercury removal was observed in the control bioreactors.

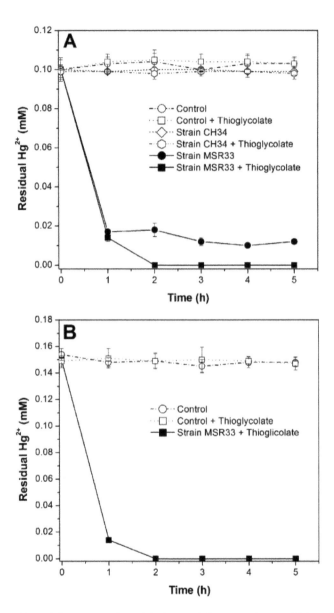

FIGURE 6: Effect of bioaugmentation with *C. metallidurans* strain MSR33 in Hg²⁺-polluted waters. A, removal of Hg²⁺ from water polluted with Hg²⁺ (0.10 mM) by strain MSR33 or strain CH34 in presence or absence of thioglycolate (5 mM). B, removal by strain MSR33 of Hg²⁺ from water polluted with Hg²⁺ (0.15 mM) in presence of thioglycolate (5 mM). Control assays without cells were incubated in presence or absence of thioglycolate (5 mM).

13.4 DISCUSSION

Mercury-polluted sites often contain other heavy metals. Therefore, for the bioremediation of mercury-polluted sites broad range heavy metal-resistant bacteria are required. *C. metallidurans* strain CH34 is a model heavy metal-resistant bacterium [27], [28]. The main goal of this work was to generate a heavy metal-resistant bacterial strain with resistance to inorganic and organic mercury species. Therefore, the heavy metal-resistant strain *C. metallidurans* CH34 with low resistance to mercury (II) and sensitive to organic mercury was modified by introducing the plasmid pTP6 that has been directly isolated from the environment and that encodes Mer proteins that confer a broad-range mercury resistance [26]. The transconjugant *C. metallidurans* strain MSR33 was characterized and applied for remediation of mercury-polluted waters.

In the present study, improved mercury resistance of the transconjugant strain MSR33 was obtained by incorporation of broad-host range plasmid pTP6 with a complex mer system as sole accessory element into *C. metallidurans* CH34. To our knowledge, this is the first study to use an IncP-1β plasmid directly isolated from the environment to generate a novel bacterial strain with improved mercury resistance. Therefore, and in contrast to genetically engineered bacteria, environmental application of transconjugant strain MSR33 is not subjected to biosafety regulation, which is an important advantage for bioremediation of mercury-polluted sites. This study showed that pTP6 stably replicated even without selective pressure as other IncP-1 plasmids that are tightly controlled by korA and korB genes [26]. Interestingly, IncP-1 plasmids have an extremely broad-host range in Gram-negative bacteria and are efficiently transferred from the host to soil bacteria [43], [44]. The natural plasmid pTP6 provided novel merB and merG genes and conferred a broad-spectrum mercury resistance to *C. metallidurans* strain MSR33. The presence of plasmid pTP6 in strain MSR33 provided also additional mer gene copies (merR, merT, merP, merA, merD and merE genes) to the host strain, improving also inorganic mercury resistance. The genetic redundancy may increase the robustness of organisms to mercury stress [45]. The MerA protein encoded by plasmid pTP6 belonged to different subclasses than MerA proteins of pMOL28 and pMOL30. On the other hand, plasmid pTP6 provides two copies of merB genes encoding MerB1 and MerB2 enzymes.

The mechanisms for mercury detoxification in strain MSR33 are shown in Figure 1. The expression of mer genes of pTP6 in strain MSR33 support growth of strain MSR33 with Hg^{2+} at high concentrations and in presence of CH_3Hg^+. Strain MSR33 possesses a high organic mercury resistance (CH_3Hg^+, 0.08 mM) and a high inorganic mercury resistance (Hg^{2+}, 0.12 mM) (Table 1). The novel resistance to methylmercury is conferred by the organomercurial lyase MerB. The organomercurial lyase catalyzes the protonolysis of the carbon mercury bond, removing the organic ligand and releasing Hg^{2+}. Inorganic mercury is 100 times less toxic than the CH_3Hg^+ form [9], [24]. An additional MerA enzyme improved strain MSR33 detoxification of Hg2+, and additional MerT and MerP proteins in strain MSR33 probably allow a rapid uptake of mercury avoiding an extracellular accumulation. The mercury resistance of strain MSR33 is higher than that of other recombinant strains such as *Pseudomonas putida* F1::mer [46]. The addition of merTPAB genes in *P. putida* F1 increased broad-spectrum mercury resistance in the range of 14–86%.

Strain MSR33 showed a similar growth rate in absence and presence of Hg^{2+} (Fig. 4). As expected, the presence of Hg^{2+} decreased the growth rate of strain CH34. Noteworthy, the growth of MSR33 was not affected by the addition of mercury (II) at exponential phase (Fig. 4C). On the other hand, an immediate cessation of growth of strain CH34 was observed after the addition of Hg^{2+} at exponential phase. The additional mer genes incorporated into strain MSR33 probably improve the mercury detoxification system. The effects of Hg^{2+} on growth are in accordance with the effects on cell morphology (Fig. 5). Cell membrane of strain MSR33 was not affected by the presence of mercury. In contrast, after the exposure to Hg^{2+} strain CH34 showed a fuzzy outer membrane. Enterobacter exposed to mercury also showed a fuzzy cell membrane [47]. After exposure to mercury, both strains MSR33 and CH34 contained electron-dense granules in the cytoplasm that probably are polyhydroxybutyrate. *C. metallidurans* CH34 harbors the genes for polyhydroxybutyrate synthesis [48] and probably accumulates these intracellular storage granules during stress.

The bioremediation potential of strain MSR33 of mercury-polluted aqueous solutions was evaluated in a bioreactor. Strain MSR33 completely removed Hg^{2+} (0.10 mM and 0.15 mM) from polluted aqueous solutions. The presence of thioglycolate was required for complete mercury removal in water. The presence of thiol molecules ensures that Hg^{2+} will be present as the dimercaptide, RS-Hg-SR. The dimercaptide form of Hg^{2+} is the substrate of the NADPH-depending mercuric reductase MerA [42]. Thiol compounds such as 2-mercaptoethanol, cysteine and thioglycolate improved mercury volatilization by *Escherichia coli, Pseudomonas sp.* strain K-62, bacterial strain M-1 and *Pseudomonas putida* strain PpY101 (pSR134) [21], [41], [49]–[53].

In the present study, strain MSR33 showed a high mercury volatilization rate, 6.8×10^{-3} ng Hg^{2+} $cell^{-1}$ h^{-1}. Saouter et al. [54] studied mercury removal from freshwater pond by resting cells of mercury-resistant *Aeromonas hydrophila* strain KT20 and *Pseudomonas aeruginosa* PAO1 derivative containing a plasmid carrying mer genes. The mercury volatilization rates of *A. hydrophila* KT20 and *P. aeruginosa* PAO1 derivative were 1.0×10^{-3} ng Hg^{2+} $cell^{-1}$ h^{-1} and 2.4×10^{-4} ng Hg^{2+} $cell^{-1}$ h^{-1}, respectively. Okino et al. [53] studied mercury removal from a mercury-polluted aqueous solution with the genetically engineered bacterium, *P. putida* strain PpY101 (pSR134) showing a volatilization rate of 6.7×10^{-3} ng Hg^{2+} $cell^{-1}$ h^{-1}.

C. metallidurans MSR33 is a stable transconjugant strain carrying plasmid pTP6, which conferred a novel organomercurial resistance and improved significantly the resistance to Hg^{2+}. Strain MSR33 was able to cleave the organic moiety from methylmercury and to reduce Hg^{2+} into a less toxic Hg elemental form. Strain MSR33 efficiently removed Hg^{2+} (0.10 mM and 0.15 mM) through mercury volatilization from mercury-contaminated waters. This study suggests that strain MSR33 may be useful for mercury bioremediation of contaminated water bodies and industrial wastewater. In addition, strain MSR33 is an interesting biocatalyst for bioremediation of mercury-polluted soils.

REFERENCES

1.   von Canstein H, Li Y, Timmis KN, Deckwer WD, Wagner-Döbler I (1999) Removal of mercury from chloralkali electrolysis wastewater by a mercury-resistant Pseudomonas putida strain. Appl Environ Microbiol 65: 5279–5284.

2. Nealson KH, Belz A, McKee B (2002) Breathing metals as a way of life: geobiology in action. Antonie Van Leeuwenhoek 81: 215–222. doi: 10.1023/a:1020518818647.

3. Valls M, de Lorenzo V (2002) Exploiting the genetic and biochemical capacities of bacteria for the remediation of heavy metal pollution. FEMS Microbiol Rev 26: 327–338. doi: 10.1111/j.1574-6976.2002.tb00618.x.

4. Pieper DH, Seeger M (2008) Bacterial metabolism of polychlorinated biphenyls. J Mol Microbiol Biotechnol 15: 121–138. doi: 10.1159/000121325.

5. Morgante V, López-López A, Flores C, González M, González B, et al. (2010) Bio-augmentation with Pseudomonas sp. strain MHP41 promotes simazine attenuation and bacterial community changes in agricultural soils. FEMS Microbiol Ecol 71: 114–126. Erratum in FEMS Microbiol Ecol (2010) 72: 152. doi: 10.1111/j.1574-6941.2009.00790.x.

6. Saavedra JM, Acevedo F, González M, Seeger M (2010) Mineralization of PCBs by the genetically modified strain Cupriavidus necator JMS34 and its application for bioremediation of PCBs in soil. Appl Microbiol Biotechnol 87: 1543–1554. doi: 10.1007/s00253-010-2575-6.

7. Nascimento AM, Chartone-Souze E (2003) Operon mer: bacterial resistance to mercury and potential for bioremediation of contaminated environments. Genet Mol Res 2: 92–101.

8. Oehmen A, Fradinho J, Serra S, Carvalho G, Capelo JL, et al. (2009) The effect of carbon source on the biological reduction of ionic mercury. J Hazard Mater 165: 1040–1048. doi: 10.1016/j.jhazmat.2008.10.094.

9. Barkay T, Miller SM, Summers AO (2003) Bacterial mercury resistance from atoms to ecosystems. FEMS Microbiol Rev 27: 355–384. doi: 10.1016/s0168-6445(03)00046-9.

10. Wagner-Döbler I (2003) Pilot plant for bioremediation of mercury-containing industrial wastewater. Appl Microbiol Biotechnol 62: 124–133. doi: 10.1007/s00253-003-1322-7.

11. Fatta D, Canna-Michaelidou S, Michael C, Demetriou Georgiou E, Christodoulidou M, et al. (2007) Organochlorine and organophosphoric insecticides, herbicides and heavy metals residue in industrial wastewaters in Cyprus. J Hazard Mater 145: 169–179. doi: 10.1016/j.jhazmat.2006.11.009.

12. Ritter JA, Bibler JP (1992) Removal of mercury from wastewater: large-scale performance of an ion exchange process. Wat Sci Technol 25: 165–172.

13. Chang JS, Hong J (1994) Biosorption of mercury by the inactivated cells of Pseudomonas aeruginosa PU21 (Rip64). Biotechnol Bioeng 44: 999–1006. doi: 10.1002/bit.260440817.

14. Deckwer WD, Becker FU, Ledakowicz S, Wagner-Döbler I (2004) Microbial removal of ionic mercury in a three-phase fluidized bed reactor. Environ Sci Technol 38: 1858–1865. doi: 10.1021/es0300517.

15. Baldrian P, in der Wiesche C, Gabriel J, Nerud F, Zadrazil F (2000) Influence of cadmium and mercury on activities of ligninolytic enzymes and degradation of polycyclic aromatic hydrocarbons by Pleurotus ostreatus in soil. Appl Environ Microbiol 66: 2471–2478. doi: 10.1128/aem.66.6.2471-2478.2000.

16. Yurieva O, Kholodii G, Minakhin L, Gorlenko Z, Kalyaeva E, et al. (1997) Intercontinental spread of promiscuous mercury-resistance transposons in environmental bacteria. Mol Microbiol 24: 321–329. doi: 10.1046/j.1365-2958.1997.3261688.x.

17. Silver S, Phung LT (2005) A bacterial view of the periodic table: genes and proteins for toxic inorganic ions. J Ind Microbiol Biotechnol 32: 587–605. doi: 10.1007/s10295-005-0019-6.

18. Kiyono M, Sone Y, Nakamura R, Pan-Hou H, Sakabe K (2009) The MerE protein encoded by transposon Tn21 is a broad mercury transporter in Escherichia coli. FEBS Lett 583: 1127–1131. doi: 10.1016/j.febslet.2009.02.039.

19. Moore MJ, Distefano MD, Zydowsky LD, Cummings RT, Walsh CT (1990) Organomercurial lyase and mercuric ion reductase: nature's mercury detoxification catalysts. Acc Chem Res 23: 301–308. doi: 10.1021/ar00177a006.

20. Misra TK (1992) Bacterial resistance to inorganic mercury salts and organomercurials. Plasmid 27: 4–16. doi: 10.1016/0147-619x(92)90002-r.

21. Kiyono M, Pan-Hou H (1999) The merG gene product is involved in phenylmercury resistance in Pseudomonas strain K-62. J Bacteriol 181: 762–730.

22. Champier L, Duarte V, Michaud-Soret I, Covès J (2004) Characterization of the MerD protein from Ralstonia metallidurans CH34: a possible role in bacterial mercury resistance by switching off the induction of the mer operon. Mol Microbiol 52: 1475–1485. doi: 10.1111/j.1365-2958.2004.04071.x.

23. Ni'Bhriain NN, Silver S, Foster TJ (1983) Tn5 insertion mutations in the mercuric ion resistance genes derived from plasmid R100. J Bacteriol 155: 690–703.

24. Permina EA, Kazakov AE, Kalinina OV, Gelfand MS (2006) Comparative genomics of regulation of heavy metal resistance in Eubacteria. BMC Microbiol 6: 49–60.

25. Brown NL, Stoyanov JV, Kidd SP, Hobman JL (2003) The MerR family of transcriptional regulators. FEMS Microbiol Rev 27: 145–163. doi: 10.1016/s0168-6445(03)00051-2.

26. Smalla K, Haines AS, Jones K, Krögerrecklenfort E, Heuer H, et al. (2006) Increased abundance of IncP-1β plasmids and mercury resistance genes in mercury-polluted river sediments: first discovery of IncP-1β plasmids with a complex mer transposon as the sole accessory element. Appl Environ Microbiol 72: 7253–7259.

27. Mergeay M, Monchy S, Vallaeys T, Auquier V, Benotmane A, et al. (2003) Ralstonia metallidurans, a bacterium specifically adapted to toxic metals: towards a catalogue of metal-responsive genes. FEMS Microbiol Rev 27: 385–410. doi: 10.1016/s0168-6445(03)00045-7.

28. Mergeay M, Nies D, Schlegel HG, Gerits J, Charles P, et al. (1985) Alcaligenes eutrophus CH34 is a facultative chemolithotroph with plasmid-bound resistance to heavy metals. J Bacteriol 162: 328–334.

29. Monchy S, Benotmane MA, Janssen P, Vallaeys T, Taghavi S, et al. (2007) Plasmids pMOL28 and pMOL30 of Cupriavidus metallidurans are specialized in the maximal response to heavy metals. J Bacteriol 189: 7417–7425. doi: 10.1128/jb.00375-07.

30. Don RH, Pemberton JM (1981) Properties of six pesticide degradation plasmids isolated from Alcaligenes paradoxus and Alcaligenes eutrophus. J Bacteriol 145: 681–686.

31. Sambrook J, Fritsch EF, Maniatis T (1989) Molecular Cloning: A Laboratory Manual, 2nd Ed. Cold Spring Harbor, New York: Cold Spring Harbor Laboratory Press.

32. Kado CI, Liu ST (1981) Rapid procedure for detection and isolation of large and small plasmids. J Bacteriol 145: 1365–1373.

33. Top E, Mergeay M, Springael D, Verstraete W (1990) Gene escape model: transfer of heavy metal resistances genes from Escherichia coli to Alcaligenes eutrophus on agar plates and in soil samples. Appl Environ Microbiol 56: 2471–2479.

34. Liebert CA, Wireman J, Smith T, Summers AO (1997) Phylogeny of mercury resistance (mer) operons of gram-negative bacteria isolated from the fecal flora of primates. Appl Environ Microbiol 63: 1066–1076.

35. Nies A, Nies DH, Silver S (1990) Nucleotide sequence and expression of a plasmid-encoded chromate resistance determinant from Alcaligenes eutrophus. J Biol Chem 265: 5648–5653.

36. Abou-Shanab RA, van Berkum P, Angle JS (2007) Heavy metal resistance and genotypic analysis of metal resistances genes in gram-positive and gram-negative bacteria present in Ni-rich serpentine soil and in the rhizosphere of Alyssum murale. Chemosphere 68: 360–367. doi: 10.1016/j.chemosphere.2006.12.051.

37. Lejon DP, Nowak V, Bouko S, Pascault N, Mougel C, et al. (2007) Fingerprinting and diversity of bacterial copA genes in response to soil types, soil organic status and copper contamination. FEMS Microbiol Ecol 61: 424–437. doi: 10.1111/j.1574-6941.2007.00365.x.

38. Larkin MA, Blackshields G, Brown NP, Chenna R, McGettigan PA, et al. (2007) Clustal W and Clustal X version 2.0. Bioinformatics 23: 2947–2948. doi: 10.1093/bioinformatics/btm404.

39. Cámara B, Herrera C, González M, Couve E, Hofer B, et al. (2004) From PCBs to highly toxic metabolites by the biphenyl pathway. Environ Microbiol 6: 842–850. doi: 10.1111/j.1462-2920.2004.00630.x.

40. Seeger M, Jerez CA (1993) Phosphate-starvation induced changes in Thiobacillus ferrooxidans. FEMS Microbiol Lett 108: 35–42. doi: 10.1016/0378-1097(93)90484-j.

41. Summers AO, Sugarman LI (1974) Cell-free mercury (II)-reducing activity in a plasmid-bearing strain of Escherichia coli. J Bacteriol 119: 242–249.

42. Fox B, Walsh CT (1982) Mercuric reductase. Purification and characterization of a transposon-encoded flavoprotein containing an oxidation-reduction-active disulfide. J Biol Chem 257: 2498–2503.

43. Pukall R, Tschäpe H, Smalla K (1996) Monitoring the spread of broad host and narrow host range plasmids in soil microcosms. FEMS Microbiol Ecol 20: 53–66. doi: 10.1016/0168-6496(96)00021-9.

44. Schlüter A, Szczepanowski R, Pühler A, Top EM (2007) Genomics of IncP-1 antibiotic resistance plasmids isolated from wastewater treatment plants provides evidence for a widely accessible drug resistance gene pool. FEMS Microbiol Rev 31: 449–477. doi: 10.1111/j.1574-6976.2007.00074.x.

45. Kafri R, Levy M, Pilpel Y (2006) The regulatory utilization of genetic redundancy through responsive backup circuits. Proc Natl Acad Sci USA 103: 11653–11658. doi: 10.1073/pnas.0604883103.

46. Horn JM, Brunke M, Deckwer WD, Timmis KN (1994) Pseudomonas putida strains which constitutively overexpress mercury resistance for biodetoxification of organomercurial pollutants. Appl Environ Microbiol 60: 357–362.

47. Vaituzis Z, Nelson JD Jr, Wan LW, Colwell RR (1975) Effects of mercuric chloride on growth and morphology of selected strains of mercury-resistant bacteria. Appl Microbiol 29: 275–286.

48. Janssen PJ, van Houdt R, Moors H, Monsieurs P, Morin N, et al. (2010) The complete genome sequence of Cupriavidus metallidurans strain CH34, a master survivalist in harsh and anthropogenic environments. PLoS ONE 5: e10433. doi:10.1371/journal.pone.0010433.

49. Schottel JL (1978) The mercuric and organomercurial detoxifying enzymes from a plasmid-bearing strain of Escherichia coli. J Biol Chem 253: 4341–4349.

50. Nakamura K, Nakahara H (1988) Simplified X-ray film method for detection of bacterial volatilization of mercury chloride by Escherichia coli. Appl Environ Microbiol 54: 2871–2873.

51. Ray S, Gachhui R, Pahan K, Chaudhury J, Mandal A (1989) Detoxification of mercury and organomercurials by nitrogen-fixing soil bacteria. J Biosci 14: 173–182. doi: 10.1007/bf02703169.

52. Nakamura K, Hagimine M, Sakai M, Furukawa K (1999) Removal of mercury from mercury-contaminated sediments using a combined method of chemical leaching and volatilization of mercury by bacteria. Biodegradation 10: 443–447.

53. Okino S, Iwasaki K, Yagi O, Tanaka H (2000) Development of a biological mercury removal-recovery system. Biotechnol Lett 22: 783–788.

54. Saouter E, Gillman M, Barkay T (1995) An evaluation of mer-specified reduction of ionic mercury as a remedial tool of mercury-contaminated freshwater pond. J Ind Microbiol 14: 343–348. doi: 10.1007/bf01569949.

CHAPTER 14

A FERRITIN FROM *Dendrorhynchus Zhejiangensis* WITH HEAVY METALS DETOXIFICATION ACTIVITY

CHENGHUA LI, ZHEN LI, YE LI, JUN ZHOU, CHUNDAN ZHANG, XIURONG SU, AND TAIWU LI

14.1 INTRODUCTION

As one of a major member of iron homeostasis proteins, ferritin plays an important role in storage and detoxification of excess iron in living cells. Structure analysis indicates the protein complex usually composed of 24 subunits, which surrounds an inorganic microcrystalline hollow capable of accommodating up to 4500 Fe^{3+} [1], [2]. In vertebrates, two types of subunits called heavy (H) and light (L) chains are identified and demonstrated to be encoded by separate genes. The H subunit has been studied in a variety of species including vertebrate and invertebrate animals, plants and bacteria [3], [4]. The L subunit, however, has been only found in vertebrates [5], [6]. The H subunits from different species contain seven conserved residues that confer ferroxidase activity for converting Fe^{2+} to Fe^{3+} allowing rapid detoxification of iron cations. The L subunit does not have

This chapter was originally published under the Creative Commons Attribution License. Li C, Li Z, Li Y, Zhou J, Zhang C, Su X, and Li T. A Ferritin from Dendrorhynchus zhejiangensis with Heavy Metals Detoxification Activity. PLoS ONE 7,12 (2012), doi:10.1371/journal.pone.0051428.

ferroxidase activity, but serves as a salt bridge that stabilizes the ferritin structure, thus playing a role in iron nucleation and long-term storage [7].

As environmental issue attracted much attention in recent times, development and implication pollutant binding or degradation related genes were considered to be promising way, especially for heavy metals. Accumulative results showed that natural ferritin could store multiple toxic metal ions (Zn^{2+}, Pb^{2+}, Ni^{2+}) by its great storage capacity, resulting to detoxification of heavy metals in vivo [8]. Additionally, ferritin had also potential application for chemotherapy and cancer treatment [9].

Dendrorhynchus zhejiangensis are one of marine animals belonged to *Neatinea, Heteronemertea, Lineidae, Dendrorhynchus*. The worm lives in the bottom of shrimp ponds, in which the concentration of heavy metal usually higher than that in seawater. Therefore, the worm was considered to be a promising material to study heavy metal detoxification. In the present study, we firstly isolated and characterized a new source of ferritin from *D. zhejiangensis*. Then, its enrichment capacity for different metals in vitro was further addressed by atomic force microscopy to fully elucidate the roles of worm ferritin.

14.2 RESULTS

14.2.1 CDNA LIBRARY ANNOTATION

The QC procedure was performed to evaluate the quality of the cDNA library, and the titer of the cDNA library was 1.7×10^6 cfu/mL. Colony PCR found that the recombinant rate was 87.5% with average size of 800 bp, indicating that the inserted fragment length was ideal and the library quality was good.

Random sequencing of 458 clones using T3 primer yielded 428 effective sequences. After removing low quality sequences, adaptor and vector sequences, the remaining were subjected for BLASTx analysis with 349 sequences successfully matched. These ESTs were classified into seven categories, including 152 unknown genes, 49 genes related to protein expression, 50 mitochondrial genes, 35 ribosomal structural genes, 19 metabolism-related genes, 11 immune-related genes, and 33 genes related to translation, intercellular material transport, and endocrine.

14.2.2 CLONING THE FULL-LENGTH CDNA OF WORM FERRITIN

A 1000 bp fragment containing polyA tail was cloned from the *D. zhejiangensis* cDNA library using gene specific primer and T7. Blastx analysis indicated the fragment was similar to the other reported ferritin. The 5'end was obtained with gene specific primer and T3 to get a 750 bp product. By overlapping the three fragments together, an 1179 bp nucleotide sequence representing the full-length cDNA of worm ferritin was assembled. The complete nucleotide and deduced amino acid sequence were shown in Fig. 1. The sequence consists of a 5'-UTR of 104 bp, a 3'-UTR of 565 bp with a poly(A) tail and polyadenylation signal AATAAA, and an ORF of 510 bp encoding a polypeptide of 169 amino acid residues. The three typical ferritin domains for ferritin, ferroxidase diiron center (Glu-23, Tyr-30, Glu-57, Glu-58, His-61, Glu-103, Gln-137), a ferrihydrite nucleation center (Lys-53, Ser-56, Glu-57, Glu-59) and an iron ion channel (His-114, Asp-127, Glu-130), were conserved in worm ferritin (Fig. 2). Blast analysis revealed worm ferritin showed higher simmilarity to other registered counterparts. For example, it shared 71% identities with *Dermacentor variabilis* (AF467696) and *Holothuria glaberrima* (ABS29643), 68% with *Phascolosoma esculenta* (ABW75858), 65% with *Crassostrea ariakensis* (ABE99842), 63% with *Macrobrachium rosenbergii* (ABY75225), and 60% with *Xenopus laevis* (CAA35760) (Fig. 2).

14.2.3 TRANSCRIPTIONAL LEVEL EXPRESSION OF FERRITIN UNDER HEAVY METALS EXPOSURE

Temporal effect of heavy metals exposure on the transcriptional activities of worm ferritin were investigated by qPCR (Fig. 3). In Fe^{2+} groups, an increase expression profile was detected at the first 36 h, and the peak

```
   1  GGCACGAGGGTTTTGCTGCGTCAGTGAACGTACGGACAAATCGGCGCTAGACCAGAAATT
  61  TACTCCCTTTTCTTTCTAAGTTGATACAGTAACTGAAATCCAAGATGTCACTCTGTCGTC
   1                                              M  S  L  C  R
 121  AGAACTACCATGAGGAATGCGAGGCTGGTGTTAATAAGCAGATTAACATGGAATTTTACG
   6   Q  N  Y  H  E  E  C  E  A  G  V  N  K  Q  I  N  M  E  G  Y
 181  CCAGCTACGTCTACATGTCGATGGCCTCCCACTTCGACAGGGATGATGTAGCTTTGAAAG
  26   A  S  Y  V  Y  M  S  M  A  S  H  F  D  R  D  D  V  A  L  K
 241  GCGCCCATGAGTTCTTCCTCAAGAGTTCAAGTGAGGAACGTGAACATGCCATGCGCCTTA
  46   G  A  H  E  F  F  L  K  S  S  S  E  E  R  E  H  A  M  R  L
 301  TCAAGTTCCAGAATCAGCGTGGTGGCCGTGTTGTCTACCAAGACATCAAGAAGCCAGAAA
  66   I  K  F  Q  N  Q  R  G  G  R  V  V  Y  Q  D  I  K  K  P  E
 361  AGGATGCCTGGGGCACCCTTACTGATGCCATGCAAGCTGCCTTGGACCTGGAGAAACATG
  86   K  D  A  W  G  T  L  T  D  A  M  Q  A  A  L  D  L  E  K  H
 421  TCAACCAGGCCCTCTTGGACCTTCATGCCCTTGCTAGCAAACACAATGATCCCCAGATGT
 106   V  N  Q  A  L  L  D  L  H  A  L  A  S  K  H  N  D  P  Q  M
 481  GTGACTTCATTGAGAATCACTACCTGACTGAACAGGTGGAAGCCATCAGGGAAATTTCTG
 126   C  D  F  I  E  N  H  Y  L  T  E  Q  V  E  A  I  R  E  I  S
 541  GATACCTGACCAACCTGAAGCGCTGTGGTCCTGGGCTTGGTGAATTCCTGTTCGACAAGG
 146   G  Y  L  T  N  L  K  R  C  G  P  G  L  G  E  F  L  F  D  K
 601  AACTCAATTCTTAATCTGCTTATGAACACCAAAGGATACCACACATTGCTTATAGGAATA
 166   E  L  N  S  *
 661  AAGAACTTTCTCATAATGTTAAACATCTTTTAGACTGTGCTCACTACAGTTCATTTTCAG
 721  TGTTCACATATTCCATGGACTAAACGTTTTGTCACAACCAAAGCAAACAGTACAAGACTG
 781  GGATTGAACTTGTGGTAACCAGAGAAAATGTCTAAAAAGGAAAAAAAAATGTAGTGTGTG
 841  CCGAAGTTGTATCTGATTTGCTGAACAAGTGGTATTCCTATATTAACATCTTCTCTGCAT
 901  GAAGAAAATCTTGGTTCATTTATTGTATCAAAGGCATCAACATTCCCAGGTCTCCTGACT
 961  ATTTTCCTTAACTTGGTATCCCACATCTTGAATCTTGCAGTCATATCTTCTAGGGTTATT
1021  CCTGCTTGATATTCTTATTTCTTAAAAGTATTTTAATTTCTGTAACTGAAACAATGTAAA
1081  TAAAAGTCTCCCAAAGTACAATTTATTTTGTAGTCCAGTGAGGCTTGGATGACAAGGGAT
1141  GTACAATAAATGCAATTAAACAAAAAAAAAAAAAAAAAAAA
```

FIGURE 1: Nucleotide and deduced amino acid sequences of ferritin cDNA of *D. zhejiangensis*. The underlined sequences represent start codons. The polyadenylation signal sequence is shown in box.

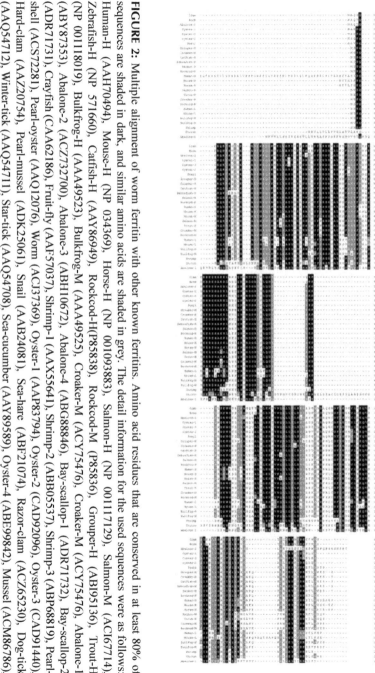

FIGURE 2: Multiple alignment of worm ferritin with other known ferritins. Amino acid residues that are conserved in at least 80% of sequences are shaded in dark, and similar amino acids are shaded in grey. The detail information for the used sequences were as follows: Human-H (AAH70494), Mouse-H (NP 034369), Horse-H (NP 001093883), Salmon-H (NP 001117129), Salmon-M (ACI67714), Zebrafish-H (NP 571660), Catfish-H (AAY86949), Rockcod-H(P85838), Rockcod-M (P85836), Grouper-H (ABI95136), Trout-H (NP 001118019), Bulkfrog-H (AAA49523), Bulkfrog-M (AAA49525), Croaker-M (ACY75476), Croaker-M (ACY75476), Abalone-1 (ABY87353), Abalone-2 (ACZ732700), Abalone-3 (ABH10672), Abalone-4 (ABG88846), Bay-scallop-1 (ADR71732), Bay-scallop-2 (ADR71731), Crayfish (CAA62186), Fruit-fly (AAF57037), Shrimp-1 (AAX55641), Shrimp-2 (ABB05537), Shrimp-3 (ABP68819), Pearl-shell (ACS72281), Pearl-oyster (AAQ12076), Worm (ACI37369), Oyster-1 (AAP83794), Oyster-2 (CAD92096), Oyster-3 (CAD91440), Hard-clam (AAZ20754), Pearl-mussel (ADK25061), Snail (AAB24081), Sea-hare (ABF21074), Razor-clam (ACZ65230), Dog-tick (AAQ54712), Winter-tick (AAQ54711), Star-tick (AAQ54708), Sea-cucumber (AAY89589), Oyster-4 (ABE99842), Mussel (ACM86786), Zhikong-scallop (AAV66904) and Crab (ADD17345).

expression was reached at 36 h with 18.5-fold increase compared to the control group (P<0.05). In Pb^{2+} group, the ferritin mRNA level increased and reached maximal level at 12 h with 7-fold increase compared to the control group (P<0.05). In Cd^{2+} group, the ferritin mRNA expression was sharply induced during the first 6 h, then slowly decreased with time going on (Fig. 3).

14.2.4 RECOMBIANT EXPRESSION OF WORM FERRITIN

The positive transformants were analyzed by SDS-PAGE to characterize the recombinant product of worm ferritin. After IPTG induction, an obvious protein band with molecular weight between 20.1 kDa and 29 kDa was detected in the positive transformants (Fig. 4), which could be purified to homogeneity by HiTrap Chelating Columns. The molecular mass of the purified product was in good agreement with the Predicted MW of worm ferritin. Moreover, the intensity of the recombinant protein band was increased with time went on. The peak expression level of recombinant protein was observed at 6 h after IPTG was introduced into the culture.

14.2.5 WESTERN BLOT ANALYSIS OF WORM FERRITIN EXPRESSION

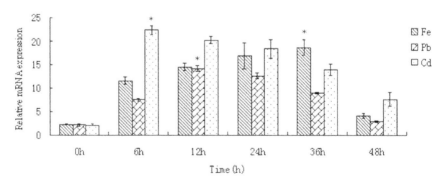

FIGURE 3: Quantification of worm ferritin mRNA expression under Fe^{2+}, Pb^{2+} and Cd^{2+} exposure by qPCR.

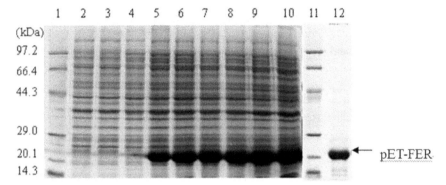

FIGURE 4: Ferritin expression and purification revealed by SDS-PAGE. lane 1 and11: low molecular marker; lane 2, 3 and 4: negative control; lane 5–10: induce expression at 1 h, 2 h, 3 h, 4 h,5 h and 6 h; lane 12: purified expression product.

With the purified recombinant worm ferritin, polyclonal antibodies were generated for Western blot (Fig. 5). The results showed that antiserum could be specifically identified by not only the recombinant protein, but also the native protein from haemocytes. No signals were detected in other control samples. The size difference between the pET-FER and native ferritin was contributed to the extra fusion expression tag in the pET28 vector.

14.2.6 CHARACTERIZATION OF RECOMBIANT WORM FERRITIN BINDING BY ATOMIC FORCE MICROSCOPY

The majority of ferritin molecules were well dispersed and had height of 6~12 nm in cross-section analysis before refolding (Fig 6aa"). In addition to a small amount of protein aggregates, most of the ferritin molecules formed a chain of beads with height of 5~10 nm and diameter of about 90~100 nm (Fig. 6bb"). After adding Fe^{2+}, Pb^{2+}, and Cd^{2+}, respectively, ferritin molecules aggregated as protein complexes with height of 9~13 nm and diameter of about 300~700 nm (Fig 6cc"), height of 35~40 nm and diameter of 400~1000 nm (Fig. 6dd"), and height of 95~100 nm and diameter of about 400~1100 nm (Fig. 6ee"), respectively.

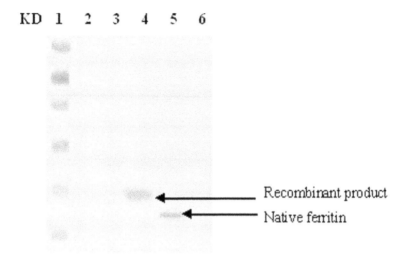

FIGURE 5: Specificity of worm ferritin polyclonal antibody determined by Western-blot. Lane 1: protein marker; lane 2 and 3: negative control; lane 4: recombinant product; lane 5: total protein extraction from worm; lane 6: antiserum control.

14.3 DISCUSSION

Atomic force microscopy (AFM) has become an important tool for studying the interaction between biological molecules based on its higher resolution and capability. The technique had been suceessfully utilized in antigen and antibody, DNA and protein complex, and DNA conformation [10]. In this study, a full-length cDNA of worm ferritin was firstly cloned by cDNA library and RACE, then, its expression pro-files were characterized at mRNA level under heavy metal exposure. Its binding activity to heavy metals was further investigated by atomic force microscopy (AFM).

In order to better understand the role of the worm ferritin in response to heavy metals challenge, temporal expression levels of worm ferritin were analyzed by real-time PCR. We showed that the expression of the ferritin

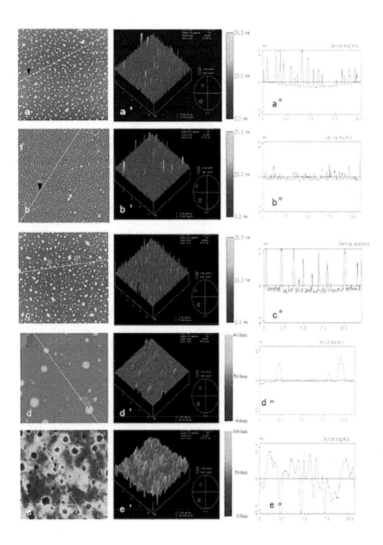

FIGURE 6: Interaction between recombiant ferritin and metal ions by AFM with 9×9 μm² scanning range. a: Ferritin training complex topography image of the front surface, a': three-dimensional topography map from the a, a": relative height of the peak figure from the a; b: Ferritin culture surface topography map refolded, b': three-dimensional topography map from the b, b": Relative height of the peak figure from the b; c: Incubation of ferritin and Fe^{2+} at 37°C for 30 min, c': three-dimensional topography map from the c, c": Relative height of the peak figure from the c; d: Incubation of ferritin and Cu^{2+}, at 37°Cfor 30 min a, d': three-dimensional topography map from the c, d": Relative height of the peak figure from the d; e: Incubation of ferritin and Cd^{2+} at 37°C for 30 min, e': three-dimensional topography map from the e, e": Relative height of the peak figure from the e;.

in all treated groups was significantly increased after 6 h of exposure to three heavy metals, indicating that worm ferritin could be induced by Cd^{2+}, Pb^{2+} and Fe^{2+}. The ferritin expression level of Cd^{2+} at same dose was about three times higher than that of Pb^{2+}, suggesting that worm ferritin may be the least sensitive to Pb^{2+}, compared with Cd^{2+} and Fe^{2+}. The similar patterns of ferritin toward heavy metals exposure was also reported in giant prawn [11], clam [12], fruit fly [13] and amphioxus [14]. Given these facts, our results suggested that the enhanced expression of worm ferritin by heavy metals treatments was probably a protective mechanism of the cell to environmental stress [15].

In order to utilize the gene in practice, knowledge on how worm ferritin binding with heavy metals should be further investigated. The dynamic process between worm ferritin and three heavy metals was observed with AFM. The size of recombinant ferritin was very similar to that of the native ferritin protein after refolding [16], with small amount of irregularly aggregated molecules and fragments. However, when divalent ions was introduced in the culture, the subunit number of ferritin shell gradually increases, and ferritin shell becomes thicker, resulting in the size of ferritin structure was greater than that of native ferritin [3]. The interactions between metal and its binding ligands, such as hydrophobic bonds, hydrogen bonds, electrostatic interaction and van der Waals force are changed, which significantly impacts protein folding. Harrison et al [17] classified ferritin's iron storage procedure into three stages, namely, the oxidation of Fe^{2+} to Fe^{3+}, movement of Fe^{3+} and the nucleation of iron core. Once the small organic molecules occupied part of the iron core, unstable iron ions in ferritin were released, resulting in decrease of electron density within the iron core [18]. As a continuation of storage reaction, the amount of released iron increased. The three-phase material exchange tunnel was narrowed by ferrtin shell through a flexible regulation. Thus, the width of ferritin was decreased. Our result was in agreement with the theory that free or denatured ferritin subunits could quickly wrap bare iron core to form native structure.

Regarding to Cd^{2+} challenge, the dramatic conformation changes of ferritin was also observed through forming a large hollow, which was speculated to be a cadmium core formed by a lot of Cd^{2+} gathering and surrounded by many ferritin molecules. Previous reports had shown Cd^{2+}

was capable of binding to both the inside and outside of horse spleen ferritin [19]. All these evidences indicated worm ferritin might be promising candidates for Fe^{2+} and Cd^{2+} detoxification.

The introduce of heavy metal into recombinant ferritin was also changed its size. The diameter of ferritin increased from 35~40 nm to 400~1000 nm when added with Pb^{2+}. It was hypothesized that the hydration of metal ions destroyed the hydration layer surrounding ferritin molecules, making it easier to form folding intermediates. The stronger the metal ion hydration was, the greater role it played in promoting ferritin aggregation, leading to the formation of large ferritin aggregates. Therefore, metals with large electron could be more likely to promote ferritin aggregation because they had higher hydration free energy [20]. In addition to metal hydration, the electrostatic interaction among metal ions shielded electrostatic repulsion among protein molecules, the so-called Debye screening. In alkaline solution, ferritin molecules had negative charges, thus the electrostatic repulsion among them made aggregation difficult. These phnomena was also expained by the fact that metal ions could accelerate protein nucleation. Metal ions with high positive charges exerted much stronger Debye screening effect. Therefore, they were more likely to enhance nucleation than metal ions with lower positive charges. Nucleation rate could also affect the protein shape, thus exposure to different metal ions had the potential to form different protein shapes [21].

In summary, we cloned the full-length cDNA of *D. zhejiangensis* ferritin and characterized its expression patterns during exposure to iron ion. Subsequently, the recombinant ferritin was produced to investigate its binding activity to Fe^{2+}, Pb^{2+}, and Cd^{2+} by AFM. All our results supported that worm ferritin was a promising candidate for heavy metals detoxification. In order to utilize the protein in practice, its binding capcity to different heavy metals should be elucidated in our further work.

14.4 MATERIALS AND METHODS

14.4.1 WORMS

Dendrorhynchus zhejiangensis was collected from shrimp ponds along the coast of Hudoudu (29.6° N, 121.6° E), Fenghua City, Zhejiang Province,

China in May 2011. The animals were acclimated in our laboratory for two days prior to the experiment. No specific permits were required for the described field studied, and no specific permissions were required for these locations.

14.4.2 CDNA LIBRARY CONSTRUCTION AND EST ANALYSIS

Eight *D. zhejiangensis* were randomly collected and homogenized with liquid nitrogen. The lysate was used for cDNA library construction. The cDNA library was constructed using the ZAP-cDNA synthesis kit and ZAP-cDNA GigapackIII Gold cloning kit (Stratagene). Random sequencing of the library using T3 primer yielded 458 successful sequencing reactions. BLAST analyses of all the 458 EST sequences revealed that one EST of 387 bp was highly similar to the previously identified ferritin. Therefore, the EST sequence was selected for the full-length cDNA cloning.

14.4.3 CLONING THE FULL-LENGTH CDNA OF WORM FERRITIN

Two specific primers, Fer-R: 5'-GTGGAAGCCATCAGGGAA-3' and Fer-F: 5'-AAGGGCATGAAGGTCCAAGAGGG-3', were designed based on the EST to amplify the 3' and 5' ends of worm ferritin combined with vector prime T7 or T3, respectively. PCR was performed at 94°C for 5 min followed by 35 cycles of 94°C for 30 s, 58°C for 30 s, and 72°C for 1 min and final extension at 72°C for 10 min. The obtained PCR amplicons were gel-purified and cloned into pMD18-T (Takara, China). After transforming into competent *E. coli*, the positive colonies were identified by colony PCR. Three positive clones were sequenced at Invitrogen (Shanghai, China) and were analyzed using BLAST (http://www.ncbi.nlm.nih.Gov/BLAST/).

14.4.4 QUANTITATIVE EXPRESSION ANALYSIS OF FERRITIN

For heavy metals challenge experiment, worms were divided in to four tanks and were exposed to three metals of Fe^{2+}, Pb^{2+} and Cd^{2+} with the final concentration of 10 µM. The fourth tank was served as control group. The seawater was changed daily and metal stock solution was added to the seawater everyday. After 6 h, 12 h, 24 h and 48 h exposure, the haemocytes from each group were collected from the control and the treated groups for RNA extraction and cDNA synthesis.

For expression analysis, the ferritin specific primers, FER-F (5'-GAG-GAATGCGAGGCTGGT-3') and FER-R (5'-CACG CTGATTCTG-GAACTTG-3'), were designed to amplify a product of 205 bp. For normalizing the ferritin transcripts, two β-actin specific primers actin-F (5'-GGACCTCTACGCCAACACTG-3') and actin-R (5'- ATG-CAAGGATGGAGCCAC-3') were used to amplify an 170 bp fragment. The fluorescent real-time PCR assay was carried out performed in Rotor-Gene 6000real-time PCR detection system and the condition was according to our previous work [12]. The 2−∆∆CT method was used to analyze the expression level of worm ferritin. All data were given in terms of relative mRNA expressed as mean ± S.D. The data were then subjected to analysis by one-way analysis of variance (ANOVA). Differences were considered significant at $P<0.05$.

14.4.5 EXPRESSION AND PURIFICATION OF WORM FERRITIN

PCR product corresponding to ferritin ORF was amplified by two gene specific primers with BamH I and Hind III sites at 5'ends. The PCR amplicons were digested with BamH I and Hind III, and then inserted into digested pET-28a(+) vector. After transforming into competent E. coli BL21, positive clones were identified by kanamycin screening and sequencing. After sequencing to ensure in-frame insertion, positive clones were incubated at 37°C at 220 rpm. When the culture reached OD600 nm = 0.5 to 0.7, 1 mM isopropyl-β-D-thiogalacto- pyranoside (IPTG) were added to induce the ferritin expression for additional 6 h. Cells were harvested at 1 h, 2 h, 3 h, 4 h, 5 h and 6 h by centrifugation at 4000 r/min for 15 min and resuspended in PBS for sonication. The expressed ferritin was purified using Ni-NTA affinity column (GE healthcare) according to the manufacturer's instruction.The expression and purified products were separated by SDS-PAGE and stained by Coomassie brilliant G250.

14.4.6 WESTERN BLOT ANALYSIS

Protein samples from positive bacteria and *D. zhejiangensis* tissue were separated by SDS-PAGE. Proteins were electronically transferred onto nitrocellulose membrane, and blocked with 5% non-fat milk in TBST. Then the membrane was incubated with anti-ferritin serum for 1 h and washed three times with TBST. The membrane was then incubated with 1:10000 diluted rabbit anti-mouse secondary antibody for 1 h and washed three times with TBST. Signals were visualized using NBT/BCIP method and the reaction was terminated with distilled water. The anti-ferritin serum was prepared according to our previous work [4].

14.4.7 FUNCTIONAL CHARACTERIZATION OF WORM FERRITIN

To achieve recombinant ferritin with biological activity, the purified protein was refolded against stepwise decrease of urea concentration in GSH/GSSG buffer (50 mM Tris–HCl, 1 mM EDTA, 50 mM NaCl, 10% glycerol, 1% glycine, 2 Mm reduced glutathione, 0.2 Mm oxide glutathione, pH 8.0) overnight at 4°C. The concentration of purified protein was quantified by BCA method.

To determine heavy metal enrichment capacity, 1 mL purified ferritin (20 μg/mL) was incubated with 20 μL of 2 mM $FeCl_2$, $PbCl_2$ and $CdCl_2$ at 22°C for 10 min, respectively. Then, 5 μL of each mixture was evenly distributed on the surface of pretreated mica sheet and dried for 3 h at room temperature at environmental humidity of 5%. The samples were observed with a Nanoscope IIIa multimode scanning probe microscopy (Veeco Instruments, USA) using Tapping mode. The NSC11 triangular cantilever probe (MikroMasch, USA) was used for scanning at a resonance frequency of 330 kHz, scan rate of 1–1.5 Hz under spring constant of 50 N/m.

REFERENCES

1. Arosio P, Levi S (2002) Ferrtin, iron homeostasis, and oxidative damage. Free Radic Biol Med 33(4): 457–463.
2. Su X, Du L, Li Y, Li T, Li D, et al. (2009) Production of recombinant protein and polyclonal mouse antiserum for ferritin from Sipuncula Phascolosoma esculenta. Fish Shellfish Immunol 27: 466–468. doi: 10.1016/j.fsi.2009.06.014.
3. Khare G, Gupta V, Nangpal P, Gupta RK, Sauter NK, et al. (2011) Ferritin structure from Mycobacterium tuberculosis: comparative study with homologues identifies extended C-terminus involved in ferroxidase activity. PLoS One 6(4): e18570. doi: 10.1371/journal.pone.0018570.
4. Jin C, Li C, Su X, Li T (2011) Identification and characterization of a Tegillarca granosa ferritin regulated by iron ion exposure and thermal stress. Dev Comp Immunol 35: 745–751. doi: 10.1016/j.dci.2011.02.006.
5. Huang H, Lin Q, Xiao Z, Tong L (2000) Studies on storing capacity of organic small molecules by pig spleen ferritin reactor. Acta Biophysica Sinica 16(1): 39–47.
6. Zhu B, Lin Q, Ke CH, Huang HQ (2011) Single subunit type of ferritin from visceral mass of Saccostrea cucullata: Cloning, expression and cisplatin-subunit analysis. Fish Shellfish Immunol 31: 453–461. doi: 10.1016/j.fsi.2011.06.020.
7. Zoysa D, Lee M J (2007) Two ferritin subunits from disk abalone (Haliotis discus): cloning, characterization and expression analysis. Fish Shellfish Immunol 23: 624–635. doi: 10.1016/j.fsi.2007.01.013.
8. Pead S, Durrant E, Webb B, Larsen C, Heaton D, et al (1995) Metal ion binding to apo, holo and reconstituted horse spleen ferritin, J Inorgan Biochem. 59: 15–27. doi: 10.1016/0162-0134(94)00050-K.
9. Hainfeld JF (1992) Uranium-loaded apoferritin with antibodies attached: molecular design for uranium neutroncapture therapy. Proc Nat Acad Sci USA 89(22): 11064–11068. doi: 10.1073/pnas.89.22.11064.
10. Stephanie LD, Johnpeter NN, Jayne CG (2009) Investigation of the magnetic properties of ferritin by AFM imaging with magnetic sample modulation. Anal Bioanal Chem 394: 215–223.
11. Qiu GF, Zheng L, Liu P (2008) Transcriptional regulation of ferritin mRNA levels by iron in the freshwater giant prawn, Macrobrachium rosenbergii. Comp Biochem Physiol B 10: 320–325. doi: 10.1016/j.cbpb.2008.03.016.
12. Li C, Li H, Su X, Li T (2011) Identification and characterization of a clam ferritin from Sinonovacula constricta. Fish Shellfish Immunol 30: 1147–1151. doi: 10.1016/j.fsi.2011.02.017.
13. Georgieva T, Dunkov BC, Dimov S, Ralchev K, Law JH (2002) Drosophila melanogaster ferritin: cDNA encoding a light chain homologue, temporal and tissue specific expression of both subunit types. Insect Biochem Mol Biol 32(3): 295–302. doi: 10.1016/S0965-1748(01)00090-X.
14. Li M, Saren G, Zhang S (2008) Identification and expression of a ferritin homolog in amphioxus Branchiostoma belcheri evidence for its dual role in immune response and iron metabolism. Comp Biochem Physiol B 150: 263–270. doi: 10.1016/j.cbpb.2008.03.014.

15. Zheng W, Hu Y, Sun L (2010) Identification and analysis of a Scophthalmus maximus ferritin that is regulated at transcription level by oxidative stress and bacterial infection. Comp Biochem Physiol B 156: 222–228. doi: 10.1016/j.cbpb.2010.03.012.

16. Zhang J, Cui C, Zhou X (2009) Study on elastic modulus of individual ferritin. Chinese Sci Bull 54(5): 723–726. doi: 10.1007/s11434-008-0544-6.

17. Harrison PM, Arosio P (1996) The ferritins: molecular properties, iron storage function and cellular regulation. Biochim Biophys Acta 1275: 161–203. doi: 10.1016/0005-2728(96)00022-9.

18. Meldrum FC, Douglas T, Levi S, Arosio P, Stephen M (1991) Reconstituted of manganese oxide cores in horse spleen and recombinant ferritins. J Inorgan Biochem 58: 59–68. doi: 10.1016/0162-0134(94)00037-B.

19. Jung JH, Eom TW, Lee YP (2011) Rhee JY, Choi EH (2011) Magnetic model for a horse spleen ferritin with a three-phasecore structure. J Magn Magn Mater 323(23): 3077–3080. doi: 10.1016/j.jmmm.2011.06.060.

20. Haikarainen T, Thanassoulas A, Stavros P, Nounesis G, Haataja S, et al. (2011) Structural and thermodynamic characterization of metal ion binding in Streptococcus suis Dpr. J Mol Biol 405: 448–460. doi: 10.1016/j.jmb.2010.10.058.

21. Margarita EA, Catalina CP, Camilo LA, Claudio OA, Hernán S (2011) Superoxide dependent reduction of free Fe^{3+} and release of Fe^{2+} from ferritin by the physiologically occurring Cu(I)–glutathione complex. Bioorgan Med Chem 19: 534–541.

AUTHOR NOTES

CHAPTER 2

Acknowledgments

The authors are especially grateful to Dr. Yirong Guo for critical review and perceptive comments on the manuscript. The comments and suggestions of the editor and two anonymous referees are appreciated.

Author Contributions

Conceived and designed the experiments: SZ. Performed the experiments: SZ CC. Analyzed the data: SZ XZ. Contributed reagents/materials/analysis tools: XZ. Wrote the paper: SZ.

CHAPTER 3

Acknowledgments

The authors gratefully thank Cécile Grand from the Agence De l'Environnement et de la Maîtrise de l'Energie (ADEME) for many fruitful scientific discussions. We warmly thank Nadia Crini, Elie Dhivert, Alice Labourier, Jean-Claude Lambert, Christiane Lovy, Jonathan Paris and Pierre-Yves Peseux for their technical assistance. The authors are indebted to three anonymous reviewers for their very helpful remarks.

Author Contributions

Conceived and designed the experiments: CF RS MC PG FR AdV. Performed the experiments: CF RS MC FD DR. Analyzed the data: CF PG MC RS. Contributed reagents/materials/analysis tools: CF DR FD. Wrote the paper: CF RS MC PG FR AdV.

CHAPTER 4

Acknowledgments
This study was partly supported by Grants-in-Aid for Scientific Research from the Ministry of Education, Culture, Sports, Science, and Technology of Japan (M. Ishizuka, No. 19671001 and Y. Ikenaka, 21810001), by JSPS AA Science Platform Program and by The Sumitomo Foundation.

CHAPTER 5

Author Contributions
Conceived and designed the experiments: CSQ LH JB. Performed the experiments: CSQ JY ZWM. Analyzed the data: CSQ ZWM. Contributed reagents/materials/analysis tools: CSQ YL. Wrote the paper: CSQ LH.

CHAPTER 6

Acknowledgments
Many thanks to the three anonymous reviews and the Academic Editor whose pertinent comments have greatly improved the quality of this paper and to Dr. V. Achal and Dr. X. Pan for lingual edit.

Author Contributions
Conceived and designed the experiments: XL XH. Performed the experiments: XL LL BG. Analyzed the data: XL YW GL. Contributed reagents/materials/analysis tools: YW XC. Wrote the paper: XL XC XY.

CHAPTER 7

Acknowledgments
This experimental work was supported by grant MSMT INCHEMBIOL 0021622412.

CHAPTER 8

Competing Interests
The authors declare that they have no competing interests.

Authors' Contributions
This work is part of the PhD thesis of AAZ where MRY and AHP, supervised the thesis, suggested the problem, participated in determination of sample points, sample preparation procedure, and wrote and edited the manuscript. All authors read and approved the final manuscript.

Acknowledgments
Sincere gratitude to the Industrial Parks Co. (Zanjan- Iran) for partial financial support (The grant number: 8/270).

CHAPTER 10

Acknowledgement
The authors are grateful to the University of Putra Malaysia for providing financial assistance.

CHAPTER 11

Acknowledgments
This research was made possible through support provided by NASA to Jackson State University through the University of Mississippi under the term of Grant No. NNG05G572H/08-08-012/300112306A. Partial research support to J.N. was provided by the U.S. Department of Education (Title III Program Grant No. P031B040101-07). Undergraduate fellowship to O.H. was provided by the Science and Technology Access to Research and Graduate Education (STARGE) Program. The opinions expressed herein are those of the authors and do not necessarily reflect the views of NASA or The University of Mississippi. Thanks to Christian Rogers, CSET Environmental Toxicology Core Laboratory for his technical assistance with metal analyses.

CHAPTER 12

Acknowledgments
This research is part of the MBD Energy Research and Development program for Biological Carbon Capture and Storage. The research was conducted with the support and co-operation of Stanwell Energy Corporation and Tarong Power Station. We thank Ian Tuart (JCU-AMCRC), Michelle Tink (JCU-Australian Centre for Tropical Freshwater Research), Richard Keane and Murray Davies (JCU-School of Chemistry) for technical support and advice. The manuscript has also benefited from the comments of three anonymous reviewers.

Author Contributions
Conceived and designed the experiments: RdN NP RJS. Analyzed the data: RJS NP. Contributed reagents/materials/analysis tools: YH RJS. Wrote the paper: RJS NP RdN.

CHAPTER 13

Acknowledgments
The authors thank Max Mergeay for providing strain *C. metallidurans* CH34 and Gabriela Lobos for analytical support.

Author Contributions
Conceived and designed the experiments: LAR CY MG SL KS MS. Performed the experiments: LAR MG. Analyzed the data: LAR MG SL MS. Contributed reagents/materials/analysis tools: MG SL KS MS. Wrote the paper: LAR MS.

CHAPTER 14

Author Contributions
Conceived and designed the experiments: CL XS. Performed the experiments: ZL YL JZ. Analyzed the data: CL JZ CZ. Contributed reagents/materials/analysis tools: TL. Wrote the paper: CL XS.

INDEX

A

acid rain, xxi–xxii, 15, 53–56, 60, 66–68
    simulated acid rain (SAR), xxi–xxii, 15, 53–57, 60–62, 64, 66–68
adsorption, 23, 68, 282, 286, 298, 301
Africa, 126–127
agricultural, xxi, xxvi–xxvii, 3, 6, 15, 34, 42–43, 54, 68, 74–75, 82, 86–90, 92, 98, 107, 128, 150–151, 182, 205, 221, 223, 225, 227, 229, 231–233, 235, 237, 239, 241, 243–245, 249, 265, 324
air pollution, 147–148, 195
anemia, 236
anthropogenic, xxvi–xxvii, 66, 159, 168, 177–178, 180–181, 217, 226, 233, 249, 251, 327
aquifer, 205
arsenic (Ar), xxi, xxv, 12, 44, 46, 127, 149–151, 153, 180, 219, 232, 243, 262, 165–266, 286–287, 292, 294, 303
Ash Dam, xxix, 286, 288–289, 291–292, 297–299
atmospheric, 15, 126, 147, 180–181, 302
atomic absorption spectrometry (AAS), xxvi, 77, 114, 202
atomic force microscopy (AFM), xxx, 330, 335, 337–339, 343

B

Bangladesh, xiii, xv, xvii, 221–222, 243–245

bank vole, xxii, 72, 77, 82–83, 85–86, 88–89, 92–93, 95, 97–98, 104–105, 107
battery, 10, 14, 127, 234
best demonstrated available technologies (BDATs), xxi, 2, 22
bioaccumulation, xxii–xxiii, 22, 70–71, 92, 99, 102, 105–106, 177, 234, 256, 262, 266, 286, 301, 303, 349
bioaugmentation, 320, 324
bioavailability, xxvii–xxviii, 2–3, 15, 41–45, 66, 70–71, 94, 99, 101, 103, 106, 270, 282–283, 299, 303
bioconcentration, xxix, 265, 281, 286–287, 291–292, 295, 298–299
    bioconcentration factor (BCF), 260, 295, 298
biomagnification, 18, 287, 303, 349
bioreactor, 322
biosolids, 5–6, 15, 43
biosorption, xxviii, 219, 286–287, 299–301, 303, 324
Brassica juncea, 266, 279
Brdicka reaction, viii, xxv, 183–185, 187, 189, 191, 193, 195, 197, 199–200

C

cadmium, xxi, xxv, 14–15, 42–45, 48, 67–68, 71, 103, 105–106, 127–128, 149, 151, 153, 191–192, 197–198, 202–203, 219, 262–267, 282–283, 286, 292, 301, 324, 338
calcium chloride (CaCl$_2$), xxii–xxiii, 71, 79–80, 86, 93, 99

carbon capture and storage (CCS), 287

cation exchange capacity (CEC), 54, 275, 278

cDNA library, xxx, 330, 340

cell morphology, 316–317, 322

cement, 23–26, 46, 181

Cepaea (grove snail), xxii–xxiii, 72, 76–77, 79, 82–83, 85–86, 91–93, 104–105

China, xiv–xx, xxiii–xxiv, 41, 54, 67–68, 107, 116, 128–129, 131, 133, 147–150, 152, 157–159, 167, 169, 175–180, 182, 231–232, 235, 244–245

Chironomus riparius (freshwater midge), xxv, 183–185, 191–192

chromium (Cr), xxi, xxiii, xxv–xxvii, 3, 6–7, 11, 20, 25, 28, 36, 44, 114, 120, 122–123, 131, 136, 139, 145, 150–151, 193, 196, 220, 223, 231, 233–234, 237, 241–242, 245, 260, 286, 291, 303

cinnabar, 18

citric acid, 32–33, 263, 276

coal, xxix, 25, 152, 285, 289, 302

coal combustion, 25

contamination

contamination factor, xxvi, 223, 226–228, 238

degree of contamination, xxiv–xxvii, 2–3, 6, 16, 18–20, 22, 24, 27, 30, 32, 34, 42, 44, 47–48, 51, 60, 66, 69–72, 88–89, 92–93, 95, 98–99, 101–105, 109, 111, 113, 115, 117, 119, 121, 123, 125–128, 146, 150, 152, 155, 157–159, 161, 163, 165, 167, 169, 171, 173, 175, 177–182, 184, 195, 199, 202–203, 205, 210, 216–217, 219–221, 223, 225–235, 237–239, 241, 243–247, 249–250, 266, 301, 326

copper (Cu), xxi–xxvii, 3, 6, 16, 20, 25, 32–33, 42, 44, 48, 50, 54–56, 60–63, 65–66, 68, 93, 106, 110, 114, 120, 122–128, 131, 136–137, 139, 145–146, 149, 159, 162, 165–173, 176, 180, 193, 196, 202–203, 208, 211, 213–217, 223, 231–233, 237, 241–242, 244–245, 249, 257, 260–261, 264–267, 282–283, 286, 291, 297–298, 302, 326, 344, 350

correlation analyses, xxiv, 66

Crocidura russula, xxii, 72, 82–83, 85, 105

crop yield, 17

Cupriavidus metallidurans, xxix–xxx, 305, 307–311, 313–323, 325, 327

D

Debye screening, 339

Dendrorhynchus zhejiangensis, xxx, 329–330, 339

detoxification, xxx, 307, 322, 325, 327, 329–331, 333, 335, 337–339, 341, 343

differential pulse polarography (DPP), xxv, 202–203

dissolved oxygen (DO), 11, 24, 72, 74, 201, 206, 214, 276–278

E

eco-toxic, 235

ecosystems, xxiii, 2, 54, 67, 71, 95, 100, 110, 159, 244, 302, 324

soil ecosystems, 2, 54

ecotoxicology, xv, xviii, xxii–xxiii, 70, 95, 100–103, 105–106, 177–178, 180, 265

effluent, xxvii, 43, 127, 167, 222–223, 244–245, 263, 302

electrical conductivity (EC), 57, 206

emission, xxiv, xxviii, 104–105, 125, 146, 162, 178, 196, 202, 243, 294
environmental risk assessment, 130
estuary, xxiv, 158–159, 161, 166–167, 171, 173, 177–178, 180–181, 245
exposure, xxiv–xxv, xxix–xxx, 10–11, 15, 70–71, 94–95, 98–102, 105, 127–131, 133–145, 147–153, 195, 198–199, 216, 219, 233–236, 271, 281, 303, 317–318, 322, 331, 334, 336, 338–339, 343
   exposure pathways, xxiv, 129–131, 133, 135, 137, 139, 141, 143, 145, 147, 149, 151, 153

F

ferritin, xxx, 329–344, 351
fertilizers, 4, 6, 15, 222, 249–250
food, xxi, xxiv, 2–3, 11, 15–18, 41, 71, 93–94, 100–101, 103–106, 127, 130, 133–135, 140, 144, 146–147, 152, 202, 233, 235–236, 243, 245, 300, 351, 353
   food chain, 2, 11, 15–16, 18, 93, 101, 105–106, 130, 202, 233, 235
   food safety, 351
   food security, xxi, 3
   food web, 71, 94, 100, 103
France, xiv–xv, xviii, 77, 104–106

G

gasoline, 10
geoaccumulation, xxvi, 159, 223–224, 226, 237, 245
geostatistics, 179
graphite furnace atomic absorption spectrometry (GFAAS), 202
greater white-toothed shrew, xxii, 72, 77, 83, 86, 88, 93, 97–98, 105, 107

growth rate (GR), xxix, 106, 151, 267 298, 322

H

habitat, 70, 72–73, 80, 95, 97–102, 104–105
hair, xxiv, 130–131, 134, 144–145, 148–149, 151, 153
hazard quotient (HQ), 130, 136–137, 139–144, 146–148, 343
heavy metal speciation, 2–3
haemoglobin, 16
human exposure, 129, 131, 133–135, 137, 139, 141, 143, 145, 147–149, 151, 153
hypertension, 233, 236

I

India, 50, 169, 178, 180–181, 199, 219–220, 231–232, 235, 244–245, 265
inductively coupled plasma-atomic emission spectrometry (ICP-AES), 162, 202
industrial, xiii, xv, xxiv–xxv, xxvii, xxix, 6–7, 15, 23, 46, 111, 114, 157–159, 167, 171, 176, 178, 181–182, 196, 202–203, 205, 213, 219–220, 222, 232–234, 236, 243–246, 249–251, 301, 323–324
ingestion, xxiv, 10–11, 130, 135–136, 138–140, 144, 147–148
inhalation, xxiv, 10, 130–131, 136, 139–140, 144, 147–149
intervention value, 20
intervention values, 20
inverse distance weighting, 182
Iran, xiii, xvii–xix, xxv, 199, 203, 213, 219–220
irrigation, 203, 244

Itai-itai disease, 119

J

Japan, xv, xvii, xix, 15, 114, 126, 220
Jinzhou Bay, xxiv, 158–159, 161, 166, 168–171, 173–176, 179, 182

K

kidney, 10–11, 15–16, 71, 105, 124, 127, 149, 233
kriging, xxiii, 163, 182

L

Lake Victoria, 131, 151
leachate, xxii, 19, 57, 59–60, 62
leaching, vii, xxi–xxii, 18–19, 30, 34, 43, 46, 48–49, 53–55, 57–68, 202, 264, 302, 327
lead (Pb), xxi–xxviii, 3, 6–7, 9–11, 15–17, 20, 25, 28, 32–33, 36, 44–45, 48–50, 54–56, 60, 64–66, 68, 71–72, 77, 80–81, 83–86, 88–93, 96–97, 100–101, 104–106, 110, 114–116, 119–120, 122–123, 125–129, 131, 133, 135–141, 143–153, 159, 162, 165–167, 169–173, 176–179, 182, 193, 197, 201–203, 205, 211, 213–217, 219, 223, 231, 233–237, 241–242, 244–245, 256, 260–267, 269–271, 273, 275–283, 286, 291, 297–198
    lead solubility, 276
legislation, 2
liver, 16, 76, 83–84, 87, 97, 105, 124, 196, 198, 233
low-molecular-weight organic acids, 32

M

manufacturing, 233
mercury (Hg), xxi, xxiii–xxiv, xxvi, xxix–xxx, 3, 17–18, 20, 24–25, 49, 105, 114, 120, 122–123, 126, 131, 133, 136, 139–140, 143–145, 147–149, 151–153, 177, 188, 197–198, 223, 226, 231, 234–235, 237, 241–242, 256, 260, 286, 298, 301, 305–310, 312–314, 316–317, 321–327
metal toxicity monitoring, 130
metallothionein (MT), xxv, 101–102, 105, 183–184, 187, 190–192, 194–200, 220, 281, 283
methylmercury, 301, 306, 322–323
mining, vii, xxiii–xxiv, 7, 18, 37, 110–111, 114, 116, 119, 124, 126–129, 131, 133, 135, 137, 139, 141, 143, 145–147, 149, 151, 153, 178, 196, 199, 234, 249, 301
    metal mining, 7
moisture, 19, 271
Monte Carlo simulation, 130, 137, 140

N

neutron activation analysis (NAA), 202
nickel (Ni), xxi, xxiii–xxiv, xxvi–xxvii, 3, 6, 14, 18, 20, 32–33, 36, 44–45, 68, 106, 114, 120, 122–123, 131, 136, 139, 145, 162, 165–166, 169–173, 176, 193, 196, 202–203, 211, 213–217, 219, 223, 231, 235–237, 241–242, 260–261, 264, 286, 291, 297–299, 301, 325–326
Nigeria, xvii, xix, 20, 45, 48, 127
nitric acid, 19, 77, 205

O

Oxychilus draparnaudi (glass snail), xxii, 72, 76–77, 82–83, 85–87, 89, 92–93 104

P

paint, 10, 146, 230, 236
pesticides, 5, 7, 15, 150–151, 250, 325
petrochemicals, 7
pH, xxii, xxvii, 10, 15–16, 44–45, 55–57, 60, 62–63, 66, 102, 205–206, 208, 233, 275, 302–303, 310
phytodegradation, xxvii, 255
phytoextraction, xxvii, 22, 34–36, 38–39, 43, 49–50, 255–257, 263–266, 269–271, 273, 275–277, 279, 281–283
   chelate-assisted phytoextraction, viii, 269, 271, 273, 275, 277, 279, 281, 283
   chemically enhanced phytoextraction, 49, 283
   natural phytoextraction, 35
phytofiltration, xxvii, 34, 40, 253, 262
phytoremediation, xxi, xxvii, 2–3, 22, 33–34, 44–45, 49–50, 220, 249, 251, 253, 255–259, 261–267, 270, 281–283
phytostabilization, xxvii, 22, 34, 39, 254, 262
pollution load, xxvi, 223, 229, 238, 241
   pollution load index (PLI), xxvi, 223, 229, 238, 241
polymerase chain reaction (PCR), xxx, 198, 310, 315, 336, 341
polyvinyl chloride (PVC), 15
population density, 3

precipitation, xxvii, 18–19, 23–24, 30, 67, 270
Principal Component Analysis (PCA), 209–210, 214, 216, 310
public health, xvi, 2, 149, 202, 269, 282
purple soil, xxi–xxii, 54–55, 60, 62, 64, 66–67, 257, 266

R

red mud, 23
reference concentration (RfC), 136
reference dose (RfD), 130, 136, 140, 144
relative errors (REs), xxii, 56, 64, 67, 126, 128, 150–151, 199, 219–220, 244–245, 263–265, 282, 324–325
remediation, i, iii, vii–viii, xxi, xxvii, 1–2, 20–22, 29, 31, 34, 41–50, 126, 223, 247, 252, 264, 267, 269–270, 287, 298–300, 324
   extraction, xxi, 19, 22, 30, 33, 41–42, 44–45, 47–48, 50, 56–57, 68, 112, 218, 273, 276, 282, 289, 303, 309, 336
   immobilization, xxi, 2–3, 22–23, 25, 28, 46–47
   isolation, 22, 269, 326
   physical separation, 22, 30
   phytoremediation, viii, xxi, xxvii, 2–3, 22, 33–34, 44–45, 49–50, 220, 249, 251, 253, 255–259, 261–267, 270, 281–283, 356
   toxicity reduction, 22
Risk Index (RI), 101, 159, 164, 173209–210, 174, 180, 245

S

sediment, xxiv–xxv, 18, 48, 67–68, 74, 80, 109–111, 113–115, 117, 119, 121, 123, 125–127, 150, 158–159, 164, 167, 171, 174–178, 180–181, 195, 225, 245, 281, 306
Sesbania exaltata (coffeeweed), xxvii, 269–270, 277–279, 281–282
sewage, 6, 15, 25, 42–43, 233–234, 244, 287
shale, 225–226
smelter, xxii, 42, 68, 71–73, 75, 93, 101–102, 104–105, 126–128, 152, 159, 161, 165, 167, 171, 176, 249
soil
    soil washing, xxi, 2–3, 22, 29–32, 47–48, 50
    topsoil, 101, 127, 233
solidification/stabilization (S/S), 22, 24–26
standard deviation (SD), 63, 68, 77, 114, 127, 138–139, 145, 150, 165–166, 219, 231, 318
steel, 18, 233–235

T

thioglycolate, xxx, 320, 323
toxicity characteristic leaching procedure (TCLP), 19
trace metal (TM), xxii–xxiii, 11, 16 42–47, 68, 70–72, 76, 78–80, 83–100, 106, 150, 153, 178–179, 181, 199, 221, 244, 246, 303
transmission electron microscopy, 301, 311, 317
Turkey, 169, 181, 196, 219, 232, 235, 245–246

U

US Environmental Protection Agency (EPA), 19, 22, 43–47, 50, 130–131, 136, 152, 219, 263, 266

V

vegetables, xxiv, 11, 15, 130, 133, 138, 144, 146–150, 153, 234, 243–244, 265
vertical engineered barriers (VEB), 22

W

water
    contamination, 16, 202
    drinking, 16, 131, 135, 201, 203, 205, 216, 219
    groundwater, xxii, xxv, 6, 10, 18, 41, 46, 60, 66, 201–203, 205, 216, 219–220, 244, 264–265
    quality, xxii, xxvi, 179, 220
    surface, 18, 202, 214, 220, 244, 246
    wastewater, xxvii, 6–7, 18, 25, 41, 167, 176, 219, 222, 235, 244, 301–302, 323–324, 326
Western blot, 334, 342
wetlands, 196, 267

X

x-ray fluorescence (XRF), 127, 202

Z

Zambia, vii, xiv, xvii, xxiii, 109–111, 113, 115–117, 119, 121, 123, 125, 127–128
Zea mays L.(corn), 11, 279